Trigonometric Sums and Their Applications

Andrei Raigorodskii • Michael Th. Rassias

Editors

Trigonometric Sums and Their Applications

 Springer

Editors
Andrei Raigorodskii
Moscow Institute of Physics and
Technology
Dolgoprudny, Russia

Moscow State University
Moscow, Russia

Buryat State University
Ulan-Ude Russia

Caucasus Mathematical Center
Adyghe State University, Maykop, Russia

Michael Th. Rassias
Institute of Mathematics
University of Zurich
Zurich, Switzerland

Moscow Institute of Physics and
Technology
Dolgoprudny, Russia

Institute for Advanced Study
Program in Interdisciplinary Studies
Princeton, NJ, USA

ISBN 978-3-030-37906-3   ISBN 978-3-030-37904-9   (eBook)
https://doi.org/10.1007/978-3-030-37904-9

Mathematics Subject Classification (2010): 42-XX, 43-XX, 44-XX, 26-XX, 30-XX, 32-XX, 33-XX, 35-XX, 40-XX, 41-XX, 46-XX, 47-XX, 65-XX

This Springer imprint is published by the registered company Springer Nature Switzerland AG.
The registered company address is: Gewerbestrasse 11, 6330 Cham, Switzerland

Preface

This volume is devoted to the study of trigonometric and exponential sums with their various applications and interconnections with other mathematical objects. These types of sums play an essential role in a broad variety of mathematical areas of research, including real and complex analysis, analytic number theory, functional analysis, approximation theory, harmonic analysis, and analytic inequalities with best constants. More specifically, the volume deals with Marcinkiewicz-Zygmund inequalities at Hermite zeros and their airy function cousins, polynomials with constrained coefficients, nonnegative sine polynomials, inequalities for weighted trigonometric sums, Dedekind- and Hardy-type sums and trigonometric sums induced by quadrature formulas, best orthogonal trigonometric approximations of classes of infinitely differentiable functions, trigonometric functions related to the Riemann zeta function and the Nyman-Beurling criterion for the Riemann hypothesis, half-discrete Hilbert-type inequalities in the whole plane with the kernel of hyperbolic secant function, reverse Hilbert-type integral inequalities with the kernel of hyperbolic cotangent function, sets with small Wiener norm, double-sided Taylor's approximations and their applications in the theory of trigonometric inequalities, norm inequalities for generalized Laplace transforms, and the first derivative of Hardy's Z-function.

The papers have been contributed by eminent experts in the corresponding domains, who have presented the state of the art in the problems treated. The present volume is expected to be a valuable source for both graduate students and research mathematicians as well as physicists and engineers. We would like to express our warmest thanks to all the authors of papers in this volume who contributed in this collective effort. Last but not least, we would like to extend our appreciation to the Springer staff for their valuable help throughout the publication process of this work.

Moscow, Russia
Zurich, Switzerland

Andrei Raigorodskii
Michael Th. Rassias

Contents

On a Category of Cotangent Sums Related to the Nyman-Beurling Criterion for the Riemann Hypothesis

Nikita Derevyanko, Kirill Kovalenko, and Maksim Zhukovskii

Abstract The purpose of the present paper is to provide a general overview of a variety of results related to a category of cotangent sums which have been proven to be associated to the so-called Nyman-Beurling criterion for the Riemann Hypothesis. These sums are also related to the Estermann Zeta function.

Keywords Cotangent sums · Riemann zeta function · Vasyunin sums · Estermann zeta function · Riemann Hypothesis · Dedekind sums · Nyman-Beurling-Báez-Duarte criterio

1 Introduction

This paper is focused on applications of certain cotangent sums to various problems related to the Riemann Hypothesis. The expression for the trigonometric sums in question is the following

Definition 1.1

$$c_0\left(\frac{r}{b}\right) := -\sum_{m=1}^{b-1} \frac{m}{b} \cot\left(\frac{\pi m r}{b}\right), \tag{1}$$

where $r, b \in \mathbb{N}$, $b \geq 2$, $1 \leq r \leq b$ and $(r, b) = 1$.

N. Derevyanko
Moscow Institute of Physics and Technology, Dolgoprudny, Russia
e-mail: nikita.derevyanko@phystech.edu

K. Kovalenko
National Research University Higher School of Economics, Moscow, Russia
e-mail: kdkovalenko@edu.hse.ru

M. Zhukovskii (✉)
Laboratory of Advanced Combinatorics and Network Applications, Moscow Institute of Physics and Technology, Dolgoprudny, Russia

© Springer Nature Switzerland AG 2020
A. Raigorodskii, M. T. Rassias (eds.), *Trigonometric Sums and Their Applications*,
https://doi.org/10.1007/978-3-030-37904-9_1

·These sums were thoroughly studied for the first time in the important paper of S. Bettin and J. Conrey [7] in which they establish a reciprocity formula for these sums among other properties. We will provide more details about their work – among other results – in Section 2.

1.1 Nyman–Beurling Criterion for the Riemann Hypothesis

There are many interesting results concerning these cotangent sums, but initially we will present some general information about the Riemann Hypothesis and some related problems. Moreover, our aim is to provide motivation for the use of cotangent sums in these problems.

In this paper we shall denote a complex variable by $s = \sigma + it$, where σ and t are the real and imaginary part of s respectively.

Definition 1.2 The Riemann zeta function is a function of the complex variable s defined in the half-plane $\{\sigma > 1\}$ by the absolutely convergent series

$$\zeta(s) := \sum_{n=1}^{\infty} \frac{1}{n^s}. \tag{2}$$

As shown by B. Riemann, $\zeta(s)$ extends to $\mathbb{C}$ as a meromorphic function with only a simple pole at $s = 1$, with the residue 1, and satisfies the functional equation

$$\zeta(s) = 2^s \pi^{s-1} \sin\left(\frac{\pi s}{2}\right) \Gamma(1-s)\zeta(1-s). \tag{3}$$

For negative integers, one has a convenient representation of the Riemann zeta function in terms of Bernoulli numbers:

$$\zeta(-n) = (-1)^n \frac{B_{n+1}}{n+1}, \text{ for } n \geq 0.$$

By the above formula one can easily deduce that $\zeta(s)$ vanishes when s is a negative even integer because $B_m = 0$ for all odd m other than 1. The negative even integers are called trivial zeros of the Riemann zeta function. All other complex points where $\zeta(s)$ vanishes are called non-trivial zeros of the Riemann zeta function, and they play a significant role in the distribution of primes.

The actual connection with the distribution of prime numbers was observed in Riemann's 1859 paper. It is in this paper that Riemann proposed his well known hypothesis.

Hypothesis 1.1 (Riemann) *The Riemann zeta function $\zeta(s)$ attains its non-trivial zeros only in complex points with $\sigma = \frac{1}{2}$. The line on the complex plane given by the equation $\sigma = \frac{1}{2}$ is usually called "critical".*

The Nyman–Beurling–Baez–Duarte–Vasyunin (also simply known as Nyman-Beurling) approach to the Riemann Hypothesis enables us to associate the study of cotangent sums and the Riemann Hypothesis through the following theorem:

Theorem 1.1 *The Riemann Hypothesis is true if and only if*

$$\lim_{N \to +\infty} d_N = 0,$$

where

$$d_N^2 = \inf_{D_N} \frac{1}{2\pi} \int_{-\infty}^{+\infty} \left| 1 - \zeta\left(\frac{1}{2} + it\right) D_N\left(\frac{1}{2} + it\right) \right|^2 \frac{dt}{\frac{1}{4} + t^2} \tag{4}$$

and the infimum is taken over all Dirichlet polynomials

$$D_N(s) = \sum_{n=1}^{N} \frac{a_n}{n^s}, \ a_n \in \mathbb{C}. \tag{5}$$

In his paper [2], B. Bagchi used a slightly different formulation of Theorem 1.1. In order to state it, we have to introduce some definitions.

Definition 1.3 The Hardy space $H^2(\Omega)$ is the Hilbert space of all analytic functions F on the half-plane Ω (we define it for a right half-plane $\{\sigma > \sigma_0\}$ of the complex plane) such that

$$\|F\|^2 := \sup_{\sigma > \sigma_0} \frac{1}{2\pi} \int_{-\infty}^{+\infty} |F(\sigma + it)|^2 dt < \infty.$$

Everywhere in this section we will use $\Omega = \{\sigma > \frac{1}{2}\}$.

Definition 1.4 For $0 \leq \lambda \leq 1$, let $F_\lambda \in H^2(\Omega)$ be defined by

$$F_\lambda(s) = (\lambda^s - \lambda)\frac{\zeta(s)}{s}, s \in \Omega,$$

and for $l = 1, 2, 3, \ldots$, let $G_l \in H^2(\Omega)$ be defined by $G_l = F_{\frac{1}{l}}$, i.e.

$$G_l(s) = (l^{-s} - l^{-1})\frac{\zeta(s)}{s}, s \in \Omega.$$

Also, let $E \in H^2(\Omega)$ be defined by

$$E(s) = \frac{1}{s}, s \in \Omega.$$

Now we can state the reformulation of Theorem 1.1 which was used in paper [2].

Theorem 1.2 *The following statements are equivalent:*

(1) *The Riemann Hypothesis is true;*
(2) *E belongs to the closed linear span of the set $\{G_l : l = 1, 2, 3 \ldots\}$;*
(3) *E belongs to the closed linear span of the set $\{F_\lambda : 0 \le \lambda \le 1\}$.*

The plan of the proof is to verify three implications: $1 \to 2, 2 \to 3$ and $3 \to 1$.

The first implication is the most challenging of all three. It is proven using some famous results obtained under the assumption that the Riemann Hypothesis is true, among which are Littlewood's Theorem 1.3 and the Lindelöf Hypothesis 1.2, and some standard techniques of functional analysis, particularly concerning convergence in the norm. More details can be found in the original paper by B. Bagchi [2].

Hypothesis 1.2 (Lindelöf) *If the Riemann Hypothesis is true, then*

$$\forall \varepsilon > 0, \ \zeta \left(\frac{1}{2} + it \right) = O(t^\varepsilon).$$

Remark A very interesting and novel approach to the Lindelöf hypothesis is presented by A. Fokas [11].

Theorem 1.3 (Littlewood) *If the following conditions are satisfied:*

- $\lim_{r \to 1^-} \sum_{n=0}^{\infty} r^n c_n = a, \ \forall i \ c_i \in \mathbb{C}, \ a \in \mathbb{C},$
- $c_n = O(\frac{1}{n}),$

   then

$$\sum_{n=0}^{\infty} c_n = a.$$

The second implication follows from the embedding

$$\{G_l : l = 1, 2, 3 \ldots\} \subset \{F_\lambda : 0 \le \lambda \le 1\}.$$

To prove the third implication ($3 \to 1$), suppose that the Riemann Hypothesis is false. Then $\exists s_0 = \sigma_0 + it_0 : \zeta(s_0) = 0$ and $\sigma_0 \ne \frac{1}{2}$, which implies that $\forall 0 \le \lambda \le 1$ $F_\lambda(s_0) = 0$. That together with statement 3 gives that $E(s_0) = \frac{1}{s_0} = 0$, i.e. $0 = 1$. This contradiction completes the proof.

1.2 The Cotangent Sum's Applications to Problems Related to the Riemann Hypothesis

The main motivation behind the study of the cotangent sum (1) follows from Theorem 1.1, which constitutes an equivalent form of the Riemann Hypothesis.

Asymptotics for d_N (4) under the assumption that the Riemann Hypothesis is true have been studied in several papers. S. Bettin, J. Conrey and D. Farmer in [9] obtained the following result.

Theorem 1.4 *If the Riemann Hypothesis is true and if*

$$\sum_{|Im(\rho)| \leq T} \frac{1}{|\zeta'(\rho)|^2} \ll T^{\frac{3}{2} - \delta}$$

for some $\delta > 0$, with the sum on the left hand side taken over all distinct zeros ρ of the Riemann zeta function with imaginery part less than or equal to T, then

$$\frac{1}{2\pi} \int_{-\infty}^{+\infty} \left| 1 - \zeta\left(\frac{1}{2} + it\right) V_N\left(\frac{1}{2} + it\right) \right|^2 \frac{dt}{\frac{1}{4} + t^2} \sim \frac{2 + \gamma - \log 4\pi}{\log N}$$

for

$$V_N(s) := \sum_{n=1}^{N} \left(1 - \frac{\log n}{\log N} \right) \frac{\mu(n)}{n^s}. \tag{6}$$

We should mention that in the sequel γ stands for the Euler–Mascheroni constant. Also, here μ is the Möbius function.

Also, from results of [9] it follows that under some restrictions, the infimum from (4) is attained for $D_N = V_N$.

Nevertheless, it is interesting to obtain an unconditional estimate for d_N.

In order to proceed further, we shall study equation (4) in more detail. In particular, we can expand the square in the integral:

$$\begin{aligned}
d_N = \inf_{D_N} \Bigg(&\int_{-\infty}^{+\infty} \left(1 - \zeta\left(\frac{1}{2} + it\right) D_N\left(\frac{1}{2} + it\right) \right. \\
&\left. - \zeta\left(\frac{1}{2} - it\right) \overline{D_N}\left(\frac{1}{2} + it\right) \right) \frac{dt}{\frac{1}{4} + t^2} \\
&+ \int_{-\infty}^{+\infty} \left| \zeta\left(\frac{1}{2} + it\right) \right|^2 \left| D_N\left(\frac{1}{2} + it\right) \right|^2 \frac{dt}{\frac{1}{4} + t^2} \Bigg).
\end{aligned}$$

The integral in the second summand can be expressed as

$$\sum_{1 \le r, b \le N} a_r \bar{a}_b r^{-\frac{1}{2}} b^{-\frac{1}{2}} \int_{-\infty}^{+\infty} \left| \zeta \left(\frac{1}{2} + it \right) \right|^2 \left(\frac{r}{b} \right)^{it} \frac{dt}{\frac{1}{4} + t^2}, \tag{7}$$

where a_i are from the Definition (5) of D_N.

Therefore, the integral

$$\int_{-\infty}^{\infty} \left| \zeta \left(\frac{1}{2} + it \right) \right|^2 \left(\frac{r}{b} \right)^{it} \frac{dt}{\frac{1}{4} + t^2}$$

plays an important role in the Nyman-Beurling criterion for the Riemann Hypothesis. Moreover, one can prove that this integral can be expressed via the so-called Vasyunin sum.

Definition 1.5 The Vasyunin sum is defined as follows:

$$V \left(\frac{r}{b} \right) := \sum_{m=1}^{b-1} \left\{ \frac{mr}{b} \right\} \cot \left(\frac{\pi mr}{b} \right), \tag{8}$$

where $\{x\} = x - \lfloor x \rfloor, x \in \mathbb{R}$.

The following proposition holds true:

Proposition 1.1

$$\frac{1}{2\pi (rb)^{1/2}} \int_{-\infty}^{\infty} \left| \zeta \left(\frac{1}{2} + it \right) \right|^2 \left(\frac{r}{b} \right)^{it} \frac{dt}{\frac{1}{4} + t^2}$$

$$= \frac{\log 2\pi - \gamma}{2} \left(\frac{1}{r} + \frac{1}{b} \right) + \frac{b-r}{2rb} \log \frac{r}{b} - \frac{\pi}{2rb} \left(V \left(\frac{r}{b} \right) + V \left(\frac{b}{r} \right) \right). \tag{9}$$

One can note that the only non-explicit function on the right hand side of this formula is the Vasyunin sum.

The next equation connects this result with the cotangent sums in question.

Proposition 1.2 *It holds that*

$$V \left(\frac{r}{b} \right) = -c_0 \left(\frac{\bar{r}}{b} \right), \tag{10}$$

where $\bar{r}$ is such that $\bar{r}r \equiv 1 \pmod{b}$.

The cotangent sum c_0 can also be used to describe some special values of the Estermann zeta function.

Definition 1.6 The Estermann zeta function $E(s, \frac{r}{b}, \alpha)$ is defined by the Dirichlet series

$$E\left(s, \frac{r}{b}, \alpha\right) = \sum_{n \geq 1} \frac{\sigma_\alpha(n) \exp(\frac{2\pi i n r}{b})}{n^s}, \tag{11}$$

where $Re\, s > Re\, \alpha + 1$, $b \geq 1$, $(r, b) = 1$, and

$$\sigma_\alpha(n) = \sum_{d|n} d^\alpha. \tag{12}$$

One can show that the Estermann zeta function $E(s, \frac{r}{b}, \alpha)$ satisfies the following functional equation:

$$E\left(s, \frac{r}{b}, \alpha\right) = \frac{1}{\pi} \left(\frac{b}{2\pi}\right)^{1+\alpha-2s} \Gamma(1-s)\Gamma(1+\alpha-s)$$

$$\times \left(\cos\left(\frac{\pi\alpha}{2}\right) E\left(1+\alpha-s, \frac{\bar{r}}{b}, \alpha\right)\right.$$

$$\left. - \cos\left(\pi s - \frac{\pi\alpha}{2}\right) E\left(1+\alpha-s, -\frac{\bar{r}}{b}, \alpha\right)\right), \tag{13}$$

where r is such that $\bar{r}r \equiv 1 (mod\ b)$.

Properties of $E(0, \frac{r}{b}, 0)$ were used by R. Balasubramanian, J. Conrey, and D. Heath-Brown [3] to prove an asymptotic formula for

$$I = \int_0^T \left|\zeta\left(\frac{1}{2} + it\right)\right|^2 \left|A\left(\frac{1}{2} + it\right)\right|^2 dt, \tag{14}$$

where $A(s)$ is a Dirichlet polynomial.

Asymptotic results for the function I as well as other functions of this type, are useful for estimating a lower bound for the portion of zeros of the Riemann zeta function $\zeta(s)$ on the critical line.

The following result of M. Ishibashi from [13] concerning $E(s, \frac{r}{b}, \alpha)$ for $s = 0$, provides the connection of the Estermann zeta function with the cotangent sum $c_0(\frac{r}{b})$, simply by setting $\alpha = 0$.

Theorem 1.5 (Ishibashi) *Let* $b \geq 2$, $1 \leq r \leq b$, $(r, b) = 1$, $\alpha \in \mathbb{N} \cup \{0\}$. *Then for even* α, *it holds that*

$$E\left(0, \frac{r}{b}, \alpha\right) = \left(-\frac{i}{2}\right)^{\alpha+1} \sum_{m=1}^{b-1} \frac{m}{b} \cot^{(\alpha)}\left(\frac{\pi m r}{b}\right) + \frac{1}{4}\delta_{\alpha,0}, \tag{15}$$

where $\delta_{\alpha,0}$ *is the Kronecker delta function.*

For odd α, it holds that

$$E\left(0, \frac{r}{b}, \alpha\right) = \frac{B_{\alpha+1}}{2(\alpha+1)}.$$
(16)

In the special case when $r = b = 1$, we have

$$E(0, 1, \alpha) = \frac{(-1)^{\alpha+1} B_{\alpha+1}}{2(\alpha+1)},$$

where B_m is the m-th Bernoulli number and $B_{2m+1} = 0$,

$$B_{2m} = \frac{(-1)^{m+1} 2(2m)!}{(2\pi)^{2m}} \zeta(2m), \quad for \ m \geq 1.$$

Thus for $b \geq 2$, $1 \leq r \leq b$, $(r, b) = 1$, it follows that

$$E\left(0, \frac{r}{b}, 0\right) = \frac{1}{4} + \frac{i}{2} c_0\left(\frac{r}{b}\right),$$
(17)

where $c_0(\frac{r}{b})$ is our cotangent sum from (1).

2 Central Properties of the Cotangent Sum c_0

The function c_0 is thoroughly studied in the papers of S. Bettin and J. Conrey [7, 8], where they have established a reciprocity formula for it, which encapsulates important information about the behaviour of these sums. However, before we state the formula itself, we shall give several definitions.

For $a \in \mathbb{C}$ and $\mathrm{Im}\,(s) > 0$, consider

$$\mathscr{S}_a(s) := \sum_{n=1}^{\infty} \sigma_a(n) e(ns) e^{2\pi i n s},$$

$$E_a(s) := 1 + \frac{2}{\zeta(-a)} \mathscr{S}_a(s).$$

It is worth mentioning that for $a = 2k+1, k \in \mathbb{Z}_{\geq 1}$, E_a is the well known Eisenstein series of weight $2k + 2$.

Definition 2.1 For $a \in \mathbb{C}$ and $Im(s) > 0$, define the function

$$\psi_a(s) := E_{a+1}(s) - \frac{1}{s^{a+1}} E_{a+1}\left(-\frac{1}{s}\right). \tag{18}$$

For $a = 2k$, $k \geq 2$, the function $\psi_a(s)$ is equal to zero, because of the modularity property of the Eisenstein series. Unfortunately, it is not true for other values of a, but the functions $\psi_a(s)$ have some remarkable properties, which were described in detail by S. Bettin and J. Conrey.

Now we can state the theorem proven in paper [7].

Theorem 2.1 *The function c_0 satisfies the following reciprocity formula:*

$$c_0\left(\frac{r}{b}\right) + \frac{b}{r} c_0\left(\frac{b}{r}\right) - \frac{1}{2\pi r} = \frac{i}{2} \psi_0\left(\frac{r}{b}\right). \tag{19}$$

This result implies that the value of $c_0(\frac{r}{b})$ can be computed within a prescribed accuracy in a polynomial of $\log b$.

S. Bettin and J. Conrey highlighted that the reciprocity formula from 2.1 is very similar to that of the Dedekind sum 5.1. We will consider Dedekind sums in more detail in section 5 of this paper.

In [8] the result for c_0 was generalized for the sums

$$c_a\left(\frac{r}{b}\right) := b^a \sum_{m=1}^{b-1} \cot\left(\frac{\pi m r}{b}\right) \zeta\left(-a, \frac{m}{b}\right), \tag{20}$$

where $\zeta(s, x)$ is the Hurwitz zeta function.

After the introduction of the sum c_0 by Bettin and Conrey and the proof of their reciprocity formula, these sums have also been very extensively studied in a number of works by H. Maier and M. Th. Rassias by the use of different techniques.

We shall now state some crucial results concerning the cotangent sum c_0.

In [22], M. Th. Rassias proved – using elementary techniques – the following asymptotic formula:

Theorem 2.2 *For $b \geq 2$, $b \in \mathbb{N}$, we have*

$$c_0\left(\frac{1}{b}\right) = \frac{1}{\pi} b \log b - \frac{b}{\pi}(\log 2\pi - \gamma) + O(1). \tag{21}$$

Subsequently in [18], M. Th. Rassias and H. Maier established an improvement, or rather an asymptotic expansion, of Theorem 2.2.

Theorem 2.3 *Let $b, n \in \mathbb{N}$, $b \geq 6N$, with $N = \lfloor \frac{n}{2} \rfloor + 1$. There exist absolute real constants $A_1, A_2 \geq 1$ and absolute real constants E_l, where $l \in \mathbb{N}$, with $|E_l| \leq (A_1 l)^{2l}$, such that for each $n \in \mathbb{N}$ we have*

$$c_0\left(\frac{1}{b}\right) = \frac{1}{\pi} b \log b - \frac{b}{\pi}(\log 2\pi - \gamma) - \frac{1}{\pi} + \sum_{l=1}^{n} E_l b^{-l} + R_n^*(b), \qquad (22)$$

where

$$|R_n^*| \le (A_2 n)^{4n} b^{-(n-1)}.$$

One could obtain Theorems 2.2 and 2.3 by the use of the reciprocity formula of Bettin and Conrey, but the proofs of Maier and Rassias follow a different method. The proof of Theorems 2.2 and 2.3 in [22] and [18], respectively, were obtained using a common underlying idea proposed in [22]. First of all, one can obtain the following relation between sums of cotangents and sums with fractional parts:

$$\sum_{\substack{a \ge 1 \\ b \nmid a}} \frac{b(1 - 2\{a/b\})}{a} = \sum_{\substack{a \ge 1 \\ b \nmid a}} \sum_{m=1}^{b} \cot\left(\frac{\pi m}{b}\right) \frac{\sin(\frac{2\pi m}{b} a)}{a}. \qquad (23)$$

This relation provided the following proposition:

Proposition 2.1 *For every positive integer $b \ge 2$, we have*

$$c_0\left(\frac{1}{b}\right) = \frac{1}{\pi} \sum_{\substack{a \ge 1 \\ b \nmid a}} \frac{b(1 - 2\{a/b\})}{a}. \qquad (24)$$

Then the difficulty lies in obtaining a good approximation of the sum $S(L; b)$ defined by

$$S(L; b) := 2b \sum_{1 \le a \le L} \frac{1}{a} \left\lfloor \frac{a}{b} \right\rfloor. \qquad (25)$$

The difference between the estimates from Theorems 2.2 and 2.3 is that stronger approximation techniques were applied in [18] to obtain more information about $S(L; b)$. Namely, the generalized Euler summation formula (26) was used to improve the result of Theorem 2.2.

Definition 2.2 If f is a function that is differentiable at least $(2N + 1)$ times in $[0, Z]$, let

$$r_N(f, Z) = \frac{1}{(2N + 1)!} \int_0^Z (u - \lfloor u \rfloor + B)^{(2N+1)} f^{(2N+1)}(u) du,$$

with the following notation:

$$(u - \lfloor u \rfloor + B)^{(2N+1)} = ((u - \lfloor u \rfloor) + B)^{(2N+1)} := \sum_{j=0}^{2N+1} \binom{2N+1}{j} (u - \lfloor u \rfloor)^j B_{2N+1-j},$$

where B_{2j} are the Bernoulli numbers.

Theorem 2.4 (Generalized Euler summation formula) *Let f be $(2N+1)$ times differentiable in the interval $[0, Z]$. Then*

$$\sum_{\nu=0}^{Z} f(\nu) = \frac{f(0) + f(Z)}{2} + \int_0^Z f(u) du$$

$$+ \sum_{j=1}^{N} \frac{B_{2j}}{(2j)!} (f^{(2j-1)}(Z) - f^{(2j-1)}(0)) + r_N(f, Z). \tag{26}$$

Particularly, H. Maier and M. Th. Rassias used Theorem 2.4 to obtain the following new representation for $S(L; b)$.

Theorem 2.5 *For $N \in \mathbb{N}$, we have*

$$S(L; b) = 2b \sum_{k \le L/b} k \left(\log \frac{(k+1)b - 1}{kb - 1} + \frac{1}{2} F_1(k, b) \right)$$

$$+ 2b \sum_{j=1}^{N} \frac{B_{2j}}{2j} \sum_{k \le L/b} k F_{2j}(k, b) + 2br_N \left(f, \frac{L}{b} \right), \tag{27}$$

where the function f satisfies

$$f(u) = \begin{cases} \frac{1}{u}, & \text{if } u \ge 1 \\ 0, & \text{if } u = 0 \end{cases}$$

and $f \in C^\infty([0, \infty))$ with $f^{(j)}(0) = 0$ for $j \le 2N + 1$.

The new form of $S(L; b)$ from (27) leads essentially to the proof of Theorem 2.3.

Furthermore, H. Maier and M. Th. Rassias obtained even more interesting results concerning $c_0(\frac{r}{b})$ for a fixed arbitrary positive integer value of r and for large integer values of b, which give us a deeper understanding of our cotangent sum for almost all values of r and b.

Proposition 2.2 *For $r, b \in \mathbb{N}$ with $(r, b) = 1$, it holds that*

$$c_0 \left(\frac{r}{b} \right) = \frac{1}{r} c_0 \left(\frac{1}{b} \right) - \frac{1}{r} Q \left(\frac{r}{b} \right), \tag{28}$$

where

$$Q\left(\frac{r}{b}\right) = \sum_{m=1}^{b-1} \cot\left(\frac{\pi m r}{b}\right)\left\lfloor\frac{rm}{b}\right\rfloor.$$

Theorem 2.6 *Let $r, b_0 \in \mathbb{N}$ be fixed, with $(b_0, r) = 1$. Let b denote a positive integer with $b \equiv b_0 \pmod r$. Then, there exists a constant $C_1 = C_1(r, b_0)$, with $C_1(1, b_0) = 0$, such that*

$$c_0\left(\frac{r}{b}\right) = \frac{1}{\pi r} b \log b - \frac{b}{\pi r}(\log 2\pi - \gamma) + C_1 b + O(1) \tag{29}$$

for large integer values of b.

We would like to mention that one could also prove the result of Theorem 2.6 via the techniques introduced by Bettin and Conrey.

2.1 Ellipse

It is interesting to mention that if one examines the graph of $c_0\left(\frac{r}{b}\right)$ for hundreds of integer values of b by the use of MATLAB, the resulting Figs. 1 and 2 always have a shape similar to an ellipse.

In 2014 H. Maier and M. Th. Rassias (and M. Th Rassias in his PhD thesis [23]) [18] tried to explain this phenomenon and obtained an important result (later

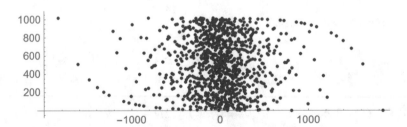

Fig. 1 axis $Ox : r$, axis $Oy : c_0, b = 1021$

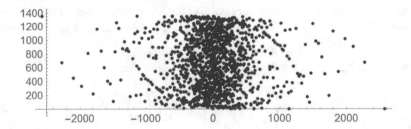

Fig. 2 axis $Ox : r$, axis $Oy : c_0, b = 1357$

generalized by S. Bettin in [6]), which establishes the equidistribution of certain normalized cotangent sums with respect to a positive measure. The result of Maier and Rassias is presented in the following theorem.

Definition 2.3 For $z \in \mathbb{R}$, let

$$F(z) = meas\{x \in [0, 1] : g(x) \le z\},$$

where "meas" denotes the Lebesgue measure,

$$g(x) := \sum_{l=1}^{+\infty} \frac{1 - 2\{lx\}}{l},$$

and

$$C_0(\mathbb{R}) = \{f \in C(\mathbb{R}) : \forall \varepsilon > 0, \exists \text{ a compact set } \mathcal{K} \subset \mathbb{R}, \text{ such that } |f(x)| < \varepsilon, \forall x \notin \mathcal{K}\}.$$

Remark The convergence of the above series has been investigated by R.Bretèche and G.Tenenbaum (see Theorem 4.2). It depends on the partial fraction expansion of the number x.

Theorem 2.7

(i) *F is a continuous function of z.*
(ii) *Let A_0, A_1 be fixed constants, such that $1/2 < A_0 < A_1 < 1$. Let also*

$$H_k = \int_0^1 \left(\frac{g(x)}{\pi}\right)^{2k} dx,$$

so H_k is a positive constant depending only on k, $k \in \mathbb{N}$.

There is a unique positive measure μ on $\mathbb{R}$ with the following properties:

(a) *For $\alpha < \beta \in \mathbb{R}$ we have*

$$\mu([\alpha, \beta]) = (A_1 - A_0)(F(\beta) - F(\alpha)).$$

(b)

$$\int x^k d\mu = \begin{cases} (A_1 - A_0)H_{k/2}, & \text{for even } k \\ 0, & \text{otherwise.} \end{cases}$$

(c) *For all $f \in C_0(\mathbb{R})$, we have*

$$\lim_{b \to +\infty} \frac{1}{\phi(b)} \sum_{\substack{r:(r,b)=1, \\ A_0 b \le r \le A_1 b}} f\left(\frac{1}{b} c_0\left(\frac{r}{b}\right)\right) = \int f d\mu,$$

where $\phi(\cdot)$ denotes the Euler phi-function.

As mentioned above, this result was later generalized by the use of a different method by S. Bettin in [6].

Outline of the proof of Theorem 2.7. In [18] H. Maier and M. Th. Rassias proved Theorem 2.3, which constitutes an improvement of their earlier Theorem 2.2.

Additionally, they investigated the cotangent sum $c_0\left(\frac{r}{b}\right)$ for a fixed arbitrary positive integer value of r and for large integer values of b and proved Theorem 2.6 as well as the following results

Theorem 2.8 *Let $k \in \mathbb{N}$ be fixed. Let also A_0, A_1 be fixed constants such that $\frac{1}{2} < A_0 < A_1 < 1$. Then there exists a constant $E_k > 0$, depending only on k, such that*

(a)

$$\sum_{\substack{r:(r,b)=1 \\ A_0 b = r = A_1 b}} Q\left(\frac{r}{b}\right)^{2k} = E_k(A_1^{2k+1} - A_0^{2k+1})b^{4k}\phi(b)(1 + o(1)) \quad (b \to +\infty),$$

(b)

$$\sum_{\substack{r:(r,b)=1 \\ A_0 b = r = A_1 b}} Q\left(\frac{r}{b}\right)^{2k-1} = o(b^{4k-2}\phi(b)) \quad (b \to +\infty),$$

(c)

$$\sum_{\substack{r:(r,b)=1 \\ A_0 b = r = A_1 b}} c_0\left(\frac{r}{b}\right)^{2k} = H_k(A_1 - A_0)b^{2k}\phi(b)(1 + o(1)) \quad (b \to +\infty),$$

(d)

$$\sum_{\substack{r:(r,b)=1 \\ A_0 b = r = A_1 b}} c_0\left(\frac{r}{b}\right)^{2k-1} = o(b^{2k-1}\phi(b)) \quad (b \to +\infty),$$

with

$$E_k = \frac{H_k}{(2k+1)}.$$

Using the method of moments, one can deduce detailed information about the distribution of the values of $c_0\left(\frac{r}{b}\right)$, where $A_0 b \leq r \leq A_1 b$ and $b \to +\infty$. Namely, one can prove Theorem 2.7.

3 The Maximum of c_0 in Rational Numbers in Short Intervals

In this section we consider some results about the maximum of c_0 in rational numbers in short intervals. More precicely, consider the following definition:

Definition 3.1 Let $0 < A_0 < 1, 0 < C < 1/2$. For $b \in \mathbb{N}$ we set

$$\Delta := \Delta(b, C) = b^{-C}.$$

We define

$$M(b, C, A_0) := \max_{A_0 b \leq r < (A_0 + \Delta) b} \left| c_0\left(\frac{r}{b}\right) \right|.$$

In [19] H. Maier and M. Th. Rassias proved the following theorems.

Theorem 3.1 Let D satisfy $0 < D < \frac{1}{2} - C$. Then we have for sufficiently large b:

$$M(b, C, A_0) \geq \frac{D}{\pi} b \log b.$$

An important part of the proof is the following key proposition.

Proposition 3.1 Let $\langle a_0; a_1, a_2, \ldots, a_n \rangle$ be the continued fraction expansion of $\frac{\bar{r}}{b} \in \mathbb{Q}$. Moreover, let $\frac{u_l}{v_l}$ be the l-th partial quotient of $\frac{\bar{r}}{b}$. Then

$$c_0\left(\frac{r}{b}\right) = -b \sum_{1 \leq l \leq n} \frac{(-1)^l}{v_l} \left(\left(\frac{1}{\pi v_l}\right) + \psi\left(\frac{v_{l-1}}{v_l}\right) \right).$$

Here ψ is an analytic function satisfying

$$\psi(x) = -\frac{\log(2\pi x) - \gamma}{\pi x} + O(\log x), \quad (x \to 0).$$

The proposition was proven in [6] by S. Bettin.

Definition 3.2 Let Δ be as in Definition 3.1 and $\Omega > 0$. We set

$$N(b, \Delta, \Omega) := \#\{r : A_0 b \leq r < (A_0 + \Delta)b, \ |\bar{r}| = \Omega b\}.$$

Another key proposition, proven in [19], is the following:

Proposition 3.2 *Let $\varepsilon < 0$ be such that*

$$D + \varepsilon < \frac{1}{2} - C.$$

Set

$$\Omega := b^{-(D+\varepsilon)}.$$

Then for sufficiently large b it holds

$$N(b, \Delta, \Omega) > 0.$$

Let $\left\{ \frac{u_i}{v_i} \right\}_{i=1}^{s}$ be the sequence of partial fractions of such $\frac{\bar{r}}{b}$. From

$$\Omega \geq \frac{\bar{r}}{b} \geq \frac{1}{v_1 + 1}$$

we obtain

$$v_1 + 1 \geq \Omega^{-1}.$$

Then by Proposition 3.2 we have

$$\sum_{l>1} \left(\left(\frac{1}{\pi v_l} \right) + \psi \left(\frac{v_{l-1}}{v_l} \right) \right) < 2\varepsilon \log b, \ \text{ for } b \geq b_0(\varepsilon).$$

Therefore,

$$\left| c_0 \left(\frac{r}{b} \right) \right| \geq \frac{2}{\pi} \log(\Omega^{-1}(1 + o(1))) \ \ b \to +\infty.$$

This proves Theorem 3.1.

Theorem 3.2 *Let C be as in Theorem 3.1 and let D satisfy $0 < D < 2 - C - E$, where $E \geq 0$ is a fixed constant. Let B be sufficiently large. Then we have*

$$M(b, C, A_0) \leq \frac{D}{\pi} b \log b$$

for all b with $B \leq b < 2B$, with at most B^E exceptions.

Proof We will need the following proposition.

Proposition 3.3 *Let $\varepsilon > 0$, $B \geq B\varepsilon$, $B < b \leq 2B$. For $1 \leq r < b$, $(r, b) = 1$, let $\{\frac{u_i}{v_i}\}_{i=1}^{s}$ be the sequence of partial fractions of $\frac{r}{b}$. Then there are at most 3 values of l for which*

$$\frac{1}{v_l} \psi \left(\frac{v_{l-1}}{v_l} \right) \geq \log \log b,$$

and at most one value of l for which

$$\frac{1}{v_l} \psi \left(\frac{v_{l-1}}{v_l} \right) \geq \varepsilon \log b.$$

Proof Let $l_i \, (i = 1, 2, 3, 4)$ be such that

$$\frac{1}{v_l} \psi \left(\frac{v_{l-1}}{v_l} \right) \geq \log \log b.$$

Then we have

$$v_{l_1} \geq \log \log b, \quad v_{l_2} \geq \exp(v_{l_1}) \geq \log b,$$

$$v_{l_3} \geq \exp(v_{l_2}) \geq b, \quad v_{l_4} \geq \exp(v_{l_3}) \geq \exp(b),$$

in contradiction to $v_s \leq b$.

In the same manner we obtain from $v_{l_j} \geq \varepsilon \log b$, $j = 1, 2$:

$$v_s \geq \exp(\exp((\log b)^{\varepsilon})) > b.$$

Assume $\varepsilon > 0$ to be fixed but arbitrarily small, $Z > 0$ fixed but arbitrarily large.

Definition 3.3 By Proposition 3.3 there is at most one value of l for which

$$\frac{1}{v_l} \psi \left(\frac{v_{l-1}}{v_l} \right) \geq \varepsilon \log b.$$

In case of the existence of l, we write

$$u_{l-1}(r, b) = u_{l-1}$$

and

$$v_{l-1}(r, b) = v_{l-1}.$$

Then for s, t with $1 \leq s, t = Z$, $(s, t) = 1$, and for fixed θ with $0 < \theta < 1$,

$\mathcal{F}(s, t)$:

$$= \left\{ (b, r, \bar{r}) : B \leq b < 2B, \ A_0 b \leq r = (A_0 + \Delta) b, \ \left| \frac{\bar{r}}{b} - \frac{s}{t} \right| \leq \theta, \ r\bar{r} \equiv 1 (\text{mod } b) \right\}.$$

Now we can formulate a proposition.

Proposition 3.4

$$\sum_{B \leq b < 2B} N(b, \Delta, \Omega) \leq \sum_{1 \leq s, t \leq Z} |\mathcal{F}(s, t)|$$

By Dirichlet's approximation theorem there is $(C_0, D_0) \in \mathbb{Z}^2$ with $1 \leq D_0 \leq B^2$, $(C_0, D_0) = 1$, such that

$$\left| A_0^{-1} - \frac{C_0}{D_0} \right| \leq \frac{1}{D_0 B^2}. \tag{30}$$

Let us estimate the cardinality of the set $\mathcal{F}(s, t)$.
From $A_0 b \leq r = (A_0 + \Delta) b$, the definition of $\bar{r}$, and (30), we obtain

$$r\bar{r} = y \left(\frac{C_0}{D_0} r + \frac{u}{D_0} \right), \text{ with } y \in \mathbb{Z}, \tag{31}$$

which after multiplication by $C_0 D_0$ becomes

$$(C_0 y - D_0 \bar{r})(C_0 r - u) = -D_0 (C_0 - \bar{r} u). \tag{32}$$

If $C_0 - \bar{r} u \neq 0$, one can deduce from the well-known estimate for the number of divisors of an integer that for a given pair $(\bar{r}, u)$, there are at most $O(B^\varepsilon)$ pairs (r, y) such that (32) holds.
There are at most $O(B^\varepsilon)$ pairs $(\bar{r}, u)$ such that $C_0 - \bar{r} u = 0$. Thus we obtain

$$|\mathcal{F}(s, t)| = O(B^{2 + 2\varepsilon - C} \theta). \tag{33}$$

From Proposition 3.4 and (33) we obtain for $\Omega = \theta B$

$$\sum_{B \leq b < 2B} N(b, \Delta, \Omega) = O(B^{2 + 2\varepsilon - C} \theta). \tag{34}$$

We now apply (34) with $\theta = B - D_0$, where

$$D > D' > 2 - C - E.$$

If we choose $\varepsilon > 0$ sufficiently small, then we conclude from (34) the following:
For all b with $B \leq b < 2B$ we have, with at most B^E exceptions:

$$N(b, B^{1-C}, \Omega) = 0.$$

Thus,

$$\frac{1}{v_l} \psi \left(\frac{v_{l-1}}{v_l} \right) \leq \frac{D'}{\pi v_{l-1}} b \log b (1 + o(1)), \ \forall l \leq Z. \tag{35}$$

The result of Theorem 3.2 follows now from Propositions 3.1 and 3.3.

4   The Function $g(x)$ and Moments of c_0

There is an interesting connection between the cotangent sums c_0 and the function

$$g(x) := \sum_{l=1}^{+\infty} \frac{1 - 2\{lx\}}{l}, \tag{36}$$

which, as we mentioned above, naturally appeared in the investigation of the moments of c_0 and of the sum $Q\left(\frac{r}{b}\right)$, which is related to c_0 by Proposition 2.2. To be precise, this series is related to c_0 by the important Theorem 2.8.

Later S. Bettin in his paper [6] extended the result of Theorem 2.8 and proved the following:

Theorem 4.1 *Let $b \geq 1$ and $k \geq 0$. Then*

$$\frac{1}{\phi(b)} \sum_{\substack{r=1 \\ (r,b)=1}}^{b} c_0\left(\frac{r}{b}\right)^k = H_k b^k + O_\varepsilon (b^{k-1+\varepsilon} (Ak \log b)^{2k}), \tag{37}$$

for some absolute constant $A > 0$ and any $\varepsilon > 0$.
  Moreover, if $0 \leq A_0 < A_1 \leq 1$, then we have

$$\frac{1}{\phi(b)} \sum_{\substack{(r,b)=1 \\ A_0 < \frac{r}{b} < A_1}} c_0\left(\frac{r}{b}\right)^k = (A_1 - A_0) H_k b^k + O_\varepsilon (b^{k-\frac{1}{2}+\varepsilon} (Ak \log b)^{2k}). \tag{38}$$

The function $g(x)$ is interesting not only in connection to the study of the cotangent sums c_0, but also in its own right. For example, it is also studied in [10] by R. Bretèche and G. Tenenbaum.

Theorem 4.2 *For each $x \in \mathbb{Q}$ the series $g(x)$ converges.*
  For $x \in \mathbb{R} \setminus \mathbb{Q}$, the series $g(x)$ converges if and only if the series

$$\sum_{m \geq 1} (-1)^m \frac{\log q_{m+1}}{q_m}$$

converges, where $(q_m)_{m \geq 1}$ *denotes the sequence of partial denominators of the continued fraction expansion of* x.

Proof The statement of the theorem is part of Theorem 4.4 of the paper by R. Bretèche and G. Tenenbaum in [10].

The function $g(x)$ also has the following property:

Theorem 4.3 *The series*

$$g(x) = \sum_{l=1}^{+\infty} \frac{1 - 2\{lx\}}{l}$$

converges almost everywhere in $[0, 1)$.

The function $g(x)$ was also of interest to L. Báez-Duatre, M. Balazard, B. Landreau and E. Saias. In [1] they studied the function

$$A(\lambda) := \int_0^{+\infty} \{t\}\{\lambda t\} \frac{dt}{t^2}.$$

and proved the following theorem.

Theorem 4.4 *Let* $\lambda > 0$ *be such that the series* $g(\lambda)$ *converges. Then the series* $g\left(\frac{1}{\lambda}\right)$ *converges too, and we have:*

$$A(\lambda) = \frac{1 - \lambda}{2} \log \lambda + \frac{\lambda + 1}{2}(\log 2\pi - \gamma) - g(\lambda) - \lambda g\left(\frac{1}{\lambda}\right).$$

Now let us show an important property of $g(x)$, which was proven in [20].

Theorem 4.5 *There are constants* $c_1, c_2 > 0$, *such that*

$$c_1 \Gamma(2k + 1) \leq \int_0^1 g(x)^{2k} dx \leq c_2 \Gamma(2k + 1)$$

for all $k \in \mathbb{N}$, *where* $\Gamma(\cdot)$ *stands for the gamma function.*

Sketch of a proof Let us consider the continued fraction expansion of x

$$x = [a_0(x); a_1(x), \ldots, a_k(x), \ldots].$$

The $a_k(x)$ are obtained via the Gauss map α, defined by

$$\alpha(x) = \left\{ \frac{1}{x} \right\}, \alpha_k(x) = \alpha(\alpha_{k-1}(x)), a_k(x) = \left\lfloor \frac{1}{\alpha_{k-1}(x)} \right\rfloor.$$

Definition 4.1 Let $x \in X = (0, 1) \backslash \mathbb{Q}$. Let also

$$\beta_k(x) = \alpha_0(x)\alpha_1(x)\dots\alpha_k(x), \quad \beta_{-1}(x) = 1,$$

$$\gamma_k(x) = \beta_{k-1}(x) \log \frac{1}{\alpha_k(x)}, \text{ where } k \geq 0,$$

so that $\gamma_0(x) = \log \frac{1}{x}$.

The number x is called a Wilton number if the series

$$\sum_{k \geq 0} (-1)^k \gamma_k(x)$$

converges.

Wilton's function $\mathcal{W}$ is defined by

$$\mathcal{W} = \sum_{k \geq 0} (-1)^k \gamma_k(x)$$

for each Wilton number x.

M. Balazard and B. Martin proved in [4] the following proposition:

Proposition 4.1 *There is a bounded function* $H : (0, 1) \to \mathbb{R}$*, which is continuous at every irrational number, such that*

$$g(x) = \mathcal{W}(x) + H(x) \tag{39}$$

almost everywhere. Also a number $x \in X$ *is a Wilton number if and only if* $\alpha(x)$ *is a Wilton number. In this case, we have:*

$$\mathcal{W}(x) = l(x) - T\mathcal{W}(x), \tag{40}$$

where

$$l(x) = \log \frac{1}{x}$$

and the operator T *is defined by*

$$Tf(x) = xf(\alpha(x)).$$

One can express (40) as

$$l(x) = (1 + T)\mathcal{W}(x). \tag{41}$$

The main idea in the evaluation of

$$\int_0^1 g(x)^{2k} dx$$

is to solve the operator equation (41) for $\mathcal{W}(x)$, which is:

$$\mathcal{W}(x) = (1 + T)^{-1} l(x). \tag{42}$$

An idea which has long been used in functional analysis for the case when T is a differential operator is to express the right-hand side of (42) as a Neumann series, which is obtained by the geometric series identity, i.e.

$$(1 + T)^{-1} = \sum_{k=0}^{+\infty} (-1)^k T^k.$$

Thus one can approximate $\mathcal{W}(x)$ by

$$\mathcal{L}(x, n) = \sum_{k=0}^{n} (-1)^k (T^k l)(x). \tag{43}$$

Definition 4.2 The measure m is defined by

$$m(\mathcal{E}) = \frac{1}{\log 2} \int_{\mathcal{E}} \frac{dx}{1 + x},$$

where $\mathcal{E}$ is any measurable subset of $(0, 1)$.

Proposition 4.2 *For $f \in L^p$ we have*

$$\int_0^1 |T^n f(x)|^p dm(x) \leq g^{(n-1)p} \int_0^1 |f(x)|^p dm(x),$$

where

$$g = \frac{\sqrt{5} - 1}{2} < 1.$$

Proof Marmi, Moussa, and Yoccoz, in their paper [21], consider a generalized continued fraction algorithm, depending on a parameter α, which becomes the usual continued fraction algorithm for the choice $\alpha = 1$. The operator T_v is defined in (2.5) of [21] and becomes T for $\alpha = 1$, $v = 1$. Then, Proposition 4.2 is the content of formulas (2.14), (2.15) of [21].

Using standard techniques of functional analysis one can prove that

$$\lim_{n \to \infty} \int_0^{1/2} \left| \mathcal{L}(x, n)^L - \mathcal{W}(x)^L \right| dx = 0.$$

One can eventually prove that

$$\int_0^{1/2} \mathcal{L}(x, n)^{2k} dx = \left(\int_0^{1/2} l(x)^{2k} \right) (1 + o(1)) = \Gamma(2k + 1)(1 + o(1)), \quad (k \to +\infty).$$

The order of magnitude of

$$\int_0^{1/2} g(x)^{2k} dx$$

now follows from (39), by the binomial theorem, since $H(x)$ is a bounded function.

Corollary 4.1 *The series*

$$\sum_{k \geq 0} \frac{H_k}{(2k)!} x^k$$

has radius of convergence π^2.

H. Maier and M. Th. Rassias proved in [14] an improvement of Theorem 4.5, by establishing an asymptotic result for the corresponding integral. Namely, they proved the following theorem.

Theorem 4.6 *Let*

$$A = \int_0^\infty \frac{\{t\}^2}{t^2} dt$$

and $K \in \mathbb{N}$. There is an absolute constant $C > 0$ such that

$$\int_0^1 |g(x)|^K dx = 2e^{-A}\Gamma(K+1)(1+O(\exp(-CK)))$$

for $K \to \infty$.

In [15], they improved this result settling also the general case of arbitrary exponents K.

Theorem 4.7 *Let $K \in \mathbb{R}$, $K > 0$. There is an absolute constant $C > 0$ such that*

$$\int_0^1 |g(x)|^K dx = \frac{e^{\gamma}}{\pi}\Gamma(K+1)(1+O(\exp(-CK)))$$

for $K \to \infty$, where γ is the Euler-Mascheroni constant.

5 Dedekind Sums

Dedekind sums have applications in many fields of mathematics, especially in number theory. These sums appear in R. Dedekind's study of the function

$$\eta(s) = e^{\frac{\pi i s}{12}} \prod_{m=1}^{\infty} (1 - e^{2\pi i m s}), \tag{44}$$

where Im $s > 0$.

Definition 5.1 Let r, b be integers, $(r, b) = 1, k \geq 1$. Then the Dedekind sum $s\left(\frac{r}{b}\right)$ is defined as follows

$$s\left(\frac{r}{b}\right) := \sum_{\mu=1}^{b} \left(\left(\frac{r\mu}{b}\right)\right)\left(\left(\frac{\mu}{b}\right)\right), \tag{45}$$

where $((\cdot))$ is the sawtooth function defined as follows:

$$((x)) := \begin{cases} x - [x] - \frac{1}{2} & \text{if } x \text{ is not an integer,} \\ 0 & \text{if } x \text{ is an integer.} \end{cases} \tag{46}$$

It is a fascinating fact that the Dedekind sum can also be expressed as a sum of cotangent products:

Proposition 5.1

$$s\left(\frac{r}{b}\right) = -\frac{1}{4b} \sum_{m=1}^{b-1} \cot\left(\frac{\pi m}{b}\right) \cot\left(\frac{\pi mr}{b}\right).$$

It is a well-known fact that Dedekind sums satisfy a reciprocity formula:

Theorem 5.1 (Dedekind sums' reciprocity formula)

$$s\left(\frac{r}{b}\right) + s\left(\frac{b}{r}\right) - \frac{1}{12rb} = \frac{1}{12}\left(\frac{r}{b} + \frac{b}{r} - 3\right). \tag{47}$$

We shall not present the proof of Proposition 5.1 and of Theorem 5.1. The interested reader can find these proofs, as well as more fundamental facts concerning Dedekind sums in the famous book by Rademacher and Grosswald [12].

It is interesting to study the relation between the cotangent sums c_0 and the Dedekind sums. We've already considered c_a for arbitrary $a \in \mathbb{C}$ (see (20)). S. Bettin in [5] proved the very interesting result that c_{-1} is a Dedekind sum up to a constant:

Proposition 5.2 *It holds that*

$$s\left(\frac{r}{b}\right) = \frac{1}{2\pi}c_{-1}\left(\frac{r}{b}\right). \tag{48}$$

6 Sums Appearing in the Nyman-Beurling Criterion for the Riemann Hypothesis Containing the Möbius Function

In a recent paper [16] H. Maier and M. Th. Rassias investigated the following sums (36)

$$\sum_{n\in I} \mu(n)g\left(\frac{n}{b}\right), \tag{49}$$

for a suitable interval I, where $\mu(n)$ is the Möbius function and the function $g(x)$ is as defined in Definition 2.3.

These sums appear in the study of the integral

$$\int_{-\infty}^{+\infty} \left|\zeta\left(\frac{1}{2} + it\right)\right|^2 \left|D_N\left(\frac{1}{2} + it\right)\right|^2 \frac{dt}{\frac{1}{4} + t^2} \tag{50}$$

Particularly we could express (50) using formulas (6) and (7) and Proposition 1.1, as follows:

$$\int_{-\infty}^{+\infty} \left| \zeta \left(\frac{1}{2} + it \right) \right|^2 \left| D_N \left(\frac{1}{2} + it \right) \right|^2 \frac{dt}{\frac{1}{4} + t^2}$$

$$= \sum_{1 \le r, b \le N} \mu(r)\mu(b) \left(1 - \frac{\log r}{\log N} \right) \left(1 - \frac{\log b}{\log N} \right)$$

$$\times \left[\frac{\log 2\pi - \gamma}{2} \left(\frac{1}{r} + \frac{1}{b} \right) + \frac{b-r}{2rb} \log \frac{r}{b} - \frac{\pi}{2rb} \left(V \left(\frac{r}{b} \right) + V \left(\frac{b}{r} \right) \right) \right]$$

If we expand the last equation, we will obtain the following sum for fixed b

$$\sum_{n \in I} \mu(n) \left(1 - \frac{\log n}{\log N} \right) \frac{1}{n} V \left(\frac{n}{b} \right),$$

which is equal to

$$\sum_{n \in I} \mu(n) \left(1 - \frac{\log n}{\log N} \right) g \left(\frac{n}{b} \right).$$

In [16] H. Maier and M. Th. Rassias proved the following result concerning the sums (49):

Theorem 6.1 *Let $0 \le \delta \le D/2$, $b^{2\delta} \le B \le b^D$, where $b^{-\delta} \le \eta \le 1$. Then there is a positive constant β depending only on δ and D, such that*

$$\sum_{Bb \le n \le (1+\eta)Bb} \mu(n) g \left(\frac{n}{b} \right) = O((\eta Bb)^{1-\beta}). \tag{51}$$

Finally, the above result was recently improved by the same authors in [17], by proving the following theorem:

Theorem 6.2 *Let $D \ge 2$. Let C be the number which is uniquely determined by*

$$C \ge \frac{\sqrt{5}+1}{2}, \quad 2C - \log C - 1 - 2\log 2 = \frac{1}{2}\log 2.$$

Let v_0 be determined by

$$v_0 \left(1 - \left(1 + 2\log 2 \left(C + \frac{\log 2}{2} \right)^{-1} \right)^{-1} + 2 + \frac{4}{\log 2}C \right) = 2.$$

Let $z_0 := 2 - \left(2 + \frac{4}{\log 2} C\right) v_0$. Then for all $\varepsilon > 0$ we have

$$\sum_{b^D \leq n < 2b^D} \mu(n) g\left(\frac{n}{b}\right) \ll_\varepsilon b^{D-z_0+\varepsilon} .$$

References

1. L. Báez-Duatre, M. Balazard, B. Landreau, E. Saias, Etude de l'autocorrelation multiplicative de la fonction 'partie fractionnaire'. Ramanujan J. **9**, 215–240 (2005)
2. B. Bagchi, On Nyman, Beurling and Baez-Duarte's Hilbert space reformulation of the Riemann hypothesis. Proc. Indian Acad. Sci. Math. Sci. **116**(2), 137–146 (2006)
3. R. Balasubramanian, J. Conrey, D. Heath-Brown, Asymptoticmeansquare of the product of the Riemann zeta-function and a Dirichlet polynomial. J. Reine Angew. Math. **357**, 161–181 (1985)
4. M. Balazard, B. Martin, Sur l'autocorrélation multiplicative de la fonction "partie fractionnaire" et une fonction définie par. J.R. Wilton, arXiv:1305.4395v1
5. S. Bettin, A generalization of Rademacher's reciprocity law. Acta Arithmetica **159**(4), 363–374 (2013)
6. S. Bettin, On the distribution of a cotangent sum. Int. Math. Res. Not. **2015**(21), 11419–11432 (2015)
7. S. Bettin, J. Conrey, A reciprocity formula for a cotangent sum. Int. Math. Res. Not. **2013**(24), 5709–5726 (2013)
8. S. Bettin, J. Conrey, Period functions and cotangent sums. Algebra Number Theory **7**(1), 215–242 (2013)
9. S. Bettin, J. Conrey, D. Farmer, An optimal choice of Dirichlet polynomials for the Nyman-Beurling criterion. (in memory of Prof. A. A. Karacuba), arXiv:1211.5191
10. R. de la Bretèche, G. Tenenbaum, Series trigonometriques à coefficients arithmetiques. J. Anal. Math. **92**, 1–79 (2004)
11. A. Fokas, A novel approach to the Lindelöf hypothesis. arXiv:1708.06607v4 [math.CA]
12. E. Grosswald, H. Rademacher, Dedekind sums. Mathematical Association of America (1972). https://doi.org/10.5948/UPO9781614440161
13. M. Ishibashi, The value of the Estermann zeta function at s = 0. Acta Arith. **73**(4), 357–361 (1995)
14. H. Maier, M.Th. Rassias, Asymptotics for moments of certain cotangent sums. Houst. J. Math. **43**(1), 207–222 (2017)
15. H. Maier, M.Th. Rassias, Asymptotics for moments of certain cotangent sums for arbitrary exponents. Houst. J. Math. **43**(4), 1235–1249 (2017)
16. H. Maier, M.Th. Rassias, Estimates of sums related to the Nyman-Beurling criterion for the Riemann Hypothesis. J. Number Theory **188**, 96–120 (2018)
17. H. Maier, M.Th. Rassias, Explicit estimates of sums related to the Nyman-Beurling criterion for the Riemann Hypothesis. J. Funct. Anal. (in press). https://doi.org/10.1016/j.jfa.2018.06.022
18. H. Maier, M.Th. Rassias, Generalizations of a cotangent sum associated to the Estermann zeta function. Commun. Contemp. Math. **18**(1), 1550078 (2016)
19. H. Maier, M.Th. Rassias, The maximum of cotangent sums related to Estermann's zeta function in rational numbers in short intervals. Applicable Anal. Discret. Math. **11**, 166–176 (2017)
20. H. Maier, M.Th. Rassias, The order of magnitude for moments for certain cotangent sums. J. Math. Anal. Appl. **429**, 576–590 (2015)

21. S. Marmi, P. Moussa, J.-C. Yoccoz, The Brjuno functions and their regularity properties. Commun. Math. Phys. **186**, 265–293 (1997)
22. M.Th. Rassias, On a cotangent sum related to zeros of the Estermann zeta function. Appl. Math. Comput. **240**, 161–167 (2014)
23. M.Th. Rassias, Analytic investigation of cotangent sums related to the Riemann zeta function, Doctoral Dissertation, ETH-Zürich (2014)

Recent Progress in the Study of Polynomials with Constrained Coefficients

Tamás Erdélyi

Abstract This survey gives a taste of the author's recent work on polynomials with constrained coefficients. Special attention is paid to unimodular, Littlewood, Newman, Rudin-Shapiro, and Fekete polynomials, their flatness and ultraflatness properties, their L_q norms on the unit circle including Mahler's measure, and bounds on the number of unimodular zeros of self-reciprocal polynomials with coefficients from a finite set of real numbers. Some interesting connections are explored, and a few conjectures are also made.

Keywords Unimodular · Littlewood · Newman · Rudin-Shapiro · Fekete · polynomials · L_q norms · Mahler's measure · Zeros

2010 Mathematics Subject Classifications. 11C08, 41A17, 26C10, 30C15

Notation

Let $\mathcal{P}_n$ be the set of all algebraic polynomials of degree at most n with real coefficients. Let $\mathcal{P}_n^c$ be the set of all algebraic polynomials of degree at most n with complex coefficients. Let

$$\mathcal{K}_n := \left\{ Q_n : Q_n(z) = \sum_{k=0}^{n} a_k z^k, \; a_k \in \mathbb{C}, \; |a_k| = 1 \right\}.$$

The class $\mathcal{K}_n$ is often called the collection of all (complex) unimodular polynomials of degree n. Let

T. Erdélyi (✉)
Department of Mathematics, Texas A&M University, College Station, TX, USA
e-mail: terdelyi@math.tamu.edu

© Springer Nature Switzerland AG 2020   29
A. Raigorodskii, M. T. Rassias (eds.), *Trigonometric Sums and Their Applications*,
https://doi.org/10.1007/978-3-030-37904-9_2

$$\mathcal{L}_n := \left\{ Q_n : Q_n(z) = \sum_{k=0}^{n} a_k z^k, \ a_k \in \{-1, 1\} \right\}.$$

The class $\mathcal{L}_n$ is often called the collection of all (real) unimodular polynomials of degree n. Let D denote the open unit disk of the complex plane. We will denote the unit circle of the complex plane by ∂D. We define the Mahler measure of Q (geometric mean of Q on ∂D) by

$$M_0(Q) := \exp\left(\frac{1}{2\pi} \int_0^{2\pi} \log |Q(e^{it})| \, dt \right)$$

for bounded measurable functions Q on ∂D. It is well known, see [76], for instance, that

$$M_0(Q) = \lim_{q \to 0+} M_q(Q),$$

where

$$M_q(Q) := \left(\frac{1}{2\pi} \int_0^{2\pi} \left| Q(e^{it}) \right|^q \, dt \right)^{1/q}, \qquad q > 0.$$

It is also well known that for a function Q continuous on ∂D we have

$$M_\infty(Q) := \max_{t \in [0, 2\pi]} |Q(e^{it})| = \max_{t \in \mathbb{R}} |Q(e^{it})| = \lim_{q \to \infty} M_q(Q).$$

It is a simple consequence of the Jensen formula that

$$M_0(Q) = |c| \prod_{k=1}^{n} \max\{1, |z_k|\}$$

for every polynomial of the form

$$Q(z) = c \prod_{k=1}^{n} (z - z_k), \qquad c, z_k \in \mathbb{C}.$$

We define the Mahler measure (geometric mean of Q on $[\alpha, \beta]$)

$$M_0(Q, [\alpha, \beta]) := \exp\left(\frac{1}{\beta - \alpha} \int_\alpha^\beta \log |Q(e^{it})| \, dt \right)$$

for $[\alpha, \beta] \subset \mathbb{R}$ and bounded measurable functions $Q(e^{it})$ on $[\alpha, \beta]$. It is well known, see [76], for instance, that

$$M_0(Q, [\alpha, \beta]) = \lim_{q \to 0+} M_q(Q, [\alpha, \beta]),$$

where, for $q > 0$,

$$M_q(Q, [\alpha, \beta]) := \left(\frac{1}{\beta - \alpha} \int_\alpha^\beta \left| Q(e^{it}) \right|^q dt \right)^{1/q}.$$

If $Q \in \mathcal{P}_n^c$ is of the form

$$Q(z) = \sum_{j=0}^n a_j z^j, \qquad a_j \in \mathbb{C},$$

then its conjugate polynomial is defined by

$$Q^*(z) := z^n \overline{Q}(1/z) := \sum_{j=0}^n \overline{a}_{n-j} z^j.$$

A polynomial $Q \in \mathcal{P}_n^c$ is called conjugate-reciprocal if $Q = Q^*$.

The Lebesgue measure of a measurable set $A \subset \mathbb{R}$ will be denoted by $m(A)$ throughout the paper.

1 Ultraflat Sequences of Unimodular Polynomials

By Parseval's formula,

$$\int_0^{2\pi} \left| P_n(e^{it}) \right|^2 dt = 2\pi(n+1)$$

for all $P_n \in \mathcal{K}_n$. Therefore

$$\min_{t \in \mathbb{R}} |P_n(e^{it})| \le \sqrt{n+1} \le \max_{t \in \mathbb{R}} |P_n(e^{it})|.$$

An old problem (or rather an old theme) is the following.

Problem 1.1 (Littlewood's Flatness Problem) How close can a polynomial $P_n \in \mathcal{K}_n$ or $P_n \in \mathcal{L}_n$ come to satisfying

$$|P_n(e^{it})| = \sqrt{n+1}, \qquad t \in \mathbb{R}? \tag{1}$$

Obviously (1) is impossible if $n \ge 1$. So one must look for less than (1), but then there are various ways of seeking such an "approximate situation". One way is the following. In his paper [87] Littlewood had suggested that, conceivably, there might exist a sequence (P_n) of polynomials $P_n \in \mathcal{K}_n$ (possibly even $P_n \in \mathcal{L}_n$) such that $(n + 1)^{-1/2}|P_n(e^{it})|$ converge to 1 uniformly in $t \in \mathbb{R}$. We shall call such sequences of unimodular polynomials "ultraflat". More precisely, we give the following definition.

Definition 1.2 Given a positive number ε, we say that a polynomial $P_n \in \mathcal{K}_n$ is ε-flat if

$$(1 - \varepsilon)\sqrt{n + 1} \le |P_n(e^{it})| \le (1 + \varepsilon)\sqrt{n + 1}, \qquad t \in \mathbb{R}.$$

Definition 1.3 Let (n_k) be an increasing sequence of positive integers. Given a sequence (ε_{n_k}) of positive numbers tending to 0, we say that a sequence (P_{n_k}) of polynomials $P_{n_k} \in \mathcal{K}_{n_k}$ is (ε_{n_k})-ultraflat if each P_{n_k} is (ε_{n_k})-flat. We simply say that a sequence (P_{n_k}) of polynomials $P_{n_k} \in \mathcal{K}_{n_k}$ is ultraflat if it is (ε_{n_k})-ultraflat with a suitable sequence (ε_{n_k}) of positive numbers tending to 0.

The existence of an ultraflat sequence of unimodular polynomials seemed very unlikely, in view of a 1957 conjecture of P. Erdős (Problem 22 in [68]) asserting that, for all $P_n \in \mathcal{K}_n$ with $n \ge 1$,

$$\max_{t \in \mathbb{R}} |P_n(e^{it})| \ge (1 + \varepsilon)\sqrt{n + 1}, \tag{2}$$

where $\varepsilon > 0$ is an absolute constant (independent of n). Yet, refining a method of Körner [83], Kahane [79] proved that there exists a sequence (P_n) with $P_n \in \mathcal{K}_n$ which is (ε_n)-ultraflat, where $\varepsilon_n = O\left(n^{-1/17}\sqrt{\log n}\right)$. (Kahane's paper contained though a slight error which was corrected in [100].) Thus the Erdős conjecture (2) was disproved for the classes $\mathcal{K}_n$. For the more restricted class $\mathcal{L}_n$ the analogous Erdős conjecture is unsettled to this date. It is a common belief that the analogous Erdős conjecture for $\mathcal{L}_n$ is true, and consequently there is no ultraflat sequence of polynomials $P_n \in \mathcal{L}_n$. An interesting result related to Kahane's breakthrough is given in [5]. For an account of some of the work done till the mid 1960s, see Littlewood's book [88] and [101].

Let (ε_n) be a sequence of positive numbers tending to 0. Let the sequence (P_n) of polynomials $P_n \in \mathcal{K}_n$ be (ε_n)-ultraflat. We write

$$P_n(e^{it}) = R_n(t)e^{i\alpha_n(t)}, \qquad R_n(t) = |P_n(e^{it})|, \qquad t \in \mathbb{R}. \tag{3}$$

It is simple to show that α_n can be chosen to be in $C^\infty(\mathbb{R})$. This is going to be our understanding throughout the paper. It is easy to find a formula for $\alpha_n(t)$ in terms of P_n. We have

$$\alpha_n'(t) = \text{Re}\left(\frac{e^{it}P_n'(e^{it})}{P_n(e^{it})}\right), \tag{4}$$

see formulas (7.1) and (7.2) on p. 564 and (8.2) on p. 565 in [104]. The angular function α_n^* and modulus function $R_n^* = R_n$ associated with the polynomial P_n^* are defined by

$$P_n^*(e^{it}) = R_n^*(t)e^{i\alpha_n^*(t)}, \qquad R_n^*(t) = |P_n^*(e^{it})|.$$

Similarly to α_n, the angular function α_n^* can also be chosen to be in $C^\infty(\mathbb{R})$ on $\mathbb{R}$. By applying formula (4) to P_n^*, it is easy to see that

$$\alpha_n'(t) + \alpha_n^{*\prime}(t) = n, \qquad t \in \mathbb{R}. \tag{5}$$

The structure of ultraflat sequences of unimodular polynomials is studied in [45–47], and [48], where several conjectures of Saffari are proved. These are closely related to each other.

Conjecture 1.4 (Uniform Distribution Conjecture for the Angular Speed) *Let (P_n) be a fixed ultraflat sequence of polynomials $P_n \in \mathcal{K}_n$. With the notation (3), in the interval $[0, 2\pi]$, the distribution of the normalized angular speed $\alpha_n'(t)/n$ converges to the uniform distribution as $n \to \infty$. More precisely, we have*

$$m(\{t \in [0, 2\pi] : 0 \le \alpha_n'(t) \le nx\}) = 2\pi x + o_n(x) \tag{6}$$

for every $x \in [0, 1]$, where $o_n(x)$ converges to 0 uniformly on $[0, 1]$. As a consequence, $|P_n'(e^{it})|/n^{3/2}$ also converges to the uniform distribution as $n \to \infty$. More precisely, we have

$$m(\{t \in [0, 2\pi] : 0 \le |P_n'(e^{it})| \le n^{3/2}x\}) = 2\pi x + o_n(x)$$

for every $x \in [0, 1]$, where $o_n(x)$ converges to 0 uniformly on $[0, 1]$.

The basis of this conjecture was that for the special ultraflat sequences of unimodular polynomials produced by Kahane [79], (6) is indeed true.

In Section 4 of [45] we prove this conjecture in general.

In the general case (6) can, by integration, be reformulated (equivalently) in terms of the moments of the angular speed $\alpha_n'(t)$. This was observed and recorded by Saffari [104]. For completeness the proof of this equivalence is presented in Section 4 of [45] and we settle Conjecture 1.4 by proving the following result.

Theorem 1.5 (Reformulation of the Uniform Distribution Conjecture) *Let (P_n) be a fixed ultraflat sequence of polynomials $P_n \in \mathcal{K}_n$. For any $q > 0$ we have*

$$\frac{1}{2\pi} \int_0^{2\pi} |\alpha_n'(t)|^q \, dt = \frac{n^q}{q+1} + o_{n,q}n^q. \tag{7}$$

with suitable constants $o_{n,q}$ converging to 0 for every fixed $q > 0$.

An immediate consequence of (7) is the remarkable fact that for large values of $n \in \mathbb{N}$, the $L_q(\partial D)$ Bernstein factors

$$\frac{\int_0^{2\pi} \left| P_n'(e^{it}) \right|^q \, dt}{\int_0^{2\pi} \left| P_n(e^{it}) \right|^q \, dt}$$

of the elements of ultraflat sequences (P_n) of unimodular polynomials are essentially independent of the polynomials. More precisely Theorem 1.5 implies the following result.

Theorem 1.6 (The Bernstein Factors) *Let q be an arbitrary positive real number. Let (P_n) be a fixed ultraflat sequence of polynomials $P_n \in \mathcal{K}_n$. We have*

$$\frac{\int_0^{2\pi} \left| P_n'(e^{it}) \right|^q \, , dt}{\int_0^{2\pi} \left| P_n(e^{it}) \right|^q \, dt} = \frac{n^{q+1}}{q+1} + o_{n,q} n^{q+1} \, ,$$

and as a limit case,

$$\frac{\max_{0 \le t \le 2\pi} \left| P_n'(e^{it}) \right|}{\max_{0 \le t \le 2\pi} \left| P_n(e^{it}) \right|} = n + o_n n \, .$$

with suitable constants $o_{n,q}$ and o_n converging to 0 for every fixed q.

In Section 3 of [45] we prove the following result which turns out to be stronger than Theorem 1.5.

Theorem 1.7 (Negligibility Theorem for Higher Derivatives) *Let (P_n) be a fixed ultraflat sequence of polynomials $P_n \in \mathcal{K}_n$. For every integer $r \ge 2$, we have*

$$\max_{0 \le t \le 2\pi} \left| \alpha_n^{(r)}(t) \right| = o_{n,r} n^r$$

with suitable constants $o_{n,r}$ converging to 0 for every fixed $r = 2, 3, \ldots$.

We show in Section 4 of [45] how Theorem 1.4 follows from Theorem 1.7.

In Section 4 of [45] we also prove an extension of Saffari's Uniform Distribution Conjecture 1.4 to higher derivatives.

Theorem 1.8 (Distribution of the Modulus of Higher Derivatives of Ultraflat Sequences of Unimodular Polynomials) *Let (P_n) be a fixed ultraflat sequence of polynomials $P_n \in \mathcal{K}_n$. The distribution of*

$$\left(\frac{\left| P_n^{(r)}(e^{it}) \right|}{n^{r+1/2}} \right)^{1/r}$$

converges to the uniform distribution as $n \to \infty$. More precisely, we have

$$m\left(\left\{ t \in [0, 2\pi] : 0 \le \left| P_n^{(r)}(e^{it}) \right| \le n^{r+1/2} x^r \right\} \right) = 2\pi x + o_{r,n}(x)$$

for every $x \in [0, 1]$, *where* $o_{r,n}(x)$ *converges to* 0 *uniformly for every fixed* $r = 1, 2, \ldots$.

In [51], based on the results in [45], we proved yet another conjecture of Queffelec and Saffari, see (1.30) in [101]. Namely we proved asymptotic formulas for the L_q norms of the real part and the derivative of the real part of ultraflat unimodular polynomials on the unit circle. A recent paper of Bombieri and Bourgain [9] is devoted to the construction of ultraflat sequences of unimodular polynomials. In particular, they obtained a much improved estimate for the error term. A major part of their paper deals also with the long-standing problem of the effective construction of ultraflat sequences of unimodular polynomials.

For $\lambda \geq 0$, let

$$\mathcal{K}_n^\lambda := \left\{ P_n : P_n(z) = \sum_{k=0}^n a_k k^\lambda z^k, \ a_k \in \mathbb{C}, \ |a_k| = 1 \right\}.$$

Ultraflat sequences (P_n) of polynomials $P_n \in \mathcal{K}_n^\lambda$ are defined and studied thoroughly in [67] where various extensions of Saffari's conjectures have been proved. Note that it is not yet known whether or not ultraflat sequences (P_n) of polynomials $P_n \in \mathcal{K}_n^\lambda$ exist for any $\lambda > 0$, in particular, it is not yet known for $\lambda = 1$.

In [47] we examined how far an ultraflat unimodular polynomial is from being conjugate-reciprocal, and we proved the following three theorems.

Theorem 1.9 *Let* (P_n) *be a fixed ultraflat sequence of polynomials* $P_n \in \mathcal{K}_n$. *We have*

$$\frac{1}{2\pi} \int_0^{2\pi} \left(|P_n'(e^{it})| - |P_n^{*\prime}(e^{it})| \right)^2 dt = \left(\frac{1}{3} + \gamma_n \right) n^3,$$

with some constants γ_n *converging to* 0.

Theorem 1.10 *Let* (P_n) *be a fixed ultraflat sequence of polynomials* $P_n \in \mathcal{K}_n$. *If*

$$P_n(z) = \sum_{k=0}^n a_{k,n} z^k, \qquad k = 0, 1, \ldots, n, \quad n = 1, 2, \ldots,$$

then

$$\sum_{k=0}^n k^2 |a_{k,n} - \bar{a}_{n-k,n}|^2 = \frac{1}{2\pi} \int_0^{2\pi} \left| (P_n' - P_n^{*\prime})(e^{it}) \right|^2 dt \geq \left(\frac{1}{3} + h_n \right) n^3,$$

with some constants h_n *converging to* 0.

Theorem 1.11 *Let* (P_n) *be a fixed ultraflat sequence of polynomials* $P_n \in \mathcal{K}_n$. *Using the notation of Theorem 1.10 we have*

$$\sum_{k=0}^{n} |a_{k,n} - \bar{a}_{n-k,n}|^2 = \frac{1}{2\pi} \int_0^{2\pi} \left| (P_n - P_n^*)(e^{it}) \right|^2 dt \geq \left(\frac{1}{3} + h_n \right) n,$$

with some constants h_n (the same as in Theorem 1.10) converging to 0.

There are quite a few recent publications on or related to ultraflat sequences of unimodular polynomials. Some of them (not mentioned before) are [10, 95, 100, 105], and [92].

2   More Recent Results on Ultraflat Sequences of Unimodular Polynomials

In a recent paper [64] we revisited the topic. Theorems 2.1–2.4 and 2.6 are new in [64], and Theorems 2.5 and 2.7 recapture old results.

In our results below Γ denotes the usual gamma function, and the $\sim$ symbol means that the ratio of the left and right hand sides converges to 1 as $n \to \infty$.

Theorem 2.1 *If (P_n) is an ultraflat sequence of polynomials $P_n \in \mathcal{K}_n$ and $q \in (0, \infty)$, then*

$$\frac{1}{2\pi} \int_0^{2\pi} \left| (P_n - P_n^*)(e^{it}) \right|^q dt \sim \frac{2^q \Gamma\left(\frac{q+1}{2}\right)}{\Gamma\left(\frac{q}{2} + 1\right) \sqrt{\pi}} \, n^{q/2}.$$

Our next theorem is a special case ($q = 2$) of Theorem 2.1. Compare it with Theorem 1.11.

Theorem 2.2 *Let (P_n) be an ultraflat sequence of polynomials $P_n \in \mathcal{K}_n$. If*

$$P_n(z) = \sum_{k=0}^{n} a_{k,n} z^k, \qquad k = 0, 1, \dots, n, \quad n = 1, 2, \dots,$$

then

$$\sum_{k=0}^{n} |a_{k,n} - \bar{a}_{n-k,n}|^2 = \frac{1}{2\pi} \int_0^{2\pi} \left| (P_n - P_n^*)(e^{it}) \right|^2 dt \sim 2n.$$

Our next theorem should be compared with Theorem 1.10.

Theorem 2.3 *Let (P_n) be an ultraflat sequence of polynomials $P_n \in \mathcal{K}_n$. Using the notation in Theorem 2.2 we have*

$$\sum_{k=0}^{n} k^2 \left| a_{k,n} - \bar{a}_{n-k,n} \right|^2 = \frac{1}{2\pi} \int_0^{2\pi} \left| (P_n' - P_n^{*\prime})(e^{it}) \right|^2 dt \sim \frac{2n^3}{3}.$$

We also proved the following result.

Theorem 2.4 *If* (P_n) *is an ultraflat sequence of polynomials* $P_n \in \mathcal{K}_n$ *and* $q \in (0, \infty)$, *then*

$$\frac{1}{2\pi} \int_0^{2\pi} \left| \frac{d}{dt} |(P_n - P_n^*)(e^{it})| \right|^q dt \sim \frac{\Gamma\left(\frac{q+1}{2}\right)}{(q+1)\Gamma\left(\frac{q}{2}+1\right)\sqrt{\pi}} n^{3q/2}.$$

As a Corollary of Theorem 2.2 we have recaptured Saffari's "near orthogonality conjecture" raised in [104] and proved first in [48].

Theorem 2.5 *Let* (P_n) *be a fixed ultraflat sequence of polynomials* $P_n \in \mathcal{K}_n$. *Using the notation in Theorem 2.2 we have*

$$\sum_{k=0}^{n} a_{k,n} a_{n-k,n} = o(n).$$

As a Corollary of Theorem 2.3 we have proved a new "near orthogonality" formula.

Theorem 2.6 *Let* (P_n) *be a fixed ultraflat sequence of polynomials* $P_n \in \mathcal{K}_n$. *Using the notation in Theorem 2.2 we have*

$$\sum_{k=0}^{n} k^2 a_{k,n} a_{n-k,n} = o(n^3).$$

Finally we have recaptured the asymptotic formulas for the real part and the derivative of the real part of ultraflat unimodular polynomials proved in [51] first.

Theorem 2.7 *If* (P_n) *is an ultraflat sequence of unimodular polynomials* $P_n \in \mathcal{K}_n$, *and* $q \in (0, \infty)$, *then for* $f_n(t) := \mathrm{Re}(P_n(e^{it}))$ *we have*

$$\frac{1}{2\pi} \int_0^{2\pi} |f_n(t)|^q \, dt \sim \frac{\Gamma\left(\frac{q+1}{2}\right)}{\Gamma\left(\frac{q}{2}+1\right)\sqrt{\pi}} n^{q/2}$$

and

$$\frac{1}{2\pi} \int_0^{2\pi} |f_n'(t)|^q \, dt \sim \frac{\Gamma\left(\frac{q+1}{2}\right)}{(q+1)\Gamma\left(\frac{q}{2}+1\right)\sqrt{\pi}} n^{3q/2}.$$

We remark that trivial modifications of the proof of Theorems 2.1–2.7 yield that the statement of the above theorem remains true if the ultraflat sequence (P_n) of polynomials $P_n \in \mathcal{K}_n$ is replaced by an ultraflat sequence (P_{n_k}) of polynomials $P_{n_k} \in \mathcal{K}_{n_k}$, where (n_k) is an increasing sequence of positive integers.

3 Flatness of Conjugate-Reciprocal Unimodular Polynomials

There is a beautiful short argument to see that

$$M_\infty(P) \geq \sqrt{4/3}\, n^{1/2} \tag{8}$$

for every conjugate-reciprocal unimodular polynomial $P \in \mathcal{K}_n$. Namely, Parseval's formula gives

$$M_\infty(P') \geq M_2(P') = \left(\frac{m(m+1)(2m+1)}{3} \right)^{1/2}, \qquad P \in \mathcal{K}_n .$$

Combining this with Lax's Bernstein-type inequality

$$M_\infty(P') \leq \frac{n}{2}\, M_\infty(P)$$

valid for all conjugate-reciprocal algebraic polynomials with complex coefficients (see p. 438 in [17], for instance), we obtain

$$M_\infty(P) \geq \frac{2}{n} \left(\frac{n(n+1)(2n+1)}{3} \right)^{1/2} \geq \sqrt{4/3}\, m^{1/2}$$

for all conjugate-reciprocal unimodular polynomials $P \in \mathcal{K}_n$. In [56] we prove the following results.

Theorem 3.1 *There is an absolute constant $\varepsilon > 0$ such that*

$$M_1(P') \leq (1 - \varepsilon)\sqrt{1/3}\, n^{3/2}$$

for every conjugate-reciprocal unimodular polynomial $P \in \mathcal{K}_n$.

Theorem 3.2 *There is an absolute constant $\varepsilon > 0$ such that*

$$M_\infty(P') \geq (1 + \varepsilon)\sqrt{1/3}\, n^{3/2}$$

for every conjugate-reciprocal unimodular polynomial $P \in \mathcal{K}_n$.

Theorem 3.3 *There is an absolute constant $\varepsilon > 0$ such that*

$$M_\infty(P) \geq (1+\varepsilon)\sqrt{4/3}\,n^{1/2}$$

for every conjugate-reciprocal unimodular polynomial $P \in \mathcal{K}_n$.

Theorem 3.4 *There is an absolute constant $\varepsilon > 0$ such that*

$$M_q(P') \leq \exp(\varepsilon(q-2)/q)\sqrt{1/3}\,n^{3/2}, \qquad 1 \leq q < 2,$$

and

$$M_q(P') \geq \exp(\varepsilon(q-2)/q)\sqrt{1/3}\,n^{3/2}, \qquad 2 < q,$$

for every conjugate-reciprocal unimodular polynomial $P \in \mathcal{K}_n$.

A polynomial $P \in \mathcal{P}_n^c$ is called skew-reciprocal if $P(-z) = P^*(z)$ for all $z \in \mathbb{C}$. A polynomial $P \in \mathcal{P}_n^c$ is called self-reciprocal if $P^* = \overline{P}$, that is, $P(z) = z^n P(1/z)$ for all $z \in \mathbb{C} \setminus \{0\}$.

Problem 3.5 Is there an absolute constant $\varepsilon > 0$ such that

$$M_\infty(P') \geq (1+\varepsilon)\sqrt{1/3}\,n^{3/2}$$

holds for all self-reciprocal and skew-reciprocal unimodular polynomials $P \in \mathcal{K}_n$?

Problem 3.6 Is there an absolute constant $\varepsilon > 0$ such that

$$M_\infty(P') \geq (1+\varepsilon)\sqrt{1/3}\,n^{3/2}$$

or at least

$$\max_{z \in \partial D} |P'(z)| - \min_{z \in \partial D} |P'(z)| \geq \varepsilon n^{3/2}$$

holds for all unimodular polynomials $P \in \mathcal{K}_n$?

Our method to prove Theorem 3.2 does not seem to work for all unimodular polynomials $P \in \mathcal{K}_n$. In an e-mail communication several years ago Saffari speculated that the answer to Problem 3.6 is no. However, we know the answer to neither Problem 3.6 nor Problem 3.5.

Let $\mathcal{L}_m$ be the collection of all polynomials of degree n with each of their coefficients in $\{-1, 1\}$.

Problem 3.7 Is there an absolute constant $\varepsilon > 0$ such that

$$M_\infty(P') \geq (1+\varepsilon)\sqrt{1/3}\,n^{3/2}$$

or at least

$$\max_{z \in \partial D} |P'(z)| - \min_{z \in \partial D} |P'(z)| \geq \varepsilon n^{3/2}$$

holds for all Littlewood polynomials $P \in \mathcal{L}_n$?

The following problem due to Erdős [68] is open for a long time.

Problem 3.8 Is there an absolute constant $\varepsilon > 0$ such that

$$M_\infty(P) \geq (1 + \varepsilon)n^{1/2}$$

or at least

$$\max_{z \in \partial D} |P(z)| - \min_{z \in \partial D} |P(z)| \geq \varepsilon n^{1/2}$$

holds for all Littlewood polynomials $P \in \mathcal{L}_n$?

The same problem may be raised only for all skew-reciprocal Littlewood polynomials $P \in \mathcal{L}_n$, and as far as we know, it is also open.

4 Average L_q Norm of Littlewood Polynomials on the Unit Circle

P. Borwein and Lockhart [25] investigated the asymptotic behavior of the mean value of normalized M_q norms of Littlewood polynomials for arbitrary $q > 0$. They proved the following result.

Theorem 4.1

$$\lim_{n \to \infty} \frac{1}{2^{n+1}} \sum_{f \in \mathcal{L}_n} \frac{(M_q(f))^q}{n^{q/2}} = \Gamma\left(1 + \frac{q}{2}\right).$$

In [32] we showed the following.

Theorem 4.2

$$\lim_{n \to \infty} \frac{1}{2^{n+1}} \sum_{f \in \mathcal{L}_n} \frac{M_q(f)}{n^{1/2}} = \left(\Gamma\left(1 + \frac{q}{2}\right)\right)^{1/q}$$

for every $q > 0$.

In [32] we also proved the following result on the average Mahler measure of Littlewood polynomials.

Theorem 4.3 *We have*

$$\lim_{n\to\infty} \frac{1}{2^{n+1}} \sum_{f\in\mathcal{L}_n} \frac{M_0(f)}{n^{1/2}} = e^{-\gamma/2} = 0.749306\cdots,$$

where

$$\gamma := \lim_{n\to\infty} \left(\sum_{k=1}^{n} \frac{1}{k} - \log n \right) = 0.577215\cdots$$

is the Euler constant.

These last two results are analogues of the results proved earlier by Choi and Mossinghoff [35] for polynomials in $\mathcal{K}_n$.

5 Rudin-Shapiro Polynomials

Finding polynomials with suitably restricted coefficients and maximal Mahler measure has interested many authors. The classes $\mathcal{L}_n$ and $\mathcal{K}_n$ are two of the most important classes considered. Observe that $\mathcal{L}_n \subset \mathcal{K}_n$ and

$$M_0(Q) \le M_2(Q) = \sqrt{n+1}$$

for every $Q \in \mathcal{K}_n$.

It is open whether or not for every $\varepsilon > 0$ there is a sequence (Q_n) of polynomials $Q_n \in \mathcal{L}_n$ such that

$$M_0(Q_n) \ge (1-\varepsilon)\sqrt{n}.$$

Beller and Newman [7] constructed a sequence (Q_n) of unimodular polynomials $Q_n \in \mathcal{K}_n$ such that

$$M_0(Q_n) \ge \sqrt{n} - \frac{c}{\log n}.$$

Littlewood asked if there were $Q_{n_k} \in \mathcal{L}_{n_k}$ satisfying

$$c_1\sqrt{n_k+1} \le |Q_{n_k}(z)| \le c_2\sqrt{n_k+1}, \qquad z \in \partial D,$$

with some absolute constants $c_1 > 0$ and $c_2 > 0$, see [10, p. 27] for a reference to this problem of Littlewood. No sequence (Q_{n_k}) of Littlewood polynomials $Q_{n_k} \in \mathcal{L}_{n_k}$ is known that satisfies the lower bound. A sequence of Littlewood polynomials that satisfies just the upper bound is given by the Rudin-Shapiro polynomials. The Rudin-Shapiro polynomials appear in Harold Shapiro's 1951 thesis [109] at MIT and are sometimes called just Shapiro polynomials. They also arise independently in Golay's paper [72]. They are remarkably simple to construct and are a rich source of counterexamples to possible conjectures. The Rudin-Shapiro polynomials are defined recursively as follows:

$$P_0(z) := 1, \qquad Q_0(z) := 1,$$

$$P_{k+1}(z) := P_k(z) + z^{2^k} Q_k(z),$$

$$Q_{k+1}(z) := P_k(z) - z^{2^k} Q_k(z), \qquad k = 0, 1, 2, \dots .$$

Note that both P_k and Q_k are polynomials of degree $n - 1$ with $n := 2^k$ having each of their coefficients in $\{-1, 1\}$. In signal processing, the Rudin-Shapiro polynomials have good autocorrelation properties and their values on the unit circle are small. Binary sequences with low autocorrelation coefficients are of interest in radar, sonar, and communication systems. It is well known and easy to check by using the parallelogram law that

$$|P_{k+1}(z)|^2 + |Q_{k+1}(z)|^2 = 2(|P_k(z)|^2 + |Q_k(z)|^2), \qquad z \in \partial D.$$

Hence

$$|P_k(z)|^2 + |Q_k(z)|^2 = 2^{k+1} = 2n, \qquad z \in \partial D.$$

It is also well known (see Section 4 of [10], for instance), that

$$Q_k(-z) = P_k^*(z) = z^{n-1} P_k(1/z), \qquad z \in \partial D,$$

and hence

$$|Q_k(-z)| = |P_k(z)|, \qquad z \in \partial D.$$

P. Borwein's book [10] presents a few more basic results on the Rudin-Shapiro polynomials. Various properties of the Rudin-Shapiro polynomials are discussed in [29] by Brillhart and in [30] by Brillhart, Lemont, and Morton.

As for $k \geq 1$ both P_k and Q_k have odd degree, both P_k and Q_k have at least one real zero. The fact that for $k \geq 1$ both P_k and Q_k have exactly one real zero was proved by Brillhart in [29]. Another interesting observation made in [30] is the fact that P_k and Q_k cannot vanish at root of unity different from -1 and 1.

Obviously

$$M_2(P_k) = 2^{k/2}$$

by the Parseval formula. In 1968 Littlewood [88] evaluated $M_4(P_k)$ and found that

$$M_4(P_k) \sim \left(\frac{4^{k+1}}{3}\right)^{1/4} = \left(\frac{4n^2}{3}\right)^{1/4}.$$

(9)

The M_4 norm of Rudin-Shapiro like polynomials on ∂D are studied in [26].
The merit factor of a Littlewood polynomial $f \in \mathcal{L}_{n-1}$ is defined by

$$MF(f) = \frac{M_2(f)^4}{M_4(f)^4 - M_2(f)^4} = \frac{n^2}{M_4(f)^4 - n^2}.$$

Observe that (9) implies that $MF(P_k) \sim 3$.

6 Mahler Measure and Moments of the Rudin-Shapiro Polynomials

Despite the simplicity of their definition not much is known about the Rudin-Shapiro polynomials. It has been shown in [59] fairly recently that the Mahler measure (M_0 norm) and the M_∞ norm of the Rudin-Shapiro polynomials P_k and Q_k of degree $n-1$ with $n := 2^k$ on the unit circle of the complex plane have the same size, that is, the Mahler measure of the Rudin-Shapiro polynomials of degree $n-1$ with $n := 2^k$ is bounded from below by $cn^{1/2}$, where $c > 0$ is an absolute constant.

Theorem 6.1 *Let P_k and Q_k be the k-th Rudin-Shapiro polynomials of degree $n-1$ with $n = 2^k$. There is an absolute constant $c_1 > 0$ such that*

$$M_0(P_k) = M_0(Q_k) \geq c_1\sqrt{n}, \ldots k = 1, 2, \ldots.$$

The following asymptotic formula, conjectured by Saffari in 1985, for the Mahler measure of the Rudin-Shapiro polynomials has been proved recently in [61].

Theorem 6.2 *We have*

$$\lim_{n \to \infty} \frac{M_0(P_k)}{n^{1/2}} = \lim_{n \to \infty} \frac{M_0(Q_k)}{n^{1/2}} = \left(\frac{2}{e}\right)^{1/2} = 0.857763 \cdots.$$

To formulate our next theorem we define

$$\widetilde{P}_k := 2^{-(k+1)/2} P_k \quad \text{and} \quad \widetilde{Q}_k := 2^{-(k+1)/2} Q_k.$$

By using the above normalization, we have

$$|\widetilde{P}_k(z)|^2 + |\widetilde{Q}_k(z)|^2 = 1, \qquad z \in \partial D.$$

For $q > 0$ let

$$I_q(\widetilde{P}_k) := \left(M_q(\widetilde{P}_k)\right)^q := \frac{1}{2\pi} \int_0^{2\pi} |\widetilde{P}_k(e^{it})|^q \, dt \, .$$

The following result is a simple consequence of Theorem 6.1.

Theorem 6.3 *There exists a constant $L < \infty$ independent of k such that*

$$\sum_{m=1}^{\infty} \frac{I_m(\widetilde{P}_k)}{m} < L \, , \qquad k = 0, 1, \ldots \, .$$

Consequently

$$I_m(\widetilde{P}_k) \leq \frac{L}{\log(m+1)} \, , \qquad k = 1, 2, \ldots \, , \qquad m = 1, 2, \ldots \, .$$

In [59] we also proved the following result.

Theorem 6.4 *There exists an absolute constant $c_2 > 0$ such that*

$$M_0(P_k, [\alpha, \beta]) \geq c_2\sqrt{n} \, , \qquad k = 1, 2, \ldots \, ,$$

with $n := 2^k$ for all $\alpha, \beta \in \mathbb{R}$ such that

$$\frac{12\pi}{n} \leq \frac{(\log n)^{3/2}}{n^{1/2}} \leq \beta - \alpha \leq 2\pi \, .$$

7   Lemmas for Theorem 6.1

As the proof of Theorem 6.1 is based on interesting new properties of the Rudin-Shapiro polynomials which have been observed only recently in [59], we list them in this section.

Lemma 7.1 *Let $k \geq 2$ be an integer, $n := 2^k$, and let*

$$z_j := e^{it_j} \, , \quad t_j := \frac{2\pi j}{n} \, , \quad j \in \mathbb{Z} \, .$$

We have

$$P_k(z_j) = 2P_{k-2}(z_j)$$

whenever j is even, and

$$P_k(z_j) = (-1)^{(j-1)/2} 2i \, Q_{k-2}(z_j)$$

whenever j is odd, where i is the imaginary unit.

Lemma 7.2 *If P_k and Q_k are the k-th Rudin-Shapiro polynomials of degree $n - 1$ with $n := 2^k$,*

$$\omega := \sin^2(\pi/8) = 0.146446\cdots,$$

and

$$z_j := e^{it_j}, \quad t_j := \frac{2\pi j}{n}, \quad j \in \mathbb{Z},$$

then

$$\max\{|P_k(z_j)|^2, |P_k(z_{j+1})|^2\} \geq \omega 2^{k+1} = 2\omega n, \quad j \in \mathbb{Z}.$$

Lemma 7.3 *Let $n, m \geq 1$,*

$$0 < \tau_1 \leq \tau_2 \leq \cdots \leq \tau_m \leq 2\pi, \quad \tau_0 := \tau_m - 2\pi, \quad \tau_{m+1} := \tau_1 + 2\pi,$$

$$\delta := \max\{\tau_1 - \tau_0, \tau_2 - \tau_1, \ldots, \tau_m - \tau_{m-1}\}.$$

For every $A > 0$ there is a $B > 0$ depending only on A such that

$$\sum_{j=1}^{m} \frac{\tau_{j+1} - \tau_{j-1}}{2} \log|P(e^{i\tau_j})| \leq \int_0^{2\pi} \log|P(e^{i\tau})| \, d\tau + B$$

for all $P \in \mathcal{P}_n^c$ and $\delta \leq An^{-1}$.

8 Saffari's Conjecture on the Shapiro Polynomials

In 1980 Saffari conjectured the following.

Conjecture 8.1 *Let P_k and Q_k be the Rudin-Shapiro polynomials of degree $n - 1$ with $n := 2^k$. We have*

$$M_q(P_k) = M_q(Q_k) \sim \frac{(2n)^{1/2}}{(q/2 + 1)^{1/q}}$$

for all real exponents $q > 0$.

Conjecture 8.1* *Equivalently to Conjecture 8.1, we have*

$$\lim_{k \to \infty} m\left(\left\{t \in [0, 2\pi) : \left|\frac{P_k(e^{it})}{\sqrt{2^{k+1}}}\right|^2 \in [\alpha, \beta]\right\}\right)$$

$$= \lim_{k \to \infty} m\left(\left\{t \in [0, 2\pi) : \left|\frac{Q_k(e^{it})}{\sqrt{2^{k+1}}}\right|^2 \in [\alpha, \beta]\right\}\right) = 2\pi(\beta - \alpha)$$

whenever $0 \le \alpha < \beta \le 1$.

This conjecture was proved for all even values of $q \le 52$ by Doche [39] and Doche and Habsieger [40]. Recently B. Rodgers [102] proved Saffari's Conjecture 8.1 for all $q > 0$. See also [44].

An extension of Saffari's conjecture is Montgomery's conjecture below.

Conjecture 8.2 *Let P_k and Q_k be the Rudin-Shapiro polynomials of degree $n - 1$ with $n := 2^k$. We have*

$$\lim_{k \to \infty} m\left(\left\{t \in [0, 2\pi) : \frac{P_k(e^{it})}{\sqrt{2^{k+1}}} \in E\right\}\right)$$

$$= \lim_{k \to \infty} m\left(\left\{t \in [0, 2\pi) : \frac{Q_k(e^{it})}{\sqrt{2^{k+1}}} \in E\right\}\right) = 2\mu(E),$$

where $\mu(E)$ denotes the Jordan measure of a Jordan measurable set $E \subset D$.

B. Rodgers [102] proved Montgomery's Conjecture 8.2 as well.

9   Consequences of Saffari's Conjecture

Let P_k and Q_k be the Rudin-Shapiro polynomials of degree $n - 1$ with $n := 2^k$,

$$R_k(t) := |P_k(e^{it})|^2 \quad \text{or} \quad R_k(t) := |Q_k(e^{it})|^2,$$

$$\omega := \sin^2(\pi/8) = 0.146446 \cdots.$$

In [62] we proved Theorems 9.1–9.5 below.

Theorem 9.1 *P_k and Q_k have $o(n)$ zeros on the unit circle.*

The proof of Theorem 9.1 follows by combining the recently proved Saffari's conjecture stated as Conjecture 8.1* and the theorem below.

Theorem 9.2 *If the real trigonometric polynomial R of degree n is of the form*

$$R(t) = |P(e^{it})|^2,$$

where $P \in \mathcal{P}_n^c$, and P has at least k zeros in K (counting multiplicities), then

$$m(\{t \in [0, 2\pi) : |R(t)| \leq \alpha \|R\|_K\}) \geq \frac{\sqrt{\alpha}}{e} \frac{k}{n}$$

for every $\alpha \in (0, 1)$.

Theorem 9.3 *There exists an absolute constant* $c > 0$ *such that each of the functions* $Re(P_k)$, $Re(Q_k)$, $Im(P_k)$, *and* $Im(Q_k)$ *has at least* cn *zeros on the unit circle.*

Theorem 9.4 *There exists an absolute constant* $c > 0$ *such that the equation* $R_k(t) = \eta n$ *has at most* $c\eta^{1/2}$ *solutions in* $[0, 2\pi)$ *for every* $\eta \in (0, 1]$ *and sufficiently large* $k \geq k_\eta$, *while the equation* $R_k(t) = \eta n$ *has at most* $c(2 - \eta)^{1/2}n$ *solutions in* $[0, 2\pi)$ *for every* $\eta \in [1, 2)$ *and sufficiently large* $k \geq k_\eta$.

Theorem 9.5 *The equation* $R_k(t) = \eta n$ *has at least* $(1-\varepsilon)\eta n/2$ *solutions in* $[0, 2\pi)$ *for every* $\eta \in (0, 2\omega)$, $\varepsilon > 0$, *and sufficiently large* $k \geq k_{\eta,\varepsilon}$. *The equation* $R_k(t) = \eta n$ *has at least* $(1 - \varepsilon)(2 - \eta)n/2$ *solutions in* $[0, 2\pi)$ *for every* $\eta \in (2 - 2\omega, 2)$, $\varepsilon > 0$, *and sufficiently large* $k \geq k_{\eta,\varepsilon}$.

Theorem 9.6 *There exists an absolute constant* $c > 0$ *such that the equation* $R_k(t) = (1 + \eta)n$ *has at least* $cn^{0.5394282}$ *distinct solutions in* $[0, 2\pi)$ *whenever* $\eta \in \mathbb{R}$ *with* $|\eta| < 2^{-8}$.

In [2] we combined close to sharp upper bounds for the modulus of the autocorrelation coefficients of the Rudin-Shapiro polynomials with a deep theorem of Littlewood (see Theorem 1 in [86]) to prove the above Theorem 9.6.

Theorem 9.7 *If*

$$|P_k(z)|^2 = P_k(z)P_k(1/z) = \sum_{j=-n+1}^{n-1} a_j z^j, \qquad z \in \partial D,$$

then

$$c_1 n^{0.7302852\cdots} \leq \max_{1 \leq j \leq n-1} |a_j| \leq c_2 n^{0.7302859\cdots}$$

with an absolute constants $c_1 > 0$ *and* $c_2 > 0$.

Theorem 9.7 has been recently improved by Choi in [31] by showing that

$$(0.27771487 \cdots)(1 + o(1)) |\lambda|^k \leq \max_{1 \leq j \leq n-1} |a_j| \leq (3.78207844 \cdots) |\lambda|^k,$$

where

$$\lambda := -\frac{\left(44 + 3\sqrt{177}\right)^{1/3} + \left(44 - 3\sqrt{177}\right)^{1/3} - 1}{3} = -1.658967081916\cdots$$

is the real root of the equation $x^3 - x^2 - 2x + 4 = 0$ and $|\lambda|^k = n^{07302852598\cdots}$. This settles a conjecture expected earlier by Saffari.

Theorem 9.8 (Littlewood) *If the real trigonometric polynomial of degree at most n is of the form*

$$f(t) = \sum_{m=0}^{n} a_m \cos(mt + \alpha_m), \qquad a_m, \alpha_m \in \mathbb{R},$$

satisfies $M_1(f) \geq c\mu$, $\mu := M_2(f)$, where $c > 0$ is a constant, $a_0 = 0$,

$$s_{\lfloor n/h \rfloor} = \sum_{m=1}^{\lfloor n/h \rfloor} \frac{a_m^2}{\mu^2} \leq 2^{-9} c^6$$

for some constant $h > 0$, and $v \in \mathbb{R}$ satisfies

$$|v| \leq V = 2^{-5} c^3,$$

then

$$\mathcal{N}(f, v) > Ah^{-1} c^5 n,$$

where $\mathcal{N}(f, v)$ denotes the number of real zeros of $f - v\mu$ in $(-\pi, \pi)$, and $A > 0$ is an absolute constant.

In [63] we improved Theorem 9.6 by showing the following two results.

Theorem 9.9 *The equation $R_k(t) = n$ has at least $n/4 + 1$ distinct zeros in $[0, 2\pi)$. Moreover, with the notation $t_j := 2\pi j/n$, there are at least $n/2 + 2$ values of $j \in \{0, 1 \ldots, n - 1\}$ for which the interval $[t_j, t_{j+1}]$ has at least one zero of the equation $R_k(t) = n$.*

Theorem 9.10 *The equation $R_k(t) = (1 + \eta)n$ has at least $(1/2 - |\eta| - \varepsilon)n/2$ distinct zeros in $[0, 2\pi)$ for every $\eta \in (-1/2, 1/2)$, $\varepsilon > 0$, and sufficiently large $k \geq k_{\eta, \varepsilon}$.*

In [62] we proved the theorem below.

Theorem 9.11 *There exist absolute constants $c_1 > 0$ and $c_2 > 0$ such that both P_k and Q_k have at least $c_2 n$ zeros in the annulus*

$$\left\{ z \in \mathbb{C} : 1 - \frac{c_1}{n} < |z| < 1 + \frac{c_1}{n} \right\}.$$

A key to the proof of Theorem 9.11 is the result below.

Theorem 9.12 *Let $t_0 \in K$. There is an absolute constant $c_3 > 0$ depending only on $c > 0$ such that P_k has at least one zero in the disk*

$$\left\{ z \in \mathbb{C} : |z - e^{it_0}| < \frac{c_3}{n} \right\},$$

whenever

$$T_k'(t_0) \geq cn^2, \qquad T_k(t) = P_k(e^{it}) P_k(e^{-it}).$$

We note that for every $c \in (0, 1)$ there is an absolute constant $c_4 > 0$ depending only on c such that every $U_n \in \mathcal{P}_n^c$ of the form

$$U_n(z) = \sum_{j=0}^{n} a_j z^j, \qquad |a_0| = |a_n| = 1, \quad a_j \in \mathbb{C}, \quad |a_j| \leq 1,$$

has at least cn zeros in the annulus

$$\left\{ z \in \mathbb{C} : 1 - \frac{c_4 \log n}{n} < |z| < 1 + \frac{c_4 \log n}{n} \right\}.$$

See Theorem 2.1 in [49].

On the other hand, there is an absolute constant $c_4 > 0$ such that for every $n \in \mathbb{N}$ there is a polynomial $U_n \in \mathcal{K}_n$ having no zeros in the above annulus. See Theorem 2.3 in [49].

So in Theorem 9.11 some special properties, in addition to being Littlewood polynomials, of the Rudin-Shapiro polynomials must be exploited.

10   Open Problems Related to the Rudin-Shapiro Polynomials

Problem 10.1 Is there an absolute constant $c > 0$ such that the equation $R_k(t) = \eta n$ has at least $c\eta n$ distinct solutions in K for every $\eta \in (0, 1)$ and sufficiently large $k \geq k_\eta$? In other words, can Theorem 9.5 be extended to all $\eta \in (0, 1)$?

Problem 10.2 Is there an absolute constant $c > 0$ such that P_k has at least cn zeros in the open unit disk?

Problem 10.3 Is there an absolute constant $c > 0$ such that Q_k has at least cn zeros in the open unit disk?

Recall that

$$Q_k(-z) = P_k^*(z) = z^{n-1} P_k(1/z), \qquad z \in \partial D.$$

Problem 10.4 Is it true that both P_k and Q_k have asymptotically half of their zeros in the open unit disk?

Observe that $P_k(-1) = Q_k(1) = 0$ if k is odd.

Problem 10.5 Is it true that if k is odd then P_k has a zero on the unit circle ∂D only at -1 and Q_k has a zero on the unit circle ∂D only at 1, while if k is even then neither P_k nor Q_k has a zero on the unit circle?

11   On the Size of the Fekete Polynomials on the Unit Circle

For a prime p the p-th Fekete polynomial is defined as

$$f_p(z) := \sum_{k=1}^{p-1} \left(\frac{k}{p}\right) z^k,$$

where

$$\left(\frac{k}{p}\right) = \begin{cases} 1, & \text{if } x^2 \equiv k \pmod{p} \text{ for an } x \neq 0, \\ 0, & \text{if } p \text{ divides } k, \\ -1, & \text{otherwise} \end{cases}$$

is the usual Legendre symbol. Note that $g_p(z) := f_p(z)/z$ is a Littlewood polynomial of degree $p - 2$, and has the same Mahler measure as f_p.

In 1980 Montgomery [91] proved the following fundamental result.

Theorem 11.1 *There are absolute constants $c_1 > 0$ and $c_2 > 0$ such that*

$$c_1 \sqrt{p} \log \log p \leq \max_{z \in \partial D} |f_p(z)| \leq c_2 \sqrt{p} \log p.$$

It was observed in [55] that Montgomery's approach can be used to prove that for every sufficiently large prime p and for every $8\pi p^{-1/8} \leq s \leq 2\pi$ there is a closed subset $E := E_{p,s}$ of the unit circle with linear measure $|E| = s$ such that

$$\frac{1}{|E|} \int_E |f_p(z)| \, |dz| \geq c_1 \, p^{1/2} \log \log(1/s)$$

with an absolute constant $c_1 > 0$.

In [12] the L_4 norm of the Fekete polynomials are computed.

Theorem 11.2 *We have*

$$M_4(f_p) = \left(\frac{5p^2}{3} - 3p + \frac{4}{3} - 12(h(-p))^2\right)^{1/4},$$

where $h(-q)$ is the class number of $\mathbb{Q}(\sqrt{-q})$.

In [66] we proved the following result.

Theorem 11.3 *For every $\varepsilon > 0$ there is a constant c_ε such that*

$$M_0(f_p) \geq \left(\frac{1}{2} - \varepsilon\right)\sqrt{p}$$

for all primes $p \geq c_\varepsilon$.

From Jensen's inequality,

$$M_0(f_p) \leq M_2(f_p) = \sqrt{p-1}.$$

However, as it was observed in [E-07] and [60], a result of Littlewood [86] implies that $\frac{1}{2} - \varepsilon$ in Theorem 11.2 cannot be replaced by $1 - \varepsilon$.

To prove Theorem 11.3 in [E-07] we needed to combine Theorems 11.4, 11.5 and one of Theorems 11.6 and 11.7 below. For a prime number p let

$$\zeta_p := \exp\left(\frac{2\pi i}{p}\right)$$

the first p-th root of unity. Our first lemma formulates a characteristic property of the Fekete polynomials. A simple proof is given in [10, pp. 37–38].

Theorem 11.4 (Gauss) *We have*

$$f_p(\zeta_p^j) = \sqrt{\left(\frac{-1}{p}\right)p}, \qquad j = 1, 2, \ldots, p - 1,$$

and $f_p(1) = 0$.

Theorem 11.5 *We have*

$$\left(\prod_{j=0}^{p-1} |Q(\zeta_p^j)|\right)^{1/p} \leq 2M_0(Q)$$

for all polynomials Q of degree at most p with complex coefficients.

Theorem 11.6 *There is an absolute constant $c > 0$ such that every $Q \in \mathcal{K}_n$ has at most $c\sqrt{n}$ real zeros.*

Theorem 11.7 *There is an absolute constant $c > 0$ such that every $Q \in \mathcal{L}_n$ has at most $\dfrac{c \log^2 n}{\log \log n}$ zeros at 1.*

For a proof of Theorem 11.6 see [18]. For a proof of Theorem 11.7 see [28].

In [54] Theorem 11.3 was extended to subarcs of the unit circle.

Theorem 11.8 *There exists an absolute constant $c_1 > 0$ such that*

$$M_0(f_p, [\alpha, \beta]) \geq c_1 p^{1/2}$$

for all primes p and for all $\alpha, \beta \in \mathbb{R}$ such that $(\log p)^{3/2} p^{-1/2} \leq \beta - \alpha \leq 2\pi$.

In [55] we gave an upper bound for the average value of $|f_p(z)|^q$ over any subarc I of the unit circle, valid for all sufficiently large primes p and all exponents $q > 0$.

Theorem 11.9 *There exists a constant $c_2(q, \varepsilon)$ depending only on $q > 0$ and $\varepsilon > 0$ such that*

$$M_q(f_p, [\alpha, \beta]) \leq c_2(q, \varepsilon) p^{1/2},$$

for all primes p and for all $\alpha, \beta \in \mathbb{R}$ such that $2p^{-1/2+\varepsilon} \leq \beta - \alpha \leq 2\pi$.

We remark that a combination of Theorems 11.8 and 11.9 shows that there is an absolute constant $c_1 > 0$ and a constant $c_2(q, \varepsilon) > 0$ depending only on $q > 0$ and $\varepsilon > 0$ such that

$$c_1 p^{1/2} \leq M_q(f_p, [\alpha, \beta]) \leq c_2(q, \varepsilon) p^{1/2}$$

for all primes p and for all $\alpha, \beta \in \mathbb{R}$ such that $(\log p)^{3/2} p^{-1/2} \leq 2p^{-1/2+\varepsilon} \leq \beta - \alpha \leq 2\pi$.

The L_q norm of polynomials related to Fekete polynomials were studied in several recent papers. See [10–12, 14, 73, 74], and [75], for example. An interesting extremal property of the Fekete polynomials is proved in [16].

Fekete might have been the first one to study analytic properties of the Fekete polynomials. He had an idea of proving non-existence of Siegel zeros (that is, real zeros "especially close to 1") of Dirichlet L-functions from the positivity of Fekete polynomials on the interval $(0, 1)$, where the positivity of Fekete polynomials is often referred to as the Fekete Hypothesis.

There were many mathematicians trying to understand the zeros of Fekete polynomials including Fekete and Pólya [71], Pólya [98], Chowla [36], Bateman, Purdy, and Wagstaff [4], Heilbronn [77], Montgomery [91], Baker and Montgomery [3], and Jung and Shen [78].

Baker and Montgomery [3] proved that f_p has a large number of zeros in $(0, 1)$ for almost all primes p, that is, the number of zeros of f_p in $(0, 1)$ tends to ∞ as p tends to ∞, and it seems likely that there are, in fact, about $\log \log p$ such zeros.

Conrey, Granville, Poonen, and Soundararajan [37] showed that f_p has asymptotically $\kappa_0 p$ zeros on the unit circle, where

$$0.500668 < \kappa_0 < 0.500813 .$$

An interesting recent paper [6] studies power series approximations to Fekete polynomials. In [60] we improved Theorem 11.2 by showing the following result.

Theorem 11.10 *There is an absolute constant* $c > 1/2$ *such that*

$$M_0(f_p) \geq c\sqrt{p}$$

for all sufficiently large primes.

However, there is not even a conjecture in the literature about what the asymptotically sharp constant c in Theorem 11.10 could be.

12 Unimodular Zeros of Self-Reciprocal Polynomials with Coefficients in a Finite Set

Research on the distribution of the zeros of algebraic polynomials has a long and rich history. A few papers closely related this section include [8, 13, 19, 21–24, 42, 43, 50, 52, 53, 57, 70, 84, 85, 94, 96, 97, 99, 107, 108, 110–114]. The number of real zeros trigonometric polynomials and the number of unimodular zeros (that is, zeros lying on the unit circle of the complex plane) of algebraic polynomials with various constraints on their coefficients are the subject of quite a few of these. We do not try to survey these in this section.

Let $S \subset \mathbb{C}$. Let $\mathcal{P}_n(S)$ be the set of all algebraic polynomials of degree at most n with each of their coefficients in S. An algebraic polynomial P of the form

$$P(z) = \sum_{j=0}^{n} a_j z^j , \qquad a_j \in \mathbb{C}, \tag{10}$$

is called conjugate-reciprocal if

$$\bar{a}_j = a_{n-j} , \qquad j = 0, 1, \ldots, n . \tag{11}$$

Functions T of the form

$$T(t) = \alpha_0 + \sum_{j=1}^{n} (\alpha_j \cos(jt) + \beta_j \sin(jt)) , \qquad \alpha_j, \beta_j \in \mathbb{R} ,$$

are called real trigonometric polynomials of degree at most n. It is easy to see that any real trigonometric polynomial T of degree at most n can be written as $T(t) = P(e^{it})e^{-int}$, where P is a conjugate-reciprocal algebraic polynomial of the form

$$P(z) = \sum_{j=0}^{2n} a_j z^j, \qquad a_j \in \mathbb{C}.i \qquad (12)$$

Conversely, if P is conjugate-reciprocal algebraic polynomial of the form (12), then there are $\theta_j \in \mathbb{R}$, $j = 1, 2, \ldots n$, such that

$$T(t) := P(e^{it})e^{-int} = a_n + \sum_{j=1}^{n} 2|a_{j+n}| \cos(jt + \theta_j)$$

is a real trigonometric polynomial of degree at most n. A polynomial P of the form (10) is called self-reciprocal if

$$a_j = a_{n-j}, \qquad j = 0, 1, \ldots, n. \qquad (13)$$

If a conjugate-reciprocal algebraic polynomial P has only real coefficients, then it is obviously self-reciprocal. If the algebraic polynomial P of the form (12) is self-reciprocal, then

$$T(t) := P(e^{it})e^{-int} = a_n + \sum_{j=1}^{n} 2a_{j+n} \cos(jt).$$

In this section, whenever we write "$P \in \mathcal{P}_n(S)$ is conjugate-reciprocal" we mean that P is of the form (10) with each $a_j \in S$ satisfying (11). Similarly, whenever we write "$P \in \mathcal{P}_n(S)$ is self-reciprocal" we mean that P is of the form (10) with each $a_j \in S$ satisfying (13). This is going to be our understanding even if the degree of $P \in \mathcal{P}_n(S)$ is less than n. It is easy to see that $P \in \mathcal{P}_n(S)$ is self-reciprocal and n is odd, then $P(-1) = 0$. We call any subinterval $[a, a + 2\pi)$ of the real number line $\mathbb{R}$ a period. Associated with an algebraic polynomial P of the form (10) we introduce the number

$$NC(P) := |\{j \in \{0, 1, \ldots, n\} : a_j \neq 0\}|.$$

Here, and in what follows $|A|$ denotes the number of elements of a finite set A. Let $NZ(P)$ denote the number of real zeros (by counting multiplicities) of an algebraic polynomial P on the unit circle. Associated with an even trigonometric polynomial (cosine polynomial) of the form

$$T(t) = \sum_{j=0}^{n} a_j \cos(jt)$$

we introduce the number

$$NC(T) := |\{j \in \{0, 1, \dots, n\} : a_j \neq 0\}|.$$

Let $NZ(T)$ denote the number of real zeros (by counting multiplicities) of a trigonometric polynomial T in a period. Let $NZ^*(T)$ denote the number of sign changes of a trigonometric polynomial T in a period. The quotation below is from [21].

"Let $0 \leq n_1 < n_2 < \cdots < n_N$ be integers. A cosine polynomial of the form $T(\theta) = \sum_{j=1}^{N} \cos(n_j\theta)$ must have at least one real zero in a period. This is obvious if $n_1 \neq 0$, since then the integral of the sum on a period is 0. The above statement is less obvious if $n_1 = 0$, but for sufficiently large N it follows from Littlewood's Conjecture simply. Here we mean the Littlewood's Conjecture proved by S. Konyagin [81] and independently by McGehee, Pigno, and Smith [89] in 1981. See also [38, pages 285–288] for a book proof. It is not difficult to prove the statement in general even in the case $n_1 = 0$ without using Littlewood's Conjecture. One possible way is to use the identity

$$\sum_{j=1}^{n_N} T\left(\frac{(2j-1)\pi}{n_N}\right) = 0.$$

See [82], for example. Another way is to use Theorem 2 of [90]. So there is certainly no shortage of possible approaches to prove the starting observation of this section even in the case $n_1 = 0$.

It seems likely that the number of zeros of the above sums in a period must tend to ∞ with N. In a private communication B. Conrey asked how fast the number of real zeros of the above sums in a period tends to ∞ as a function N. In [37] the authors observed that for an odd prime p the Fekete polynomial

$$f_p(z) = \sum_{k=0}^{p-1} \binom{k}{p} z^k$$

(the coefficients are Legendre symbols) has $\sim \kappa_0 p$ zeros on the unit circle, where $0.500813 > \kappa_0 > 0.500668$. Conrey's question in general does not appear to be easy. Littlewood in his 1968 monograph 'Some Problems in Real and Complex Analysis' [88, problem 22] poses the following research problem, which appears to still be open: 'If the n_m are integral and all different, what is the lower bound on the number of real zeros of $\sum_{m=1}^{N} \cos(n_m\theta)$? Possibly $N - 1$, or not much less.' Here real zeros are counted in a period. In fact no progress appears to have been made on this in the last half century. In a recent paper [21] we showed that this is false. There exist cosine polynomials $\sum_{m=1}^{N} \cos(n_m\theta)$ with the n_m integral and all different so that the number of its real zeros in a period is $O(N^{9/10}(\log N)^{1/5})$ (here the frequencies $n_m = n_m(N)$ may vary with N). However, there are reasons to believe that a cosine polynomial $\sum_{m=1}^{N} \cos(n_m\theta)$ always has many zeros in a period."

Let, as before,

$$\mathcal{L}_n := \left\{ P : P(z) = \sum_{j=0}^{n} a_j z^j, \ a_j \in \{-1, 1\} \right\}.$$

Elements of $\mathcal{L}_n$ are often called Littlewood polynomials of degree n. Let

$$\mathcal{H}_n := \left\{ P : P(z) = \sum_{j=0}^{n} a_j z^j, \ a_j \in \mathbb{C}, \ |a_0| = |a_n| = 1, \ |a_j| \le 1 \right\}.$$

Observe that $\mathcal{L}_n \subset \mathcal{H}_n$. In [24] we proved that any polynomial $P \in \mathcal{K}_n$ has at least $8n^{1/2} \log n$ zeros in any open disk centered at a point on the unit circle with radius $33n^{-1/2} \log n$. Thus polynomials in $\mathcal{H}_n$ have quite a few zeros near the unit circle. One may naturally ask how many unimodular roots a polynomial in $\mathcal{H}_n$ can have. Mercer [90] proved that if a Littlewood polynomial $P \in \mathcal{L}_n$ of the form (10) is skew-reciprocal, that is, $a_j = (-1)^j a_{n-j}$ for each $j = 0, 1, \ldots, n$, then it has no zeros on the unit circle. However, by using different elementary methods it was observed in both [50] and [90] that if a Littlewood polynomial P of the form (10) is self-reciprocal, then it has at least one zero on the unit circle. Mukunda [93] improved this result by showing that every self-reciprocal Littlewood polynomial of odd degree has at least 3 zeros on the unit circle. Drungilas [41] proved that every self-reciprocal Littlewood polynomial of odd degree $n \ge 7$ has at least 5 zeros on the unit circle and every self-reciprocal Littlewood polynomial of even degree $n \ge 14$ has at least 4 zeros on the unit circle. In [21] we proved that the average number of zeros of self-reciprocal Littlewood polynomials of degree n is at least $n/4$. However, it is much harder to give decent lower bounds for the quantities $NZ_n := \min_P NZ(P)$, where $NZ(P)$ denotes the number of zeros of a polynomial P lying on the unit circle and the minimum is taken for all self-reciprocal Littlewood polynomials $P \in \mathcal{L}_n$. It has been conjectured for a long time that $\lim_{n\to\infty} NZ_n = \infty$. In [58] we showed that $\lim_{n\to\infty} NZ(P_n) = \infty$ whenever $P_n \in \mathcal{L}_n$ is self-reciprocal and $\lim_{n\to\infty} |P_n(1)| = \infty$. This follows as a consequence of a more general result, see Corollary 2.3 in [58], stated as Corollary 12.5 here, in which the coefficients of the self-reciprocal polynomials P_n of degree at most n belong to a fixed finite set of real numbers. In [20] we proved the following result.

Theorem 12.1 *If the set $\{a_j : j \in \mathbb{N}\} \subset \mathbb{R}$ is finite, the set $\{j \in \mathbb{N} : a_j \ne 0\}$ is infinite, the sequence (a_j) is not eventually periodic, and*

$$T_n(t) = \sum_{j=0}^{n} a_j \cos(jt),$$

then $\lim_{n\to\infty} NZ(T_n) = \infty$.

In [20] Theorem 12.1 is stated without the assumption that the sequence (a_j) is not eventually periodic. However, as the following example shows, Lemma 3.4 in [20], dealing with the case of eventually periodic sequences (a_j), is incorrect. Let

$$T_n(t) := \cos t + \cos((4n+1)t) + \sum_{k=0}^{n-1} (\cos((4k+1)t) - \cos((4k+3)t))$$

$$= \frac{1 + \cos((4n+2)t)}{2\cos t} + \cos t.$$

It is easy to see that $T_n(t) \neq 0$ on $[-\pi, \pi] \setminus \{-\pi/2, \pi/2\}$ and the zeros of T_n at $-\pi/2$ and $\pi/2$ are simple. Hence T_n has only two (simple) zeros in a period. So the conclusion of Theorem 12.1 above is false for the sequence (a_j) with $a_0 := 0$, $a_1 := 2$, $a_3 := -1$, $a_{2k} := 0$, $a_{4k+1} := 1$, $a_{4k+3} := -1$ for every $k = 1, 2, \ldots$. Nevertheless, Theorem 12.1 can be saved even in the case of eventually periodic sequences (a_j) if we assume that $a_j \neq 0$ for all sufficiently large j. See Lemma 3.11 in [22] where Theorem 1 in [20] is corrected as

Theorem 12.2 *If the set $\{a_j : j \in \mathbb{N}\} \subset \mathbb{R}$ is finite, $a_j \neq 0$ for all sufficiently large j, and*

$$T_n(t) = \sum_{j=0}^{n} a_j \cos(jt),$$

then $\lim_{n\to\infty} NZ(T_n) = \infty$.

It was expected that the conclusion of the above theorem remains true even if the coefficients of T_n do not come from the same sequence, that is,

$$T_n(t) = \sum_{j=0}^{n} a_{j,n} \cos(jt),$$

where the set

$$S := \{a_{j,n} : j \in \{0, 1, \ldots, n\}, n \in \mathbb{N}\} \subset \mathbb{R}$$

is finite and

$$\lim_{n\to\infty} |\{j \in \{0, 1, \ldots, n\}, a_{j,n} \neq 0\}| = \infty.$$

Associated with an algebraic polynomial

$$P(z) = \sum_{j=0}^{n} a_j z^j, \qquad a_j \in \mathbb{C},$$

let

$$NC_k(P) := |\{u : 0 \leq u \leq n - k + 1, \ a_u + a_{u+1} + \cdots + a_{u+k-1} \neq 0\}| \,.$$

In [58] we proved the following results.

Theorem 12.3 *If* $S \subset \mathbb{R}$ *is a finite set,* $P_{2n} \in \mathcal{P}_{2n}(S)$ *are self-reciprocal polynomials,*

$$T_n(t) := P_{2n}(e^{it})e^{-int} \,,$$

and

$$\lim_{n \to \infty} NC_k(P_{2n}) = \infty$$

for every $k = 1, 2, \ldots,$ *then*

$$\lim_{n \to \infty} NZ(P_{2n}) = \lim_{n \to \infty} NZ(T_n) = \infty \,.$$

Some of the most important consequences of the above theorem obtained in [58] are stated below.

Corollary 12.4 *If* $S \subset \mathbb{R}$ *is a finite set,* $P_n \in \mathcal{P}_n(S)$ *are self-reciprocal polynomials, and*

$$\lim_{n \to \infty} |P_n(1)| = \infty \,,$$

then

$$\lim_{n \to \infty} NZ(P_n) = \infty \,.$$

Corollary 12.5 *Suppose the finite set* $S \subset \mathbb{R}$ *has the property that*

$$s_1 + s_2 + \cdots + s_k = 0, \ s_1, s_2, \ldots, s_k \in S, \ \text{implies} \ s_1 = s_2 = \cdots = s_k = 0,$$

that is, any sum of nonzero elements of S *is different from* 0. *If* $P_n \in \mathcal{P}_n(S)$ *are self-reciprocal polynomials and*

$$\lim_{n \to \infty} NC(P_n) = \infty \,,$$

then

$$\lim_{n \to \infty} NZ(P_n) = \infty .$$

J. Sahasrabudhe [106] examined the case when $S \subset \mathbb{Z}$ is finite. Exploiting the assumption that the coefficients are integer he proved that for any finite set $S \subset \mathbb{Z}$ a self-reciprocal polynomial $P \in \mathcal{P}_{2n}(S)$ has at least

$$c \ (\log \log \log |P(1)|)^{1/2-\varepsilon} - 1$$

zeros on the unit circle of $\mathbb{C}$ with a constant $c > 0$ depending only on $M = M(S) :=$ $\max\{|z| : z \in S\}$ and $\varepsilon > 0$.

Let $\varphi(n)$ denote the Euler's totient function defined as the number of integers $1 \le k \le n$ that are relative prime to n. In an earlier version of his paper Sahasrabudhe [106] used the trivial estimate $\varphi(n) \ne \sqrt{n}$ for $n \ge 3$ and he proved his result with the exponent $1/4 - \varepsilon$ rather than $1/2 - \varepsilon$. Using the nontrivial estimate $\varphi(n) \ge n/8(\log \log n)$ [103] for all $n > 3$ allowed him to prove his result with $1/2 - \varepsilon$.

In the papers [20, 58], and [106] the already mentioned Littlewood Conjecture, proved by Konyagin [81] and independently by McGehee, Pigno, and B. Smith [89], plays a key role, and we rely on it heavily in the proof of the main results of this paper as well. This states the following.

Theorem 12.6 *There is an absolute constant $c > 0$ such that*

$$\int_0^{2\pi} \left| \sum_{j=1}^m a_j e^{i\lambda_j t} \right| dt \ge c\delta \log m$$

whenever $\lambda_1, \lambda_2, \ldots, \lambda_n$ *are distinct integers and* $a_1, a_2, \ldots, a_m$ *are complex numbers of modulus at least* $\delta > 0$. *Here* $c = 1/30$ *is a suitable choice.*

This is an obvious consequence of the following result a book proof of which has been worked out by DeVore and Lorentz in [38, pages 285–288].

Theorem 12.7 *If* $\lambda_1 < \lambda_2 < \cdots < \lambda_m$ *are integers and* $a_1, a_2, \ldots, a_m$ *are complex numbers, then*

$$\int_0^{2\pi} \left| \sum_{j=1}^m a_j e^{i\lambda_j t} \right| dt \ge \frac{1}{30} \sum_{j=1}^m \frac{|a_j|}{j} .$$

In [65] we proved the following results.

Theorem 12.8 *If* $S \subset \mathbb{Z}$ *is a finite set,* $M = M(S) := \max\{|z| : z \in S\}$, $P \in$ $\mathcal{P}_{2n}(S)$ *is a self-reciprocal polynomial,*

$$T(t) := P(e^{it})e^{-int} ,$$

then

$$NZ^*(T_n) \geq \left(\frac{c}{1 + \log M}\right) \frac{\log\log\log |P(1)|}{\log\log\log\log |P(1)|} - 1$$

with an absolute constant $c > 0$, whenever $|P(1)| \geq 16$.

Corollary 12.9 *If $S \subset \mathbb{Z}$ is a finite set, $M = M(S) := \max\{|z| : z \in S\}$, $P \in \mathcal{P}_n(S)$ is a self-reciprocal polynomial, then*

$$NZ(P) \geq \left(\frac{c}{1 + \log M}\right) \frac{\log\log\log |P(1)|}{\log\log\log\log |P(1)|} - 1$$

with an absolute constant $c > 0$, whenever $|P(1)| \geq 16$.

This improves the exponent $1/2 - \varepsilon$ to $1 - \varepsilon$ in a recent breakthrough result [106] by Sahasrabudhe. We note that in both Sahasrabudhe's paper and this paper the assumption that the finite set S contains only integers is deeply exploited. Our next result is an obvious consequence of Corollary 12.9.

Corollary 12.10 *If the set $S \subset \mathbb{Z}$ is finite, $M = M(S) := \max\{|z| : z \in S\}$,*

$$T(t) = \sum_{j=0}^{n} a_j \cos(jt), \qquad a_j \in S,$$

then

$$NZ^*(T) \geq \left(\frac{c}{1 + \log M}\right) \frac{\log\log\log |T(0)|}{\log\log\log\log |T(0)|} - 1$$

with an absolute constant $c > 0$, whenever $|T(0)| \geq 16$.

13 Bourgain's L_1 Problem and Related Results

For $n \geq 1$ let

$$\mathcal{A}_n := \left\{ P : P(z) = \sum_{j=1}^{n} z^{k_j} : 0 \leq k_1 < k_2 < \cdots < k_n, \, k_j \in \mathbb{Z} \right\},$$

that is, $\mathcal{A}_n$ is the collection of all sums of n distinct monomials. For $p \geq 0$ we define

$$S_{n,p} := \sup_{Q \in \mathcal{A}_n} \frac{M_p(Q)}{\sqrt{n}} \quad \text{and} \quad S_p := \liminf_{n \to \infty} S_{n,p} \leq \Sigma_p := \limsup_{n \to \infty} S_{n,p}.$$

We also define

$$I_{n,p} := \inf_{Q \in \mathcal{A}_n} \frac{M_p(Q)}{\sqrt{n}} \qquad \text{and} \qquad I_p := \limsup_{n \to \infty} I_{n,p} \geq \Omega_p := \liminf_{n \in \to \infty} I_{n,p}.$$

Observe that Parseval's formula gives $\Omega_2 = \Sigma_2 = 1$. The problem of calculating Σ_1 appears in a paper of Bourgain [27]. Deciding whether $\Sigma_1 < 1$ or $\Sigma_1 = 1$ would be a major step toward confirming or disproving other important conjectures. Karatsuba [80] observed that $\Sigma_1 \geq 1/\sqrt{2} \geq 0.707$. Indeed, taking, for instance,

$$P_n(z) = \sum_{k=0}^{n-1} z^{2^k}, \qquad n = 1, 2, \ldots,$$

it is easy to see that

$$M_4(P_n)^4 = 2n(n-1) + n, \tag{14}$$

and as Hölder's inequality implies

$$n = M_2(P_n)^2 \leq M_1(P_n)^{2/3} M_4(P_n)^{4/3},$$

we conclude

$$M_1(P_n) \geq \frac{n^{3/2}}{(2n(n-1)+n)^{1/2}} = \frac{n}{(2n-1)^{1/2}} \geq \frac{\sqrt{n}}{\sqrt{2}}. \tag{15}$$

Similarly, if $S_n := \{a_1 < a_2 < \cdots < a_n\}$ is a Sidon set (that is, S_n is a subset of integers such that no integer has two essentially distinct representations as the sum of two elements of S_n), then the polynomials

$$P_n(z) = \sum_{a \in S_n} z^a, \qquad n = 1, 2, \ldots,$$

satisfy (14) and (15). In fact, it was observed in [15] that

$$\min_{P \in \mathcal{A}_n} M_4(P)^4 = 2n(n-1) + n,$$

and such minimal polynomials in $\mathcal{A}_n$ are precisely constructed by Sidon sets as above.

Improving Karatsuba's result, by using a probabilistic method Aistleitner [1] proved that $\Sigma_1 \geq \sqrt{\pi}/2 \geq 0.886$. We note that P. Borwein and Lockhart [25] investigated the asymptotic behavior of the mean value of normalized L_p norms of Littlewood polynomials for arbitrary $p > 0$. Using the Lindeberg Central Limit Theorem and the Dominated Convergence Theorem, they proved that

$$\lim_{n \to \infty} \frac{1}{2^{n+1}} \sum_{f \in \mathcal{L}_n} \frac{(M_p(f))^p}{n^{p/2}} = \Gamma(1 + p/2),$$

where $\mathcal{L}_n$ is, as before, the set of Littlewood polynomials of degree n. It follows simply from the case $p = 1$ of the result in [25] quoted above that $\Sigma_1 \geq \sqrt{\pi/8} \geq 0.626$. Moreover, this can be achieved by taking the sum of approximately half of the monomials of $\{x^0, x^1, \ldots, x^{2n}\}$ and letting n tend to ∞.

Observe that Parseval's formula gives $\Omega_2 = \Sigma_2 = 1$. In [33] we proved the following results.

Theorem 13.1 *Let (k_j) be a strictly increasing sequence of nonnegative integers satisfying*

$$k_{j+1} > k_j \left(1 + \frac{c_j}{j^{1/2}}\right), \qquad j = 1, 2, \ldots,$$

where $\lim_{j \to \infty} c_j = \infty$. Let

$$P_n(z) = \sum_{j=1}^{n} z^{k_j}, \qquad n = 1, 2, \ldots.$$

We have

$$\lim_{n \to \infty} \frac{M_p(P_n)}{\sqrt{n}} = \Gamma(1 + p/2)^{1/p}$$

for every $p \in (0, 2)$.

Theorem 13.2 *Let (k_j) be a strictly increasing sequence of nonnegative integers satisfying*

$$k_{j+1} > qk_j, \qquad j = 1, 2, \ldots,$$

where $q > 1$. Let

$$P_n(z) = \sum_{j=1}^{n} z^{k_j}, \qquad n = 1, 2, \ldots.$$

We have

$$\lim_{n \to \infty} \frac{M_p(P_n)}{\sqrt{n}} = \Gamma(1 + p/2)^{1/p}$$

for every $p \in [1, \infty)$.

Corollary 13.3 *We have $\Sigma_p \geq S_p \geq \Gamma(1 + p/2)^{1/p}$ for all $p \in (0, 2)$.*

The special case $p = 1$ recaptures a recent result of Aistleitner [1], the best known lower bound for Σ_1.

Corollary 13.4 *We have* $\Sigma_1 \geq S_1 \geq \sqrt{\pi}/2$.

Corollary 13.5 *We have* $\Omega_p \leq I_p \leq \Gamma(1 + p/2)^{1/p}$ *for all* $p \in (2, \infty)$.

We remark here that the same results also hold for the polynomials $\sum_{j=1}^{n} a_j z^{k_j}$ with coefficients a_j if a general form of the Salem-Zygmund theorem is used (e.g. see (2) in [69]).

Our final result in [33] shows that the upper bound $\Gamma(1 + p/2)^{1/p}$ in Corollary 13.5 is optimal at least for even integers.

Corollary 13.6 *For any even integer* $p = 2m \geq 2$, *we have*

$$\lim_{n \to \infty} \min_{P \in \mathcal{A}_n} \frac{M_p(P)}{\sqrt{n}} = \Gamma(1 + p/2)^{1/p}.$$

Observe that a standard way to prove a Nikolskii-type inequality for trigonometric polynomials [2, p. 394] applies to the classes $\mathcal{A}_n$. Indeed,

$$M_p(P)^p = \frac{1}{2\pi} \int_0^{2\pi} \left| P(e^{it}) \right|^p dt$$

$$\leq \left(\frac{1}{2\pi} \int_0^{2\pi} \left| P(e^{it}) \right|^2 dt \right) \left(\max_{t \in [0, 2\pi]} \left| P(e^{it}) \right| \right)^{p-2}$$

$$= nn^{p-2} = n^{p-1}$$

for every $P \in \mathcal{A}_n$ and $p \geq 2$, and the Dirichlet kernel $D_n(z) := 1 + z + \cdots + z^n$ shows the sharpness of this upper bound up to a multiplicative factor constant $c > 0$. So if we study the original Bourgain problem in the case of $p > 2$, we should normalize by dividing by $n^{1-1/p}$ rather than $n^{1/2}$.

In [34] we examined

$$S_{n,0}(I) := \sup_{Q \in \mathcal{A}_n} \frac{M_0(Q, I)}{\sqrt{n}} \quad \text{and} \quad S_0(I) := \liminf_{n \to \infty} S_{n,0}(I)$$

for intervals $I = [\alpha, \beta]$ with $0 < |I| := \beta - \alpha \leq 2\pi$ and proved the following results.

Theorem 13.7 *There are polynomials* $Q_n \in \mathcal{A}_n \cap \mathcal{P}_N$ *with* $N = 2n + o(n)$ *such that*

$$M_0(Q_n) \geq \left(\frac{1}{2\sqrt{2}} + o(1) \right) \sqrt{n}, \qquad n = 1, 2, \dots,$$

and hence $S_0 \geq \dfrac{1}{2\sqrt{2}}$.

Theorem 13.8 *There are polynomials $Q_n \in \mathcal{A}_n \cap \mathcal{P}_N$ with $N = 2n + o(n)$, an absolute constant $c_1 > 0$, and a constant $c_2(\varepsilon) > 0$ depending only on $\varepsilon > 0$ such that*

$$M_0(Q_n, I) \geq c_1 \sqrt{n} , \qquad n = 1, 2, \ldots ,$$

for every interval $I := [\alpha, \beta] \subset \mathbb{R}$ such that

$$\frac{4\pi}{n} \leq \frac{(\log n)^{3/2}}{n^{1/2}} \leq \beta - \alpha \leq 2\pi , \tag{16}$$

while

$$M_1(Q_n, I) \leq c_2(\varepsilon) \sqrt{n} , \qquad n = 1, 2, \ldots ,$$

for every interval $I := [\alpha, \beta] \subset \mathbb{R}$ such that

$$(n/2)^{-1/2+\varepsilon} \leq \beta - \alpha \leq 2\pi . \tag{17}$$

Note that Theorem 13.8 implies that there is an absolute constant $c_1 > 0$ such that $S_0(I) \geq c_1$ for all intervals $I := [\alpha, \beta] \subset \mathbb{R}$ satisfying (16).

Theorem 13.9 *There are polynomials $Q_n \in \mathcal{L}_n$ such that*

$$M_0(Q_n) \geq \left(\frac{1}{2} + o(1) \right) \sqrt{n} , \qquad n = 1, 2, \ldots .$$

Theorem 13.10 *There are polynomials $Q_n \in \mathcal{L}_n$, an absolute constant $c_1 > 0$, and a constant $c_2(\varepsilon) > 0$ depending only on $\varepsilon > 0$ such that*

$$M_0(Q_n, I) \geq c_1 \sqrt{n} , \qquad n = 1, 2, \ldots ,$$

for every interval $I := [\alpha, \beta] \subset \mathbb{R}$ satisfying (16), while

$$M_1(Q_n, I) \leq c_2(\varepsilon) \sqrt{n} , \qquad n = 1, 2 \ldots ,$$

for every interval $I := [\alpha, \beta] \subset \mathbb{R}$ satisfying (17).

Reference

1. C. Aistleitner, On a problem of Bourgain concerning the L_1-norm of exponential sums. Math. Z. **275**, 681–688 (2013)
2. J.-P. Allouche, K.-K.S. Choi, A. Denise, T. Erdélyi, B. Saffari, Bounds on autocorrelation coefficients of Rudin-Shapiro polynomials. Anal. Math. **45**, 705 (2019). https://doi.org/10.1007/s10476-019-0003-4

3. R.C. Baker, H.L. Montgomery, Oscillations of quadratic L-functions, in *Analytic Number Theory*, ed. by B.C. Berndt et al. (Birkhäuser, Boston, 1990), pp. 23–40
4. P.T. Bateman, G.B. Purdy, S.S Wagstaff, Jr., Some numerical results on Fekete polynomials. Math. Comput. **29**, 7–23 (1975)
5. J. Beck, "Flat" polynomials on the unit circle – note on a problem of Littlewood. Bull. Lond. Math. Soc. **23**, 269–277 (1991)
6. J. Bell, I. Shparlinski, Power series approximations to Fekete polynomials. J. Approx. Theory **222**, 132–142 (2017)
7. E. Beller, D.J. Newman, An extremal problem for the geometric mean of polynomials. Proc. Am. Math. Soc. **39**, 313–317 (1973)
8. A. Bloch, G. Pólya, On the roots of certain algebraic equations. Proc. Lond. Math. Soc. **33**, 102–114 (1932)
9. E. Bombieri, J. Bourgain, On Kahane's ultraflat polynomials. J. Eur. Math. Soc. **11**(3), 627–703 (2009)
10. P. Borwein, *Computational Excursions in Analysis and Number Theory* (Springer, New York, 2002)
11. P. Borwein, K.-K.S. Choi, Merit factors of polynomials formed by Jacobi symbols. Can. J. Math. **53**(1), 33–50 (2001)
12. P. Borwein, K.-K.S. Choi, Explicit merit factor formulae for Fekete and Turyn polynomials. Trans. Am. Math. Soc. **354**(1), 219–234 (2002)
13. P. Borwein, K.-K.S. Choi, R. Ferguson, J. Jankauskas, On Littlewood polynomials with prescribed number of zeros inside the unit disk. Can. J. Math. **67**, 507–526 (2015)
14. P. Borwein, K.-K.S. Choi, J. Jedwab, Binary sequences with merit factor greater than 6.34. IEEE Trans. Inform. Theory **50**(12), 3234–3249 (2004)
15. P. Borwein, K.-K.S. Choi, I. Mercer, Expected norms of zero-one polynomials. Can. Math. Bull. **51**, 297–507 (2008)
16. P. Borwein, K.-K.S. Choi, S. Yazdani, An extremal property of Fekete polynomials. Proc. Am. Math. Soc. **129**(1), 19–27 (2001)
17. P. Borwein, T. Erdélyi, *Polynomials and Polynomial Inequalities* (Springer, New York, 1995)
18. P. Borwein, T. Erdélyi, On the zeros of polynomials with restricted coefficients. Ill. J. Math. **41**(4), 667–675 (1997)
19. P. Borwein, T. Erdélyi, Trigonometric polynomials with many real zeros and a Littlewood-type problem. Proc. Am. Math. Soc. **129**(3), 725–730 (2001)
20. P. Borwein, T. Erdélyi, Lower bounds for the number of zeros of cosine polynomials in the period: a problem of Littlewood. Acta Arith. **128**(4), 377–384 (2007)
21. P. Borwein, T. Erdélyi, R. Ferguson, R. Lockhart, On the zeros of cosine polynomials: solution to a problem of Littlewood. Ann. Math. (2) **167**(3), 1109–1117 (2008)
22. P. Borwein, T. Erdélyi, G. Kós, Littlewood-type problems on [0, 1]. Proc. Lond. Math. Soc. **79**, 22–46 (1999)
23. P. Borwein, T. Erdélyi, G. Kós, The multiplicity of the zero at 1 of polynomials with constrained coefficients. Acta Arith. **159**(4), 387–395 (2013)
24. P. Borwein, T. Erdélyi, F. Littmann, Zeros of polynomials with finitely many different coefficients. Trans. Am. Math. Soc. **360**, 5145–5154 (2008)
25. P. Borwein, R. Lockhart, The expected L_p norm of random polynomials. Proc. Am. Math. Soc. **129**, 1463–1472 (2001)
26. P. Borwein, M.J. Mossinghoff, Rudin-Shapiro like polynomials in L_4. Math. Comput. **69**, 1157–1166 (2000)
27. J. Bourgain, On the sprectal type of Ornstein's class one transformations. Israel J. Math. **84**, 53–63 (1993)
28. D. Boyd, On a problem of Byrnes concerning polynomials with restricted coefficients. Math. Comput. **66**, 1697–1703 (1997)
29. J. Brillhart, On the Rudin-Shapiro polynomials. Duke Math. J. **40**(2), 335–353 (1973)
30. J. Brillhart, J.S. Lemont, P. Morton, Cyclotomic properties of the Rudin-Shapiro polynomials. J. Reine Angew. Math. (Crelle's J.) **288**, 37–65 (1976)

31. K.-K.S. Choi, Bounds on autocorrelation coefficients of Rudin-Shapiro polynomials II. J. Approx. Theory (2019)
32. K.-K.S. Choi, T. Erdélyi, On the average Mahler measures on Littlewood polynomials. Proc. Am. Math. Soc. Ser. B **1**, 105–120 (2015)
33. K.-K.S. Choi, T. Erdélyi, On a problem of Bourgain concerning the L_p norms of exponential sums. Math. Zeit. **279**, 577–584 (2015)
34. K.-K.S. Choi, T. Erdélyi, Sums of monomials with large Mahler measure. J. Approx. Theory **197**, 49–61 (2015)
35. K.-K.S. Choi, M.J. Mossinghoff, Average Mahler's measure and L_p norms of unimodular polynomials. Pacific J. Math. **252**(1), 31–50 (2011)
36. S. Chowla, Note on Dirichlet's L-functions. Acta Arith. **1**, 113–114 (1935)
37. B. Conrey, A. Granville, B. Poonen, K. Soundararajan, Zeros of Fekete polynomials. Ann. Inst. Fourier (Grenoble) **50**, 865–884 (2000)
38. R.A. DeVore, G.G. Lorentz, *Constructive Approximation* (Springer, Berlin, 1993)
39. Ch. Doche, Even moments of generalized Rudin-Shapiro polynomials. Math. Comput. **74**(252), 1923–1935 (2005)
40. Ch. Doche, L. Habsieger, Moments of the Rudin-Shapiro polynomials. J. Fourier Anal. Appl. **10**(5), 497–505 (2004)
41. P. Drungilas, Unimodular roots of reciprocal Littlewood polynomials. J. Korean Math. Soc. **45**(3), 835–840 (2008)
42. A. Dubickas, On the order of vanishing at 1 of a polynomial. Lith. Math. J. **39**, 365–370 (1999)
43. A. Dubickas, Polynomials with multiple roots at 1. Int. J. Number Theory **10**(2), 391–400 (2014)
44. S.B. Ekhad, D. Zeilberger, Integrals involving Rudin-Shapiro polynomials and sketch of a proof of Saffari's conjecture, in *Analytic Number Theory, Modular Forms and q-Hypergeometric Series*. Springer Proceedings in Mathematics and Statistics, vol. 221 (Springer, Cham, 2017), pp. 253–265
45. T. Erdélyi, The phase problem of ultraflat unimodular polynomials: the resolution of the conjecture of Saffari. Math. Ann. **300**, 39–60 (2000)
46. T. Erdélyi, The resolution of Saffari's Phase Problem. C. R. Acad. Sci. Paris Sér. I Math. **331**, 803–808 (2000)
47. T. Erdélyi, How far is a sequence of ultraflat unimodular polynomials from being conjugate reciprocal. Michigan Math. J. **49**, 259–264 (2001)
48. T. Erdélyi, Proof of Saffari's near-orthogonality conjecture for ultraflat sequences of unimodular polynomials. C. R. Acad. Sci. Paris Sér. I Math. **333**, 623–628 (2001)
49. T. Erdélyi, On the zeros of polynomials with Littlewood-type coefficient constraints. Michigan Math. J. **49**, 97–111 (2001)
50. T. Erdélyi, Polynomials with Littlewood-type coefficient constraints, in *Approximation Theory X: Abstract and Classical Analysis*, ed. by C.K. Chui, L.L. Schumaker, J. Stöckler (Vanderbilt University Press, Nashville, 2002), pp. 153–196
51. T. Erdélyi, On the real part of ultraflat sequences of unimodular polynomials. Math. Ann. **326**, 489–498 (2003)
52. T. Erdélyi, An improvement of the Erdős-Turán theorem on the distribution of zeros of polynomials. C. R. Acad. Sci. Paris, Ser. I **346**(5), 267–270 (2008)
53. T. Erdélyi, Extensions of the Bloch-Pólya theorem on the number of real zeros of polynomials. J. Théor. Nombres Bordeaux **20**(2), 281–287 (2008)
54. T. Erdélyi, Sieve-type lower bounds for the Mahler measure of polynomials on subarcs. Comput. Methods Funct. Theory **11**, 213–228 (2011)
55. T. Erdélyi, Upper bounds for the L_q norm of Fekete polynomials on subarcs. Acta Arith. **153**(1), 81–91 (2012)
56. T. Erdélyi, On the flatness of conjugate reciprocal unimodular polynomials. J. Math. Anal. Appl. **432**(2), 699–714 (2015)

57. T. Erdélyi, Coppersmith-Rivlin type inequalities and the order of vanishing of polynomials at 1. Acta Arith. **172**(3), 271–284 (2016)
58. T. Erdélyi, On the number of unimodular zeros of self-reciprocal polynomials with coefficients from a finite set. Acta Arith. **176**(2), 177–200 (2016)
59. T. Erdélyi, The Mahler measure of the Rudin-Shapiro polynomials. Constr. Approx. **43**(3), 357–369 (2016)
60. T. Erdélyi, Improved lower bound for the Mahler measure of the Fekete polynomials. Constr. Approx. **48**(2), 383–399 (2018)
61. T. Erdélyi, The asymptotic value of the Mahler measure of the Rudin-Shapiro polynomials. J. Anal. Math. (accepted)
62. T. Erdélyi, On the oscillation of the modulus of Rudin-Shapiro polynomials on the unit circle. Mathematika **66**, 144–160 (2020)
63. T. Erdélyi, Improved results on the oscillation of the modulus of Rudin-Shapiro polynomials on the unit circle. Proc. Am. Math. Soc. (2019, accepted)
64. T. Erdélyi, The asymptotic distance between an ultraflat unimodular polynomial and its conjugate reciprocal. https://arxiv.org/abs/1810.04287
65. T. Erdélyi, Improved lower bound for the number of unimodular zeros of self-reciprocal polynomials with coefficients from a finite set. Acta Arith. (2019). https://arxiv.org/pdf/1702.05823.pdf
66. T. Erdélyi, D. Lubinsky, Large sieve inequalities via subharmonic methods and the Mahler measure of Fekete polynomials. Can. J. Math. **59**, 730–741 (2007)
67. T. Erdélyi, P. Nevai, On the derivatives of unimodular polynomials (Russian). Mat. Sbornik **207**(4), 123–142 (2016); translation in Sbornik Math. **207**(3–4), 590–609 (2016)
68. P. Erdős, Some unsolved problems. Michigan Math. J. **4**, 291–300 (1957)
69. P. Erdős, On trigonometric sums with gaps. Publ. Math. Inst. Hung. Acad. Sci. Ser. A **7**, 37–42 (1962)
70. P. Erdős, P. Turán, On the distribution of roots of polynomials. Ann. Math. **51**, 105–119 (1950)
71. M. Fekete, G. Pólya, Über ein Problem von Laguerre. Rend. Circ. Mat. Palermo **34**, 89–120 (1912)
72. M.J. Golay, Static multislit spectrometry and its application to the panoramic display of infrared spectra. J. Opt. Soc. Am. **41**, 468–472 (1951)
73. C. Günther, K.-U. Schmidt, L_q norms of Fekete and related polynomials. Can. J. Math. **69**(4), 807–825 (2017)
74. J. Jedwab, D.J. Katz, K.-U. Schmidt, Advances in the merit factor problem for binary sequences. J. Combin. Theory Ser. A **120**(4), 882–906 (2013)
75. J. Jedwab, D.J. Katz, K.-U. Schmidt, Littlewood polynomials with small L_4 norm. Adv. Math. **241**, 127–136 (2013)
76. G.H. Hardy, J.E. Littlewood, G. Pólya, *Inequalities* (Cambridge University Press, London, 1952)
77. H. Heilbronn, On real characters. Acta Arith. **2**, 212–213 (1937)
78. J. Jung, S.W. Shin, On the sparsity of positive-definite automorphic forms within a family. J. Anal. Math. **129**(1), 105–138 (2016)
79. J.P. Kahane, Sur les polynomes a coefficient unimodulaires. Bull. Lond. Math. Soc. **12**, 321–342 (1980)
80. A.A. Karatsuba, An estimate for the L_1 -norm of an exponential sum. Mat. Zametki **64**, 465–468 (1998)
81. S.V. Konyagin, On a problem of Littlewood. Math. USSR Izvestia **18**, 205–225 (1981)
82. S.V. Konyagin, V.F. Lev, Character sums in complex half planes. J. Theor. Nombres Bordeaux **16**(3), 587–606 (2004)
83. T. Körner, On a polynomial of J.S. Byrnes. Bull. Lond. Math. Soc. **12**, 219–224 (1980)
84. J.E. Littlewood, On the mean values of certain trigonometrical polynomials. J. Lond. Math. Soc. **36**, 307–334 (1961)
85. J.E. Littlewood, On the real roots of real trigonometrical polynomials (II). J. Lond. Math. Soc. **39**, 511–552 (1964)

86. J.E. Littlewood, The real zeros and value distributions of real trigonometrical polynomials. J. Lond. Math. Soc. **41**, 336–342 (1966)

87. J.E. Littlewood, On polynomials $\sum \pm z^m$, $\sum \exp(\alpha_m i) z^m$, $z = e^{i\theta}$. J. Lond. Math. Soc. **41**, 367–376 (1966)

88. J.E. Littlewood, *Some Problems in Real and Complex Analysis* (Heath Mathematical Monographs, Lexington, 1968)

89. O.C. McGehee, L. Pigno, B. Smith, Hardy's inequality and the L_1 norm of exponential sums. Ann. Math. **113**, 613–618 (1981)

90. I.D. Mercer, Unimodular roots of special Littlewood polynomials. Can. Math. Bull. **49**(3), 438–447 (2006)

91. H.L. Montgomery, An exponential polynomial formed with the Legendre symbol. Acta Arith. **37**, 375–380 (1980)

92. H.L. Montgomery, Littlewood polynomials, in *Analytic Number Theory, Modular Forms and q-Hypergeometric Series*, ed. by G. Andrews, F. Garvan. Springer Proceedings in Mathematics and Statistics, vol. 221 (Springer, Cham, 2017), pp. 533–553

93. K. Mukunda, Littlewood Pisot numbers. J. Number Theory **117**(1), 106–121 (2006)

94. H. Nguyen, O. Nguyen, V. Vu, On the number of real roots of random polynomials. Commun. Contemp. Math. **18**, 1550052 (2016)

95. A. Odlyzko, Search for ultraflat polynomials with plus and minus one coefficients, in *Connections in Discrete Mathematics*, ed. by S. Butler, J. Cooper, G. Hurlbert (Cambridge University Press, Cambridge, 2018), pp. 39–55

96. A.M. Odlyzko, B. Poonen, Zeros of polynomials with 0, 1 coefficients. L'Enseign. Math. **39**, 317–348 (1993)

97. C. Pinner, Double roots of $[-1, 1]$ power series and related matters. Math. Comput. **68**, 1149–1178 (1999)

98. G. Pólya, Verschiedene Bemerkungen zur Zahlentheorie. Jahresber. Dtsch. Math. Ver. **28**, 31–40 (1919)

99. I.E. Pritsker, A.A. Sola, Expected discrepancy for zeros of random algebraic polynomials. Proc. Am. Math. Soc. **142**, 4251–4263 (2014)

100. H. Queffelec, B. Saffari, Unimodular polynomials and Bernstein's inequalities. C. R. Acad. Sci. Paris Sér. I Math. **321**(3), 313–318 (1995)

101. H. Queffelec, B. Saffari, On Bernstein's inequality and Kahane's ultraflat polynomials. J. Fourier Anal. Appl. **2**(6), 519–582 (1996)

102. B. Rodgers, On the distribution of Rudin-Shapiro polynomials and lacunary walks on $SU(2)$, to appear in Adv. Math. arxiv.org/abs/1606.01637

103. J.B. Rosser, L. Schoenfeld, Approximate formulas for some functions of prime numbers. Ill. J. Math. **6**, 64–94 (1962)

104. B. Saffari, The phase behavior of ultraflat unimodular polynomials, in *Probabilistic and Stochastic Methods in Analysis, with Applications*, ed. by J.S. Byrnes, Jennifer L. Byrnes, Kathryn A. Hargreaves, K. Berry (Kluwer Academic Publishers, Dordrecht, 1992), pp. 555–572.

105. B. Saffari, Some polynomial extremal problems which emerged in the twentieth century, in *Twentieth Century Harmonic Analysis – A Celebration*, ed. by J.S. Byrnes (Kluwer Academic Publishers, Dordrecht, 2001), pp. 201–233

106. J. Sahasrabudhe, Counting zeros of cosine polynomials: on a problem of Littlewood. Adv. Math. **343**, 495–521 (2019)

107. E. Schmidt, Über algebraische Gleichungen vom Pólya-Bloch-Typos. Sitz. Preuss. Akad. Wiss., Phys.-Math. Kl. 321 (1932)

108. I. Schur, Untersuchungen über algebraische Gleichungen. Sitz. Preuss. Akad. Wiss., Phys.-Math. Kl. 403–428 (1933)

109. H.S. Shapiro, Extremal problems for polynomials and power series, Master thesis, MIT (1951)

110. K. Soundararajan, Equidistribution of zeros of polynomials. Am. Math. Mon. **126**(3), 226–236 (2019, accepted)

111. G. Szegő, Bemerkungen zu einem Satz von E. Schmidt uber algebraische Gleichungen. Sitz. Preuss. Akad. Wiss., Phys.-Math. Kl. 86–98 (1934)
112. T. Tao, V. Vu, Local universality of zeros of random polynomials. Int. Math. Res. Not. **13**, 5053–5139 (2015)
113. V. Totik, Distribution of simple zeros of polynomials. Acta Math. **170**, 1–28 (1993)
114. V. Totik, P. Varjú, Polynomials with prescribed zeros and small norm. Acta Sci. Math. (Szeged) **73**, 593–612 (2007)

Classes of Nonnegative Sine Polynomials

Horst Alzer and Man Kam Kwong

Abstract We present several one-parameter classes of nonnegative sine polynomials. One of our theorems states that the inequality

$$0 \le \sum_{k=1}^{n} \left(\frac{1}{n} + \frac{1}{k} \right) (n - k + \alpha) \sin(kx) \quad (\alpha \in \mathbb{R})$$

holds for all $n \ge 1$ and $x \in [0, \pi]$ if and only if $\alpha \in [0, 3]$. This extends a result of Dimitrov and Merlo (2002), who proved the inequality for $\alpha = 1$.

Keywords Sine polynomials · Inequalities

2010 Mathematics Subject Classification 26D05

1 Introduction and Statement of Main Results

The classical Fejér-Jackson-Gronwall inequality

$$0 < \sum_{k=1}^{n} \frac{\sin(kx)}{k} \quad (n \ge 1; 0 < x < \pi) \tag{1}$$

is one of the first inequalities for sine polynomials which appeared in the literature. The validity of (1) was conjectured by Fejér in 1910 and the first proofs were published by Jackson [14] and Gronwall [13] in 1911 and 1912, respectively. This

H. Alzer (✉)
Waldbröl, Germany
e-mail: h.alzer@gmx.de

M. K. Kwong
Department of Applied Mathematics, The Hong Kong Polytechnic University, Hong Kong, China
e-mail: mankwong@connect.polyu.hk

© Springer Nature Switzerland AG 2020
A. Raigorodskii, M. T. Rassias (eds.), *Trigonometric Sums and Their Applications*,
https://doi.org/10.1007/978-3-030-37904-9_3

result has attracted the attention of many mathematicians, who discovered new proofs as well as refinements, counterparts and variants of (1). In 1958, Vietoris [24] proved a remarkable inequality for a certain class of sine polynomials which includes (1) as special case.

Vietoris' theorem *If*

$$a_1 \geq a_2 \geq \cdots \geq a_n > 0 \quad and \quad 2ka_{2k} \leq (2k-1)a_{2k-1} \quad (1 \leq k \leq n/2), \qquad (2)$$

then

$$0 < \sum_{k=1}^{n} a_k \sin(kx) \quad (n \geq 1; 0 < x < \pi).$$

For more information on this theorem we refer to Alzer, Koumandos and Lamprecht [1], Askey [3], Askey and Steinig [5], Koumandos [15], Kwong [17, 18]. Numerous related results for further classes of trigonometric sums were published by Belov [6], Brown and Hewitt [7], Brown and Wilson [8], Mondal and Swaminathan [22], Turán [23] and others.

A collection of the most important and attractive inequalities for trigonometric polynomials and its relatives as well as historical comments on this subject can be found in Askey [2], Askey and Gasper [4], Milovanović, Mitrinović and Rassias [21, chapter 4]. Information on applications of these inequalities in fields – for instance, Fourier analysis, approximation theory and geometric function theory are given in Dimitrov and Merlo [9] and Koumandos [16].

It is the aim of this paper to present new one-parameter classes of nonnegative trigonometric polynomials. More precisely, we provide necessary and sufficient conditions for the parameters such that certain sine polynomials are nonnegative for all n and $x \in [0, \pi]$. Our work has been inspired by an interesting research article published by Dimitrov and Merlo [9] in 2002. One of the inequalities given in [9] states that

$$0 \leq \sum_{k=1}^{n} \left(\frac{1}{n} + \frac{1}{k} \right)(n - k + 1) \sin(kx) \quad (n \geq 1; 0 \leq x \leq \pi). \qquad (3)$$

Our first result offers an extension of (3).

Theorem 1 *The inequality*

$$0 \leq \sum_{k=1}^{n} \left(\frac{1}{n} + \frac{1}{k} \right)(n - k + \alpha) \sin(kx) \quad (\alpha \in \mathbb{R}) \qquad (4)$$

holds for all integers $n \geq 1$ and real numbers $x \in [0, \pi]$ if and only if $\alpha \in [0, 3]$.

Inequality (4) with $\alpha = 0$ can be written in a more elegant form:

$$0 \le \sum_{k=1}^{n-1} \left(\frac{n}{k} - \frac{k}{n}\right) \sin(kx) \quad (n \ge 1; 0 \le x \le \pi). \tag{5}$$

This inequality can be generalized. In fact, an application of Vietoris' theorem reveals that (5) remains valid if we replace the factor $(n/k - k/n)$ by $(n/k - k/n)^\beta$ with $\beta \ge 1$. In view of this result it is natural to ask for *all* real parameters β such that the modified inequality holds. The following theorem gives an answer to this question.

Theorem 2 *The inequality*

$$0 \le \sum_{k=1}^{n-1} \left(\frac{n}{k} - \frac{k}{n}\right)^\beta \sin(kx) \quad (\beta \in \mathbb{R}) \tag{6}$$

holds for all integers $n \ge 2$ *and real numbers* $x \in [0, \pi]$ *if and only if* $\beta \ge \log(2)/\log(16/5) = 0.59592\ldots$..

It is well-known that (1) remains valid if the summation runs over only odd k. Indeed, we have

$$0 < \sum_{\substack{k=1 \\ k \text{ odd}}}^{n} \frac{\sin(kx)}{k} \quad (n \ge 1; 0 < x < \pi). \tag{7}$$

For an interesting extension of (7) we refer to Meynieux and Tudor [20]. With regard to this result we ask for corresponding inequalities for the sums given in (4) and (6). The following counterpart of (4) is valid.

Theorem 3 *The inequality*

$$0 \le \sum_{\substack{k=1 \\ k \text{ odd}}}^{n} \left(\frac{1}{n} + \frac{1}{k}\right)(n - k + a) \sin(kx) \quad (a \in \mathbb{R}) \tag{8}$$

holds for all integers $n \ge 1$ *and real numbers* $x \in [0, \pi]$ *if and only if* $a \ge 0$.

Our fourth theorem offers a companion to inequality (6).

Theorem 4 *The inequality*

$$0 \le \sum_{\substack{k=1 \\ k \text{ odd}}}^{n-1} \left(\frac{n}{k} - \frac{k}{n}\right)^b \sin(kx) \quad (b \in \mathbb{R}) \tag{9}$$

holds for all integers $n \ge 2$ *and real numbers* $x \in [0, \pi]$ *if and only if* $b \ge 0$.

In the next section, we collect several lemmas which we need to prove our theorems. The proofs of the four theorems are presented in Sections 3–6. We conclude the paper with a few remarks which are given in Section 7.

The numerical values have been calculated via the computer program MAPLE 13.

2 Lemmas

The first three lemmas are due to Fejér [10–12]. They present inequalities for certain sine polynomials.

Lemma 1 *If* $a_0 \geq a_1 \geq \cdots \geq a_m \geq 0$ *and* $\theta \in [0, 2\pi]$, *then*

$$\sum_{k=0}^{m} a_k \sin((k + 1/2)\theta) \geq 0. \tag{10}$$

Lemma 2 *If* $m \geq 2$ *and* $x \in [0, \pi]$, *then*

$$\sum_{k=1}^{m-1} \sin(kx) + \frac{1}{2} \sin(mx) \geq 0.$$

We recall that a sequence $\{a_1, \ldots, a_m\}$ is said to be convex, if

$$a_{k-1} - 2a_k + a_{k+1} \geq 0 \quad (k = 2, \ldots, m - 1).$$

Remark 1 Let $\gamma > \beta > 0$. If $\{a_1^\beta, \ldots, a_m^\beta\}$ is convex, then $\{a_1^\gamma, \ldots, a_m^\gamma\}$ is convex.

Lemma 3 *If* $\{a_1, \ldots, a_m, 0\}$ *is convex and* $a_m \geq 0$, *then, for* $x \in [0, \pi]$,

$$\sum_{k=1}^{m} a_k \sin(kx) \geq 0.$$

Remark 2 The condition $\{a_1, \ldots, a_m, 0\}$ is convex is equivalent to $\{a_1, \ldots, a_m\}$ is convex and $a_{m-1} \geq 2a_m$.

Lemma 4 *If* $b_3 \geq 0$ *and*

$$b_3 + 2 |b_2 - b_3| + 2 |2b_4 - b_3| \leq 1, \tag{11}$$

then, for $x \in [0, \pi]$,

$$\sin(x) + \sum_{k=2}^{4} b_k \sin(kx) \geq 0. \tag{12}$$

Proof Let $R(x)$ be the expression on the left-hand side of (12). Then,

$$R(x) = \sum_{k=1}^{4} Q_k(x),$$

where

$$Q_1(x) = \left(1 - b_3 - 2\,|b_2 - b_3| - 2\,|2b_4 - b_3|\right)\sin(x),$$

$$Q_2(x) = b_3\left(\sum_{k=1}^{3}\sin(kx) + \frac{1}{2}\sin(4x)\right),$$

$$Q_3(x) = 2\,|b_2 - b_3|\sin(x) + (b_2 - b_3)\sin(2x),$$

$$Q_4(x) = 2\,|2b_4 - b_3|\sin(x) + (b_4 - b_3/2)\sin(4x).$$

From (11) and Lemma 2 we conclude that Q_1 and Q_2 are nonnegative on $[0, \pi]$. Let $p_1 = b_2 - b_3$ and $p_2 = 2b_4 - b_3$. The representations

$$Q_3(x) = 2\sin(x)\left(|p_1| + p_1\cos(x)\right) \quad \text{and} \quad Q_4(x) = 2\sin(x)\left(|p_2| + p_2\cos(x)\cos(2x)\right)$$

show that $Q_3(x) \geq 0$ and $Q_4(x) \geq 0$ for $x \in [0, \pi]$. It follows that R is nonnegative on $[0, \pi]$.

Lemma 5 *Let*

$$J_\beta(x) = 1 - \left(\frac{7}{45}\right)^\beta + 2\left(\frac{2}{5}\right)^\beta x + 4\left(\frac{7}{45}\right)^\beta x^2. \tag{13}$$

If $\beta \in [0.5, 1]$ and $x \in \mathbb{R}$, then $J_\beta(x) > 0$.

Proof Let

$$\tau_\beta = -\frac{1}{4}\left(\frac{18}{7}\right)^\beta.$$

Then, for $\beta \in [0.5, 1]$ and $x \in \mathbb{R}$,

$$J_\beta(x) \geq J_\beta(\tau_\beta) = 1 - \left(\frac{7}{45}\right)^\beta - \frac{1}{4}\left(\frac{36}{35}\right)^\beta \geq 1 - \left(\frac{7}{45}\right)^{0.5} - \frac{9}{35} = 0.34\ldots.$$

Lemma 6 *Let*

$$y_2(\beta) = \left(\frac{7}{16}\right)^\beta, \quad y_3(\beta) = \left(\frac{2}{9}\right)^\beta, \quad y_4(\beta) = \left(\frac{3}{32}\right)^\beta. \tag{14}$$

If $\beta \in [0.8, 1]$, then the sequence $\{1, y_2(\beta), y_3(\beta), y_4(\beta)\}$ is convex.

Proof We have to show that the functions

$$\Phi(\beta) = 1 + \left(\frac{2}{9}\right)^{\beta} - 2\left(\frac{7}{16}\right)^{\beta} \quad \text{and} \quad \Lambda(\beta) = \left(\frac{7}{16}\right)^{\beta} + \left(\frac{3}{32}\right)^{\beta} - 2\left(\frac{2}{9}\right)^{\beta}$$

are nonnegative on $[0.8, 1]$. Let $0.8 \le r \le \beta \le s \le 1$. Then,

$$\Phi(\beta) \ge \Phi^*(r, s) \quad \text{and} \quad \Lambda(\beta) \ge \Lambda^*(r, s),$$

where

$$\Phi^*(r, s) = 1 + \left(\frac{2}{9}\right)^{s} - 2\left(\frac{7}{16}\right)^{r} \quad \text{and} \quad \Lambda^*(r, s) = \left(\frac{7}{16}\right)^{s} + \left(\frac{3}{32}\right)^{s} - 2\left(\frac{2}{9}\right)^{r}.$$

Since

$$\Phi^*(0.8, 1) > 0 \quad \text{and} \quad \Lambda^*\left(0.8 + \frac{k}{100}, 0.8 + \frac{k+1}{100}\right) > 0 \quad (k = 0, 1, \ldots, 19),$$

we conclude that $\Phi(\beta) > 0$ and $\Lambda(\beta) > 0$ for $\beta \in [0.8, 1]$.

Lemma 7 *Let*

$$Y(\beta) = 1 + 3\left(\frac{2}{9}\right)^{\beta} - 2\left(\frac{7}{16}\right)^{\beta} - 4\left(\frac{3}{32}\right)^{\beta}. \tag{15}$$

If $\beta \in [0.59, 0.81]$, then $Y(\beta) > 0$.

Proof Let $0.59 \le r \le \beta \le s \le 0.81$. Then,

$$Y(\beta) \ge 1 + 3\left(\frac{2}{9}\right)^{s} - 2\left(\frac{7}{16}\right)^{r} - 4\left(\frac{3}{32}\right)^{r} = Y^*(r, s), \quad \text{say.}$$

We have

$$Y^*\left(0.59 + \frac{k}{250}, 0.59 + \frac{k+1}{250}\right) > 0 \quad (k = 0, 1, \ldots, 54).$$

This yields $Y(\beta) > 0$ for $\beta \in [0.59, 0.81]$.

Lemma 8 *Let*

$$z_1(\beta) = 1 + \frac{3}{2}\left(\frac{1}{7}\right)^{\beta} - 5\left(\frac{11}{175}\right)^{\beta}, \quad z_2(\beta) = \left(\frac{16}{35}\right)^{\beta} - \left(\frac{1}{7}\right)^{\beta},$$

$$z_3(\beta) = \left(\frac{9}{35}\right)^{\beta} - \left(\frac{1}{7}\right)^{\beta}. \tag{16}$$

If $\beta \in [0.59, 0.85]$, then the sequence $\{z_1(\beta), z_2(\beta), z_3(\beta), 0\}$ is convex and $z_3(\beta) > 0$.

Proof We have to show that for $\beta \in [0.59, 0.85]$,

$$K(\beta) - L(\beta) \geq 0 \quad \text{and} \quad B(\beta) \geq 0, \tag{17}$$

where

$$K(\beta) = 1 + \frac{5}{2}\left(\frac{1}{7}\right)^{\beta} + \left(\frac{9}{35}\right)^{\beta}, \quad L(\beta) = 5\left(\frac{11}{175}\right)^{\beta} + 2\left(\frac{16}{35}\right)^{\beta}$$

and

$$B(\beta) = \left(\frac{1}{7}\right)^{\beta} + \left(\frac{16}{35}\right)^{\beta} - 2\left(\frac{9}{35}\right)^{\beta}.$$

Let $0.59 \leq r \leq \beta \leq s \leq 0.85$. Then,

$$K(\beta) - L(\beta) \geq K(s) - L(r) = M(r, s), \quad \text{say.}$$

We have

$$M\left(0.59 + \frac{k}{2500}, 0.59 + \frac{k+1}{2500}\right) > 0 \quad (k = 0, 1, \ldots, 99)$$

and

$$M\left(0.63 + \frac{k}{130}, 0.63 + \frac{k+1}{130}\right) > 0 \quad (k = 0, 1, \ldots, 28).$$

This implies that the first inequality in (17) is valid for $\beta \in [0.59, 0.85]$.
    Let

$$W(\beta) = \left(\frac{35}{9}\right)^{\beta} B(\beta).$$

Since

$$W''(\beta) \geq 0 \quad (\beta \in \mathbb{R}), \quad W'(0.59) = 0.39\ldots, \quad W(0.59) = 0.11\ldots,$$

we conclude that the second inequality in (17) holds for $\beta \geq 0.59$.

3 Proof of Theorem 1

We denote the sum in (4) by $S_{n,\alpha}(x)$. First, we assume that (4) holds for all $n \geq 1$ and $x \in [0, \pi]$. Then,

$$0 \leq S_{1,\alpha}(x) = 2\alpha \sin(x).$$

This gives $\alpha \geq 0$. Moreover, since

$$0 \leq S_{2,\alpha}(x) = \sin(x)\left(\frac{3}{2}(1 + \alpha) + 2\alpha \cos(x)\right),$$

we get

$$0 \leq \frac{3}{2}(1 + \alpha) + 2\alpha \cos(x).$$

We let $x \to \pi$ and obtain

$$0 \leq \frac{3}{2}(1 + \alpha) - 2\alpha = \frac{1}{2}(3 - \alpha).$$

Thus, $\alpha \leq 3$.

Now, we assume that $n \geq 1$ and $\alpha \in [0, 3]$. We define

$$x_k = x_k(n, \alpha) = \left(\frac{1}{n} + \frac{1}{k}\right)(n - k + \alpha) \quad (k = 1, \dots, n).$$

Since we have for $k \geq 1$,

$$x_k - x_{k+1} = \frac{k^2 + k + n^2 + n\alpha}{nk(k + 1)} > 0$$

and

$$(2k - 1)x_{2k-1} - 2kx_{2k} = \frac{4k - 1 - \alpha}{n} \geq 0,$$

we conclude that (2) holds, so that Vietoris' theorem gives $S_{n,\alpha}(x) \geq 0$ for $x \in [0, \pi]$.

4   Proof of Theorem 2

Inequality (6) is equivalent to

$$0 \leq \sum_{k=1}^{n-1}\left(\frac{n^2 - k^2}{(n^2 - 1)k}\right)^{\beta} \sin(kx) = T_{n,\beta}(x), \quad \text{say.} \tag{18}$$

If (18) is valid for all $n \geq 2$ and $x \in [0, \pi]$, then we obtain for $x \in (0, \pi)$:

$$0 \leq \frac{T_{3,\beta}(x)}{\sin(x)} = 1 + 2\left(\frac{5}{16}\right)^{\beta} \cos(x).$$

We let $x \to \pi$ and get

$$0 \leq 1 - 2\left(\frac{5}{16}\right)^{\beta} \quad \text{or} \quad \beta \geq \beta_1 = \frac{\log(2)}{\log(16/5)} = 0.59592\dots.$$

Next, let $n \geq 2$ and $x \in [0, \pi]$. As pointed out in Section 1, in order to prove (18) we may assume that $\beta \in [\beta_1, 1]$. First, we consider the cases $n = 2, 3, 4, 5, 6$.

Cases $n = 2, 3, 4$ We have

$$T_{2,\beta}(x) = \sin(x) \geq 0$$

and

$$T_{3,\beta}(x) = \sin(x)\left(1 + 2\left(\frac{5}{16}\right)^{\beta}\cos(x)\right) \geq \sin(x)\left(1 - 2\left(\frac{5}{16}\right)^{\beta}\right) \geq 0.$$

A short calculation yields that

$$T_{4,\beta}(x) = \sin(x)J_{\beta}(X),$$

where J_{β} is defined in (13) and $X = \cos(x)$. Using Lemma 5 reveals that $T_{4,\beta}(x) \geq 0$.

Case $n = 5$ We have

$$T_{5,\beta}(x) = \sin(x) + \sum_{k=2}^{4} y_k(\beta)\sin(kx)$$

where $y_2(\beta)$, $y_3(\beta)$, $y_4(\beta)$ are defined in (14). Let $\beta_2 = 0.803\ldots$ be given by

$$\left(\frac{2}{9}\right)^{\beta_2} = 2\left(\frac{3}{32}\right)^{\beta_2}.$$

Case 1: $\beta \geq \beta_2$. Then, $y_3(\beta) \geq 2y_4(\beta)$. We apply Lemmas 3 and 6 and Remark 2 and find that $T_{5,\beta}(x) \geq 0$.
Case 2: $\beta < \beta_2$. We have $2y_4(\beta) > y_3(\beta)$ and $y_2(\beta) > y_3(\beta)$. This implies

$$1 - y_3(\beta) - 2\,|y_2(\beta) - y_3(\beta)| - 2\,|2y_4(\beta) - y_3(\beta)|$$

$$= 1 - 2y_2(\beta) + 3y_3(\beta) - 4y_4(\beta) = Y(\beta),$$

where Y is defined in (15). Applying Lemmas 4 and 7 yields $T_{5,\beta}(x) \geq 0$.

Case $n = 6$ We obtain

$$T_{6,\beta}(x) = \sin(x) + \sum_{k=2}^{5} d_k(\beta)\sin(kx),$$

where

$$d_2(\beta) = \left(\frac{16}{35}\right)^\beta, \quad d_3(\beta) = \left(\frac{9}{35}\right)^\beta, \quad d_4(\beta) = \left(\frac{1}{7}\right)^\beta, \quad d_5(\beta) = \left(\frac{11}{175}\right)^\beta.$$

By direct computation we obtain that the sequence $\{1, d_2(\beta), d_3(\beta), d_4(\beta), d_5(\beta)\}$ is convex for $\beta = 1/2$, so that Remark 1 reveals that this sequence is convex for $\beta > 1/2$. Let $\beta_3 = 0.844\ldots$ be given by

$$\left(\frac{1}{7}\right)^{\beta_3} = 2\left(\frac{11}{175}\right)^{\beta_3}.$$

Case 1: $\beta \geq \beta_3$. Then, $\{1, d_2(\beta), d_3(\beta), d_4(\beta), d_5(\beta)\}$ is convex and $d_4(\beta) \geq 2d_5(\beta)$. Applying Lemma 3 and Remark 2 leads to $T_{6,\beta}(x) \geq 0$.

Case 2: $\beta < \beta_3$. Let

$$P_{1,\beta}(x) = d_4(\beta)\left(\sum_{k=1}^{4} \sin(kx) + \frac{1}{2}\sin(5x)\right),$$

$$P_{2,\beta}(x) = \left(d_5(\beta) - \frac{1}{2}d_4(\beta)\right)\left(5\sin(x) + \sin(5x)\right)$$

and

$$P_{3,\beta}(x) = z_1(\beta)\sin(x) + z_2(\beta)\sin(2x) + z_3(\beta)\sin(3x)$$

with $z_1(\beta), z_2(\beta), z_3(\beta)$ as defined in (16). Then we have the representation

$$T_{6,\beta}(x) = P_{1,\beta}(x) + P_{2,\beta}(x) + P_{3,\beta}(x).$$

Using Lemma 2 gives $P_{1,\beta}(x) \geq 0$. Since $d_5(\beta) > d_4(\beta)/2$ and

$$|\sin(mx)| \leq m\sin(x) \quad (m \in \mathbb{N}; 0 \leq x \leq \pi),$$

we obtain $P_{2,\beta}(x) \geq 0$. From Lemmas 3 and 8 we conclude that $P_{3,\beta}(x) \geq 0$. Thus, $T_{6,\beta}(x) \geq 0$.

Case $n \geq 7$ Let

$$T_{n,\beta}^*(x) = T_{n,\beta}(\pi - x).$$

The following statements are valid.

Proposition 1 $T_{n,\beta}^*$ is nonnegative on $[2.5/n, \pi]$.

Proposition 2 $T_{n,\beta}^*$ is nonnegative on $[0, \pi/n]$ if n is even.

Proposition 3 $T_{n,\beta}^*$ is nonnegative on $[0, 2.5/n]$ if n is odd.

From Propositions 1 and 2 we conclude that (18) is valid for $x \in [0, \pi]$ and even $n \geq 8$, and Propositions 1 and 3 reveal that this inequality holds for all $x \in [0, \pi]$ and odd $n \geq 7$.

Outline of the Proofs The proofs of the three propositions require numerous technical and lengthy computations which are collected in a separate paper published in arXiv; see [19].

Proposition 1 is proved by carefully applying the Comparison Principle to $T^*_{n,\beta}$, making use of the polynomials $\sum_{k=1}^{m} (-1)^{k+1} \sin(kx)$ which have known closed forms.

Let $\{a_1, a_2, \ldots, a_{n-1}\}$ denote the coefficient sequence of $T^*_{n,\beta}$. The proofs of the other two Propositions are based on the observation that there exists an integer $m \leq n$, such that the subsequence $\{a_1, a_2, \ldots, a_m\}$ is convex while the sequence $\{a_m, a_2, \ldots, a_{n-1}, 0\}$ has an odd number of terms and is concave. $T^*_{n,\beta}$ can then be decomposed into the sum of two sine polynomials, the first constructed using the first sub-sequence of coefficients and the second polynomial using the second sub-sequence. Lower bounds for these polynomials are then derived with careful, albeit rather technical, analysis. The lower bounds obtained happen to be enough to ensure that the sum is nonnegative.

5 Proof of Theorem 3

Let $S_{n,\alpha}(x)$, $H_n(x)$, $U_{n,a}(x)$ be the sums given in (4), (7), (8), respectively. If (8) holds for all $n \geq 1$ and $x \in [0, \pi]$, then

$$0 \leq U_{1,a}(x) = 2a \sin(x).$$

This yields $a \geq 0$.

Next, we assume that $n \geq 1$ and $a \geq 0$. We define

$$G_n(x) = \sum_{\substack{k=1 \\ k \text{ odd}}}^{n} \sin(kx) = \frac{\sin^2(Nx)}{\sin(x)} \quad (N = [(n+1)/2]). \tag{19}$$

From (7) and (19) we obtain for $x \in [0, \pi]$,

$$\frac{\partial}{\partial a} U_{n,a}(x) = \frac{1}{n} G_n(x) + H_n(x) \geq 0.$$

Thus,

$$U_{n,a}(x) \geq U_{n,0}(x) = \frac{1}{2} \left(S_{n,0}(x) + S_{n,0}(\pi - x) \right). \tag{20}$$

Using (4) with $\alpha = 0$ we conclude from (20) that (8) holds.

6 Proof of Theorem 4

We denote the sum in (9) by $V_{n,b}(x)$. If (9) is valid for all $n \geq 2$ and $x \in [0, \pi]$, then

$$0 \leq V_{4,b}(\pi/2) = \frac{1}{12^b}(45^b - 7^b).$$

This yields $b \geq 0$.

Next, let $n \geq 2$, $x \in [0, \pi]$ and $b \geq 0$. We set $\theta = 2x$ and $m = [(n-1)/2]$. Then, we obtain

$$V_{n,b}(x) = \sum_{k=0}^{m} c_k \sin((k+1/2)\theta),$$

where

$$c_k = c_k(n, b) = \left(\frac{n}{2k+1} - \frac{2k+1}{n}\right)^b.$$

Since $x \mapsto (n/x - x/n)^b$ is nonnegative and decreasing on $(0, n]$, we obtain $c_0 \geq c_1 \geq \cdots \geq c_m \geq 0$, so that (10) leads to $V_{n,b}(x) \geq 0$.

7 Remarks

The four theorems given in Section 1 provide lower bounds for the sine polynomials. We ask: do there exist upper bounds for these polynomials which do not depend on n and x? The answer is "no". We consider the sum $\tilde{T}_{n,\beta}(x)$ given in (6). Let $\beta \in \mathbb{R}$. Applying the arithmetic mean – geometric mean inequality yields for $n \geq 2$,

$$\frac{1}{n-1}\tilde{T}_{n,\beta}(\pi/n) = \frac{1}{n-1}\sum_{k=1}^{n-1}\left(\frac{n}{k} - \frac{k}{n}\right)^{\beta}\sin(k\pi/n)$$

$$\geq \left\{\prod_{k=1}^{n-1}\left(\frac{n}{k} - \frac{k}{n}\right)^{\beta}\sin(k\pi/n)\right\}^{1/(n-1)} = \delta_{n,\beta}, \quad \text{say.}$$

Since

$$\prod_{k=1}^{n-1}\left(\frac{n}{k} - \frac{k}{n}\right) = \frac{(2n-1)!}{(n-1)!n^n} \quad \text{and} \quad \prod_{k=1}^{n-1}\sin(k\pi/n) = \frac{n}{2^{n-1}},$$

we get

$$\delta_{n,\beta} = \frac{1}{2}\left\{\left(\frac{1}{\sqrt{2}}\right)^{1/(n-1)} \cdot \left(\frac{\sigma_{2n}}{\sigma_n}\right)^{1/(n-1)} \cdot \left(\frac{4}{e}\right)^{n/(n-1)}\right\}^{\beta} \cdot n^{1/(n-1)}, \qquad (21)$$

where

$$\sigma_n = \frac{n!e^n}{n^n\sqrt{n}}.$$

Using Stirling's formula

$$\lim_{n\to\infty} \sigma_n = \sqrt{2\pi}$$

we conclude from (21) that

$$\lim_{n\to\infty} \delta_{n,\beta} = \frac{1}{2}\left\{1 \cdot 1 \cdot \frac{4}{e}\right\}^{\beta} \cdot 1 = \frac{1}{2}\left(\frac{4}{e}\right)^{\beta}. \qquad (22)$$

Since

$$\tilde{T}_{n,\beta}(\pi/n) \geq (n-1)\delta_{n,\beta},$$

we obtain from (22) that

$$\lim_{n\to\infty} \tilde{T}_{n,\beta}(\pi/n) = \infty.$$

This reveals that there is no upper bound for $\tilde{T}_{n,\beta}(x)$ which is independent of n and x. Similarly, we can show that this is also true for the other three sine polynomials.

References

1. H. Alzer, S. Koumandos, M. Lamprecht, A refinement of Vietoris' inequality for sine polynomials. Math. Nach. **283**, 1549–1557 (2010)
2. R. Askey, *Orthogonal Polynomials and Special Functions*. Regional Conference Series in Applied Mathematics, vol. 21 (SIAM, Philadelphia, 1975)
3. R. Askey, Vietoris's inequalities and hypergeometric series, in *Recent Progress in Inequalities*, ed. by G.V. Milovanović (Kluwer, Dordrecht, 1998) pp. 63–76
4. R. Askey, G. Gasper, Inequalities for polynomials, in *The Bieberbach Conjecture*, ed. by A. Baernstein II et al. Mathematical Surveys and Monographs, vol. 2 (American Mathematical Society, Providence, 1986), pp. 7–32
5. R. Askey, J. Steinig, Some positive trigonometric sums. Trans. Am. Math. Soc. **187**, 295–307 (1974)

6. A.S. Belov, Examples of trigonometric series with non-negative partial sums. Math. USSR Sb. **186**, 21–46 (1995) (Russian); **186**, 485–510 (English translation)
7. G. Brown, E. Hewitt, A class of positive trigonometric sums. Math. Ann. **268**, 91–122 (1984)
8. G. Brown, D.C. Wilson, A class of positive trigonometric sums, II. Math. Ann. **285**, 57–74 (1989)
9. D.K. Dimitrov, C.A. Merlo, Nonnegative trigonometric polynomials. Constr. Approx. **18**, 117–143 (2002)
10. L. Fejér, Über die Positivität von Summen, die nach trigonometrischen oder Legendreschen Funktionen fortschreiten. Acta Litt. Sci. Szeged **2**, 75–86 (1925)
11. L. Fejér, Einige Sätze, die sich auf das Vorzeichen einer ganzen rationalen Funktion beziehen.... Monatsh. Math. Phys. **35**, 305–344 (1928)
12. L. Fejér, Trigonometrische Reihen und Potenzreihen mit mehrfach monotoner Koeffizienten-folge. Trans. Am. Math. Soc. **39**, 18–59 (1936)
13. T.H. Gronwall, Über die Gibbssche Erscheinung und die trigonometrischen Summen $\sin x + \frac{1}{2}\sin 2x + \cdots + \frac{1}{n}\sin nx$. Math. Ann. **72**, 228–243 (1912)
14. D. Jackson, Über eine trigonometrische Summe. Rend. Circ. Mat. Palermo **32**, 257–262 (1911)
15. S. Koumandos, An extension of Vietoris's inequalities. Ramanujan J. **14**, 1–38 (2007)
16. S. Koumandos, Inequalities for trigonometric sums, in *Nonlinear Analysis*, ed. by P.M. Pardalos et al. Optimization and Its Applications, vol. 68 (Springer, New York, 2012), pp. 387–416
17. M.K. Kwong, An improved Vietoris sine inequality. J. Approx. Theory **189**, 29–42 (2015)
18. M.K. Kwong, Improved Vietoris sine inequalities for non-monotone, non-decaying coefficients. arXiv:1504.06705 [math.CA] (2015)
19. M.K. Kwong, Technical details of the proof of the sine inequality $\sum_{k=1}^{n-1}\left(\frac{n}{k} - \frac{k}{n}\right)^{\beta}\sin(kx) \geq 0$, arXiv:1702.03387 [math.CA] (2017)
20. R. Meynieux, Gh. Tudor, Compléménts au traité de Mitrinović III: Sur un schéma général pour obtenir des inégalités. Univ. Beograd. Publ. Elektrotehn. Fak. Mat. Fiz. **412–460**, 171–174 (1973)
21. G.V. Milovanović, D.S. Mitrinović, Th.M. Rassias, *Topics in Polynomials: Extremal Problems, Inequalities, Zeros* (World Scientific, Singapore, 1994)
22. S.R. Mondal, A. Swaminathan, On the positivity of certain trigonometric sums and their applications. Comput. Math. Appl. **62**, 3871–3883 (2011)
23. P. Turán, On a trigonometrical sum. Ann. Soc. Polon. Math. **25**, 155–161 (1952)
24. L. Vietoris, Über das Vorzeichen gewisser trigonometrischer Summen, Sitzungsber. Öst. Akad. Wiss. **167**, 125–135 (1958); Anz. Öst. Akad. Wiss. **159**, 192–193

Inequalities for Weighted Trigonometric Sums

Horst Alzer and Omran Kouba

Abstract We prove that the double-inequality

$$\left(\sum_{j=1}^{n} \frac{w_j}{1-\sin^2\frac{j\pi}{n+1}}\right)^a \leq \sum_{j=1}^{n} \frac{w_j}{1-\sin\frac{j\pi}{n+1}} \cdot \sum_{j=1}^{n} \frac{w_j}{1+\sin\frac{j\pi}{n+1}} \leq \left(\sum_{j=1}^{n} \frac{w_j}{1-\sin^2\frac{j\pi}{n+1}}\right)^b$$

holds for all even integers $n \geq 2$ and positive real numbers w_j ($j = 1, \ldots, n$) with $w_1 + \cdots + w_n = 1$ if and only if $a \leq 1$ and $b \geq 2$. Moreover, we present a cosine counterpart of this result.

Keywords Inequalities · Sine sums · Cosine sums · Optimal constants

1 Introduction

In the literature, we can find many beautiful identities involving finite trigonometric sums. As an example, we mention a reciprocity theorem for the tangent sum

$$E(h, k) = \sum_{j=1}^{k-1} \frac{\tan(hj\pi/k)}{\tan(2j\pi/k)}$$

which was published by Eisenstein [5, pp. 108–110] in 1844:

H. Alzer (✉)
Waldbröl, Germany
e-mail: h.alzer@gmx.de

O. Kouba
Department of Mathematics, Higher Institute for Applied Sciences and Technology, Damascus, Syria
e-mail: omran_kouba@hiast.edu.sy

© Springer Nature Switzerland AG 2020
A. Raigorodskii, M. T. Rassias (eds.), *Trigonometric Sums and Their Applications*,
https://doi.org/10.1007/978-3-030-37904-9_4

$$hE(h, k) + kE(k, h) = -\frac{(h-k)^2}{2}.$$

Here, h and k are odd natural numbers which are relatively prime. A proof of Eisenstein's identity was given by Stern [16] in 1861.

The theory of finite trigonometric sums has attracted the attention of numerous researchers, mainly because these sums have remarkable applications in various mathematical branches, like, for example, geometry, theory of matrices, number theory, graph theory, and even in physics. Detailed information on this subject can be found in Berndt and Yeap [3], the recently published papers Fonseca et al. [6, 7], Kouba [12], Merca [13] and the references cited therein.

A proof for the elementary formula

$$\sum_{j=1}^{n} \frac{1}{1 - \cos^2\left(\frac{j\pi}{n+1}\right)} = \frac{n(n+2)}{3} \tag{1}$$

can be found in Chen [4]. From (1) we obtain

$$\sum_{j=1}^{n} \frac{1}{1 - \cos\left(\frac{j\pi}{n+1}\right)} = \sum_{j=1}^{n} \frac{1}{1 + \cos\left(\frac{j\pi}{n+1}\right)} = \frac{n(n+2)}{3}. \tag{2}$$

In this paper, we present inequalities for the weighted versions of the cosine sums given in (1) and (2),

$$C_n(\mathbf{w}) = \sum_{j=1}^{n} \frac{w_j}{1 - \cos\left(\frac{j\pi}{n+1}\right)},$$

$$C_n^*(\mathbf{w}) = \sum_{j=1}^{n} \frac{w_j}{1 + \cos\left(\frac{j\pi}{n+1}\right)}, \tag{3}$$

$$A_n(\mathbf{w}) = \frac{C_n(\mathbf{w}) + C_n^*(\mathbf{w})}{2} = \sum_{j=1}^{n} \frac{w_j}{1 - \cos^2\left(\frac{j\pi}{n+1}\right)},$$

where

$$\mathbf{w} \in \mathcal{W}_n \triangleq \left\{ (w_1, \dots, w_n) \in \mathbb{R}^n \mid w_j > 0 \text{ for } j=1, \dots, n, \text{ and } w_1 + \dots + w_n = 1 \right\}.$$

More precisely, we determine all real parameters α and β such that the inequalities

$$(A_n(\mathbf{w}))^\alpha \leq C_n(\mathbf{w}) \, C_n^*(\mathbf{w}) \leq (A_n(\mathbf{w}))^\beta$$

are valid for all $n \geq 1$ and $\mathbf{w} \in \mathcal{W}_n$.

We study also the same problem for the corresponding weighted sine sums

$$S_n(\mathbf{w}) = \sum_{j=1}^{n} \frac{w_j}{1 - \sin\left(\frac{j\pi}{n+1}\right)},$$

$$S_n^*(\mathbf{w}) = \sum_{j=1}^{n} \frac{w_j}{1 + \sin\left(\frac{j\pi}{n+1}\right)}, \tag{4}$$

$$B_n(\mathbf{w}) = \frac{S_n(\mathbf{w}) + S_n^*(\mathbf{w})}{2} = \sum_{j=1}^{n} \frac{w_j}{1 - \sin^2\left(\frac{j\pi}{n+1}\right)},$$

where $n \geq 2$ is an even integer.

A key role in the proofs of our inequalities plays the double-inequality

$$\sum_{j=1}^{n} \frac{w_j}{1 - p_j^2} \leq \sum_{j=1}^{n} \frac{w_j}{1 - p_j} \cdot \sum_{j=1}^{n} \frac{w_j}{1 + p_j} \leq \left(\sum_{j=1}^{n} \frac{w_j}{1 - p_j^2}\right)^2 \tag{5}$$

which holds for all $\mathbf{w} \in \mathcal{W}_n$ and $p_j \in (-1, 1)$, $(j = 1, \ldots, n)$. An application of the classical arithmetric mean – geometric mean inequality written in the form $xy \leq (x/2 + y/2)^2$ gives the right-hand side of (5). The left-hand side is due to Milne [14], who published an integral version in 1925. The identity

$$\sum_{j=1}^{n} \frac{w_j}{1 - p_j} \sum_{j=1}^{n} \frac{w_j}{1 + p_j} - \sum_{j=1}^{n} \frac{w_j}{1 - p_j^2} = \frac{1}{2} \sum_{i=1}^{n} \sum_{j=1}^{n} \frac{w_i w_j (p_i - p_j)^2}{(1 - p_i^2)(1 - p_j^2)}$$

leads to a short proof of Milne's result which is a special case of the well-known Chebyshev inequality; see [9, section 2.17]. For more information on Milne's inequality we refer to Alzer and Kovačec [1, 2] and Rao [15].

Throughout, we maintain the above notation. In Section 2 we state and prove a lemma and in Section 3 we present two theorems, a corollary and their proofs.

2 A Technical Lemma

We study the sums in (4) with $w_1 = \cdots = w_n = 1$. Let $n \geq 2$ be an even integer and let

$$F_n = \sum_{j=1}^{n} \frac{1}{1 - \sin^2\left(\frac{j\pi}{n+1}\right)}, \quad G_n = \sum_{j=1}^{n} \frac{1}{1 + \sin\left(\frac{j\pi}{n+1}\right)}, \quad H_n = \sum_{j=1}^{n} \frac{1}{1 - \sin\left(\frac{j\pi}{n+1}\right)}. \tag{6}$$

The following technical lemma is used in the proof of Theorem 2.

Lemma *Let $n \geq 2$ be an even integer. Then,*

$$F_n = n(n + 2), \tag{7}$$

$$G_n = \frac{2(n + 1)}{\pi} - 1 + \frac{\pi}{6(n + 1)} + O\left(\frac{1}{n^3}\right), \tag{8}$$

$$H_n = 2(n + 1)^2 - \frac{2(n + 1)}{\pi} - 1 - \frac{\pi}{6(n + 1)} + O\left(\frac{1}{n^3}\right). \tag{9}$$

Proof (i) Let $n = 2m$. We denote by T_k the Chebyshev polynomial of the first kind of degree k which is defined by $T_k(\cos(\theta)) = \cos(k\theta)$. Let

$$\theta_j = \frac{\pi}{2} - \frac{j\pi}{2m + 1} \quad \text{with} \quad j \in \{-m, \ldots, m\}.$$

Then,

$$\sin\left(\frac{j\pi}{2m + 1}\right) = \cos(\theta_j).$$

It follows that

$$T_{2m+1}\left(\sin\left(\frac{j\pi}{2m + 1}\right)\right) = T_{2m+1}(\cos(\theta_j)) = \cos\left((2m + 1)\theta_j\right)$$

$$= \cos\left(\frac{\pi}{2} - (j - m)\pi\right) = \sin((j - m)\pi) = 0$$

which implies that $(\sin(j\pi/(2m + 1)))_{-m \leq j \leq m}$ are the zeros of T_{2m+1}. Thus, there exists a constant λ such that

$$T_{2m+1}(x) = \lambda \prod_{j=-m}^{m} \left(x - \sin\left(\frac{j\pi}{2m + 1}\right)\right) = \lambda x \prod_{j=1}^{m} \left(x^2 - \sin^2\left(\frac{j\pi}{2m + 1}\right)\right).$$

Taking the logarithmic derivative gives

$$\frac{T'_{2m+1}(x)}{T_{2m+1}(x)} = \frac{1}{x} + \sum_{j=1}^{m} \frac{2x}{x^2 - \sin^2\left(\frac{j\pi}{2m+1}\right)}. \tag{10}$$

We set $x = 1$ and use $T_{2m+1}(1) = 1$, $T'_{2m+1}(1) = (2m + 1)^2$. Then, (10) yields

$$(2m + 1)^2 = 1 + 2 \sum_{j=1}^{m} \frac{1}{1 - \sin^2\left(\frac{j\pi}{2m+1}\right)}$$

$$= 1 + \sum_{j=1}^{2m} \frac{1}{1 - \sin^2\left(\frac{j\pi}{2m+1}\right)}$$

$$= 1 + F_{2m}$$

which leads to (7).

(ii) We apply the Euler-Maclaurin formula

$$\sum_{j=0}^{N} f(j) = \int_0^N f(x)dx + \frac{f(0) + f(N)}{2} + \frac{f'(N) - f'(0)}{12} - \frac{1}{2}\int_0^N P(\{x\})f''(x)dx,$$

where $\{x\}$ denotes the fractional part of x, and $P(x) = x^2 - x + 1/6$; see Knopp [10, section 64] and Kouba [11, section 8].

Taking $N = n + 1$,

$$f(x) = \frac{1}{1 + \sin\left(\frac{\pi x}{n+1}\right)},$$

and noting that

$$f(0) + f(N) = 2, \qquad \sum_{j=0}^{N} f(j) = 2 + G_n,$$

$$\int_0^N f(x)dx = \frac{2(n+1)}{\pi}, \qquad f'(N) - f'(0) = \frac{2\pi}{n+1},$$

we obtain

$$G_n = \frac{2(n+1)}{\pi} - 1 + \frac{\pi}{6(n+1)} + R_n, \qquad (11)$$

where

$$R_n = -\frac{1}{2}\int_0^{n+1} P(\{x\})f''(x)dx.$$

We have

$$-2R_n = \sum_{k=0}^{n} \int_k^{k+1} P(x - k)f''(x)dx$$

$$= \sum_{k=0}^{n} \int_{0}^{1} P(x) f''(x+k) dx$$

$$= \int_{0}^{1} P(x) \sum_{k=0}^{n} f''(x+k) dx.$$

In order to go one step further, we consider Q the Bernoulli polynomial of degree 4, that is,

$$Q(x) = x^4 - 2x^3 + x^2 - \frac{1}{30}.$$

Then,

$$Q(0) = Q(1) = -\frac{1}{30}, \quad Q'(0) = Q'(1) = 0, \quad Q''(x) = 12P(x).$$

Two integrations by parts yield

$$-24R_n = \int_{0}^{1} Q''(x) \sum_{k=0}^{n} f''(x+k) dx$$

$$= -\int_{0}^{1} Q'(x) \sum_{k=0}^{n} f'''(x+k) dx$$

$$= \frac{f'''(n+1) - f'''(0)}{30} + \int_{0}^{1} Q(x) \sum_{k=0}^{n} f^{(4)}(x+k) dx.$$

So, if g is the function defined by

$$g(x) = \frac{1}{1 + \sin(\pi x)}$$

then,

$$f^{(k)}(x) = \frac{1}{(n+1)^k} g^{(k)}\left(\frac{x}{n+1}\right) \quad (0 \le k \in \mathbb{Z}).$$

Thus,

$$-24R_n = \frac{1}{30(n+1)^3} (g'''(1) - g'''(0)) + \frac{1}{(n+1)^4} \int_{0}^{1} Q(x) \sum_{k=0}^{n} g^{(4)}\left(\frac{x+k}{n+1}\right) dx.$$

Hence, with $M \triangleq \sup_{0 \leq x \leq 1} |g^{(4)}(x)|$ we conclude that there exists a real number M^* such that

$$24|R_n| \leq \frac{M}{30(n+1)^3} + \frac{M}{(n+1)^3} \int_0^1 |Q(x)| dx \leq \frac{M^*}{(n+1)^3}, \tag{12}$$

and (8) follows from (11) and (12).

(iii) We have $H_n = 2F_n - G_n$, so that (7) and (8) lead to (9). □

Remark 1 The referee pointed out that (7) follows from the representation

$$F_n = \sum_{j=1}^{n} \left(1 + \tan^2\left(\frac{j\pi}{n+1}\right)\right)$$

and the identity

$$\sum_{j=1}^{n} \tan^2\left(\frac{j\pi}{n+1}\right) = n(n+1) \tag{13}$$

which is valid for all even integers $n \geq 2$; see Hansen [8, p. 646]. It is also worth mentioning that (13) appears in the work of Stern [16, p. 155].

3 Main Results

We are now in a position to present our main results. First, we offer inequalities involving the three cosine sums given in (3).

Theorem 1 *Let α and β be real numbers. The inequalities*

$$(A_n(\mathbf{w}))^\alpha \leq C_n(\mathbf{w}) \, C_n^*(\mathbf{w}) \leq (A_n(\mathbf{w}))^\beta \tag{14}$$

hold for all integers $n \geq 1$ and $\mathbf{w} \in \mathcal{W}_n$ if and only if $\alpha \leq 1$ and $\beta \geq 2$.

Proof We set

$$p_j = \cos\left(\frac{j\pi}{n+1}\right) \qquad \text{for } j = 1, \ldots, n,$$

then (5) gives

$$A_n(\mathbf{w}) \leq C_n(\mathbf{w}) C_n^*(\mathbf{w}) \leq (A_n(\mathbf{w}))^2. \tag{15}$$

Since $A_n(\mathbf{w}) \geq 1$, we obtain for $\alpha \leq 1$ and $\beta \geq 2$,

$$(A_n(\mathbf{w}))^\alpha \le A_n(\mathbf{w}) \quad \text{and} \quad (A_n(\mathbf{w}))^2 \le (A_n(\mathbf{w}))^\beta . \tag{16}$$

From (15) and (16) we conclude that (14) is valid.

Next, suppose that (14) holds for all $n \ge 1$ and $\mathbf{w} \in \mathscr{W}_n$. In particular, for $n = 2$ we have

$$C_2(\mathbf{w}) = \frac{2}{3} + \frac{4}{3}w_1, \quad C_2^*(\mathbf{w}) = 2 - \frac{4}{3}w_1, \quad A_2(\mathbf{w}) = \frac{4}{3},$$

so (14) leads to

$$\left(\frac{4}{3}\right)^\alpha \le \frac{4}{9}(1 + 2w_1)(3 - 2w_1) \le \left(\frac{4}{3}\right)^\beta \quad (0 < w_1 < 1).$$

Taking $w_1 = 1/2$ we obtain $\beta \ge 2$ and letting w_1 tend to 0^+ we obtain $\alpha \le 1$. □

Our second theorem provides a sine counterpart of (14) involving the sums defined in (4). It turns out that its proof is more difficult than the proof of Theorem 1. The hard part is to show that the right-hand side of (17) (given below) is in general not true if $b < 2$.

Theorem 2 *Let a and b be real numbers. The inequalities*

$$(B_n(\mathbf{w}))^a \le S_n(\mathbf{w})\, S_n^*(\mathbf{w}) \le (B_n(\mathbf{w}))^b \tag{17}$$

hold for all even integers $n \ge 2$ and $\mathbf{w} \in \mathscr{W}_n$ if and only if $a \le 1$ and $b \ge 2$.

Proof We apply (5) with

$$p_j = \sin\left(\frac{j\pi}{n+1}\right) \quad \text{for } j = 1, \dots, n$$

and use that $B_n(\mathbf{w}) \ge 1$. Then, for $a \le 1$ and $b \ge 2$,

$$(B_n(\mathbf{w}))^a \le B_n(\mathbf{w}) \le S_n(\mathbf{w})\, S_n^*(\mathbf{w}) \le (B_n(\mathbf{w}))^2 \le (B_n(\mathbf{w}))^b .$$

It follows that (17) holds if $a \le 1$ and $b \ge 2$.

Now assume that (17) is valid for all even $n \ge 2$ and $\mathbf{w} \in \mathscr{W}_n$. Then considering the case $n = 2$ we get

$$1 \le \frac{S_2(\mathbf{w}) S_2^*(\mathbf{w})}{(B_2(\mathbf{w}))^a} = 4^{1-a}.$$

Thus, $a \le 1$.

Next, we prove that $b \geq 2$. The right-hand side of (17) is equivalent to

$$\Phi_n(\mathbf{w}) = \frac{\log\left(S_n(\mathbf{w})S_n^*(\mathbf{w})\right)}{\log\left(B_n(\mathbf{w})\right)} \leq b. \tag{18}$$

Let $n \geq 4$, $t > 0$ and

$$\mathbf{w}_t = \frac{1}{1+t}\left(\frac{1}{2} + \frac{t}{n}, \frac{t}{n}, \dots, \frac{t}{n}, \frac{1}{2} + \frac{t}{n}\right).$$

Then,

$$S_n(\mathbf{w}_t) = \frac{1}{1+t}\left(\frac{1}{1-\sin\left(\frac{\pi}{n+1}\right)} + t\frac{H_n}{n}\right), \quad S_n^*(\mathbf{w}_t) = \frac{1}{1+t}\left(\frac{1}{1+\sin\left(\frac{\pi}{n+1}\right)} + t\frac{G_n}{n}\right),$$

and

$$B_n(\mathbf{w}_t) = \frac{1}{1+t}\left(\frac{1}{1-\sin^2\left(\frac{\pi}{n+1}\right)} + t\frac{F_n}{n}\right),$$

where H_n, G_n and F_n are defined in (6).

We obtain

$$\log\left(S_n(\mathbf{w}_t)S_n^*(\mathbf{w}_t)\right) = \log\left((1+tJ_n)(1+tK_n)\right) - 2\log(1+t) - \log\left(\cos^2\left(\frac{\pi}{n+1}\right)\right), \tag{19}$$

$$\log\left(B_n(\mathbf{w}_t)\right) = \log(1+tL_n) - \log(1+t) - \log\left(\cos^2\left(\frac{\pi}{n+1}\right)\right) \tag{20}$$

with

$$J_n = \left(1 - \sin\left(\frac{\pi}{n+1}\right)\right)\frac{H_n}{n},$$

$$K_n = \left(1 + \sin\left(\frac{\pi}{n+1}\right)\right)\frac{G_n}{n},$$

$$L_n = \cos^2\left(\frac{\pi}{n+1}\right)\frac{F_n}{n}.$$

Using (7), (8), (9) and the asymptotic expansions

$$\sin\left(\frac{\pi}{n+1}\right) = \frac{\pi}{n} - \frac{\pi}{n^2} + O\left(\frac{1}{n^3}\right), \quad \cos^2\left(\frac{\pi}{n+1}\right) = 1 - \frac{\pi^2}{n^2} + O\left(\frac{1}{n^3}\right)$$

gives

$$J_n = 2n - \frac{2(\pi - 1)^2}{\pi} - \frac{2\pi^2 - 3\pi + 2}{\pi n} + O\left(\frac{1}{n^2}\right),$$

$$K_n = \frac{2}{\pi} + \frac{\pi + 2}{\pi n} + O\left(\frac{1}{n^2}\right), \tag{21}$$

$$L_n = n + 2 - \frac{\pi^2}{n} + O\left(\frac{1}{n^2}\right).$$

Now, we set $t = \pi/n^2$ and apply (19), (20), (21) and

$$\log(1 + x) = x - \frac{1}{2}x^2 + O(x^3) \quad (x \to 0).$$

This yields

$$\log\left(S_n(\mathbf{w}_{\pi/n^2}) S_n^*(\mathbf{w}_{\pi/n^2})\right) = \frac{2\pi}{n} + \frac{2\pi - 3\pi^2}{n^2} + O\left(\frac{1}{n^3}\right) \tag{22}$$

and

$$\log\left(B_n(\mathbf{w}_{\pi/n^2})\right) = \frac{\pi}{n} + \frac{\pi + \pi^2/2}{n^2} + O\left(\frac{1}{n^3}\right). \tag{23}$$

From (18), (22) and (23) we conclude that

$$\Phi_n(\mathbf{w}_{\pi/n^2}) = 2 - \frac{4\pi}{n} + O\left(\frac{1}{n^2}\right).$$

Thus,

$$\lim_{n \to \infty} n\left(2 - \Phi_n(\mathbf{w}_{\pi/n^2})\right) = 4\pi.$$

In particular, we obtain

$$\lim_{n \to \infty} \Phi_n(\mathbf{w}_{\pi/n^2}) = 2. \tag{24}$$

Using (18) and (24) gives $b \geq 2$. $\qquad\square$

Remark 2 If the weight $\mathbf{w} = (w_1, \ldots, w_n)$ is symmetric, that is, $w_j = w_{n+1-j}$ for $j = 1, \ldots, n$, then we have $C_n(\mathbf{w}) = C_n^*(\mathbf{w}) = A_n(\mathbf{w})$, so that the second inequality in (14) (with $\beta = 2$) reduces to an equality. This is not true for our sine sums.

An application of the left-hand sides of (14) and (17) leads to a sharp inequality involving $C_n(\mathbf{w})$, $C_n^*(\mathbf{w})$, $S_n(\mathbf{w})$, $S_n^*(\mathbf{w})$ and the sine sum

$$D_n(\mathbf{w}) = \sum_{j=1}^{n} \frac{w_j}{\sin^2\left(\frac{2j\pi}{n+1}\right)}.$$

Corollary *For all even integers $n \geq 2$ and $\mathbf{w} \in \mathscr{W}_n$ we have*

$$4\, D_n(\mathbf{w}) \leq C_n(\mathbf{w})\, C_n^*(\mathbf{w}) + S_n(\mathbf{w})\, S_n^*(\mathbf{w}). \tag{25}$$

The constant factor 4 *is best possible.*

Proof We have

$$A_n(\mathbf{w}) + B_n(\mathbf{w}) = 4\, D_n(\mathbf{w}),$$

so that (14) and (17) with $\alpha = a = 1$ lead to (25). Moreover, since

$$\frac{C_2(\mathbf{w})\, C_2^*(\mathbf{w}) + S_2(\mathbf{w})\, S_2^*(\mathbf{w})}{D_2(\mathbf{w})} = 4 + \frac{4}{3} w_1 (1 - w_1),$$

we conclude that in (25) the factor 4 cannot be replaced by a larger number. □

Acknowledgements We thank the referee for helpful comments.

References

1. H. Alzer, A. Kovačec, The inequality of Milne and its converse. J. Inequal. Appl. **7**, 603–611 (2002)
2. H. Alzer, A. Kovačec, The inequality of Milne and its converse, II. J. Inequal. Appl. **2006**, 7, article ID 21572 (2006)
3. B.C. Berndt, B.P. Yeap, Explicit evaluations and reciprocity theorems for finite trigonometric sums. Adv. Appl. Math. **29**, 358–385 (2002)
4. H. Chen, On some trigonometric power sums. Int. J. Math. Math. Sci. **30**, 185–191 (2002)
5. G. Eisenstein, *Mathematische Werke, Band I* (Chelsea, New York, 1975)
6. C.M. da Fonseca, M.L. Glasser, V. Kowalenko, Basic trigonometric power sums with applications. Ramanujan J. **42**, 401–428 (2017)
7. C.M. da Fonseca, M.L. Glasser, V. Kowalenko, Generalized cosecant numbers and trigonometric sums. Appl. Anal. Disc. Math. **12**, 70–109 (2018)
8. E.R. Hansen, *A Table of Series and Products* (Prentice Hall, Englewood Cliffs, 1975)
9. G.H. Hardy, J.E. Littlewood, G. Pólya, *Inequalities* (Cambridge University Press, Cambridge, 1952)
10. K. Knopp, *Theorie und Anwendung der unendlichen Reihen* (Springer, Berlin, 1964)
11. O. Kouba, Lecture notes. Bernoulli polynomials and applications, arXiv:1309.7569v2 [math.CA]
12. O. Kouba, Inequalities for finite trigonometric sums. An interplay: with some series related to harmonic numbers. J. Inequal. Appl. **2016**, 15, paper no. 173 (2016)
13. M. Merca, A note on cosine power sums. J. Integer Seq. **15**, 7, article 12.5.3 (2012)

14. E.A. Milne, Note on Rosseland's integral for the stellar absorption coefficient. Mon. Not. R. Astron. Soc. **85**, 979–984 (1925)
15. C.R. Rao, Statistical proofs of some matrix inequalities. Linear Algebra Appl. **321**, 307–320 (2000)
16. M. Stern, Ueber einige Eigenschaften der Function Ex. J. Reine Angew. Math. **59**, 146–162 (1861)

Norm Inequalities for Generalized Laplace Transforms

J. C. Kuang

Abstract This paper introduced the new generalized Laplace transform. It contains the generalized Stieltjes transform and the Hankel transform etc. The corresponding new operator norm inequalities are obtained.The discrete versions of the main results are also given.As applications,a large number of known and new results have been obtained by proper choice of kernel. They are significant improvement and generalizations of many famous results.

Keywords Laplace transform · Integral operator · Norm inequality

Mathematics Subject Classification 47A30

1 Introduction

Given a function f on $(0, \infty)$ such that $e^{-\alpha y}|f(y)|$ is integrable over the interval $(0, \infty)$ for some real α, we define $F(z)$ as

$$F(z) = \int_0^\infty e^{-zy} f(y) dy, \tag{1}$$

where we require that $Re(z) > \alpha$ so that the integral in (1) converges.F is called the (one-sided)Laplace transform of f. We consider only one-sided Laplace transforms with the real parameter x, that is,

$$F(x) = \int_0^\infty e^{-xy} f(y) dy, \ x > \alpha, \tag{2}$$

J. C. Kuang (✉)
Department of Mathematics, Hunan Normal University Changsha, Hunan, P. R. China
e-mail: jckuang@163.com

© Springer Nature Switzerland AG 2020
A. Raigorodskii, M. T. Rassias (eds.), *Trigonometric Sums and Their Applications*,
https://doi.org/10.1007/978-3-030-37904-9_5

as these play the most important role in the solution of initial and boundary value problems for partial differential equations (see [1–3]). In fact, the tools we shall use for solving Cauchy and initial and boundary value problems are integral transforms. Specifically, we shall consider the Fourier transform, the Fourier sine and cosine transforms, the Hankel transform, and the Laplace transform. In asymptotic analysis, we often study functions f which have $N + 1$ continuous derivatives while f^{N+2} is piecewise continuous on $(0, \infty)$, then by [4], we have

$$F(x) = \int_0^\infty e^{-xy} f(y)dy \sim \sum_{k=0}^N x^{-(k+1)} f^{(k)}(0)$$

represents an asymptotic expansion of F, as $x \to \infty$, to $N + 1$ terms. But the following Hardy's results of being neglected [5]:

Theorem 1 *If $f \in L_\omega^p(0, \infty)$, $1 < p < \infty$, $\omega(x) = x^{p-2}$, then F is defined by (2) satisfies*

$$\|F\|_p \leq \Gamma(1/p)\|f\|_{p,\omega}, \tag{3}$$

where $\|f\|_{p,\omega} = (\int_0^\infty |f(x)|^p \omega(x)dx)^{1/p}$ and $\Gamma(\alpha)$ is the Gamma function:

$$\Gamma(\alpha) = \int_0^\infty x^{\alpha-1} e^{-x} dx \ (\alpha > 0).$$

It is important to note that x appear in (2) only through the product xy, this suggests that, as a generalization, Hardy [5] introduced the generalized Laplace transform of f:

$$T_0(f, x) = \int_0^\infty K(xy) f(y)dy, \ x \in (0, \infty), \tag{4}$$

and proved the following

Theorem 2 *Let K be a non-negative and measurable function on $(0, \infty)$. If f be a non-negative and not null, $1 < p < \infty$, $\frac{1}{p} + \frac{1}{q} = 1$, then the integral operator T_0 is defined by (4):$T_0 : L_\omega^p(0, \infty) \to L^p(0, \infty)$ and $L^p(0, \infty) \to L_\omega^p(0, \infty)$ exist as a bounded operator and*

$$\|T_0 f\|_p \leq c(\frac{1}{p})\|f\|_{p,\omega}, \ \|T_0 f\|_{p,\omega} \leq c(\frac{1}{q})\|f\|_p, \tag{5}$$

where $\omega(x) = x^{p-2}$ and

$$c(s) = \int_0^\infty K(t)t^{s-1}dt. \tag{6}$$

The constants are the best possible.

Remark 1 When $K(xy) = e^{-xy}$, Theorem 2 reduces to Theorem 1.

In 2015, the author [6] introduced the wider class of integral operators

$$T_1(f, x) = \int_{\mathbb{R}_+^n} K(\|x\|^{\lambda_1} \cdot \|y\|^{\lambda_2}) f(y) dy, \tag{7}$$

where $x \in \mathbb{R}_+^n = \{x = (x_1, x_2, \cdots, x_n) : x_k \geq 0, 1 \leq k \leq n\}$, $\|x\| = (\sum_{k=1}^n |x_k|^2)^{1/2}$, $\lambda_1 \times \lambda_2 \neq 0$, and obtained the operator norm inequalities for T_1 defined by (7) on the multiple weighted Orlicz spaces. In particular, if $n = 1, \lambda_1 = \lambda_2 = 1$, then T_1 in (7) reduces to T_0 in (4). In this paper, we introduce the new integral operator T defined by

$$T(f, x) = \int_{E_n(\alpha)} K(\|x\|_\alpha^{\lambda_1} \cdot \|y\|_\alpha^{\lambda_2}) f(y) dy, \quad x \in E_n(\alpha), \tag{8}$$

where

$$E_n(\alpha) = \{x = (x_1, x_2, \cdots, x_n) : x_k \geq 0, 1 \leq k \leq n, \|x\|_\alpha = (\sum_{k=1}^n |x_k|^\alpha)^{1/\alpha}, \alpha > 0\},$$

$\lambda_1, \lambda_2 > 0$, and the corresponding new operator norm inequalities are obtained. The discrete versions of the main results are also given. As applications, a large number of known and new results have been obtained by proper choice of kernel. They are significant improvements and generalizations of many famous results. We note that $E_n(\alpha)$ is a $n-$ dimensional vector space, when $1 \leq \alpha < \infty$, $E_n(\alpha)$ is a normed vector space. In particular, $E_n(2)$ is a $n-$ dimensional Euclidean space $\mathbb{R}_+^n$. Hence, when $\alpha = 2$, (8) reduces to (7). When $n = 1$, (7) reduces to

$$T_2(f, x) = \int_0^\infty K(x^{\lambda_1} y^{\lambda_2}) f(y) dy. \tag{9}$$

If $\lambda_1 = \lambda_2 = 1$, $K(xy) = (1 + xy)^{-\beta}$, then

$$T_3(f, x) = x^\beta \int_0^\infty \frac{1}{(1 + xy)^\beta} f(y) dy \tag{10}$$

is called the generalized Stieltjes transform of f. If $K(xy) = J_\alpha(xy)(xy)^{1/2}$, then

$$T_4(f, x) = \int_0^\infty J_\alpha(xy)(xy)^{1/2} f(y) dy \tag{11}$$

is called the Hankel transform, where $J_\alpha(t)$ is the Bessel function of the first kind of order α, that is,

$$J_\alpha(t) = (\frac{t}{2})^\alpha \sum_{k=0}^\infty \frac{(-1)^k}{k! \Gamma(\alpha + k + 1)} (\frac{t}{2})^{2k} (see [4]).$$

We use the standard notations:

$$\|f\|_{p,\omega} = (\int_{E_n(\alpha)} |f(x)|^p \omega(x)dx)^{1/p},$$

$$L^p(\omega) = \{f : f \ is \ measurable, \ and \|f\|_{p,\omega} < \infty\},$$

where, ω is a non-negative measurable function on $E_n(\alpha)$. If $\omega(x) \equiv 1$, we will denote $L^p(\omega)$ by $L^p(E_n(\alpha))$, and $\|f\|_{p,1}$ by $\|f\|_p$.

$$B(u, v) = \int_0^1 x^{u-1}(1-x)^{v-1}dx \quad (u, v > 0)$$

is the Beta function and

$$\zeta(p, \beta) = \sum_{k=0}^{\infty} \frac{1}{(k+\beta)^p}, \quad (p > 1, \beta > 0)$$

is the extended Riemann zeta function. In particular, $\zeta(p, 1) = \zeta(p) = \sum_{k=1}^{\infty} \frac{1}{k^p}$ is the Riemann zeta function.

2   Main Results

Our main results reads as follows.

Theorem 3 *Let* $1 < p, q < \infty, \frac{1}{p} + \frac{1}{q} \geq 1, 0 < \frac{\lambda}{n} = 2 - \frac{1}{p} - \frac{1}{q}$, $\omega(x) = \|x\|_{\alpha}^{n(\frac{\lambda_1}{\lambda_2}(p-1)-1)}$, $x \in E_n(\alpha)$, *and the radial kernel* $K(\|x\|_{\alpha}^{\lambda_1} \cdot \|y\|_{\alpha}^{\lambda_2})$ *be a nonnegative measurable function defined on* $E_n(\alpha) \times E_n(\alpha), \lambda_1, \lambda_2 > 0$. *(i) If*

$$c_1 = \frac{(\Gamma(1/\alpha))^n}{\alpha^{n-1}\Gamma(n/\alpha)} \int_0^{\infty} \{K(u)\}^{\frac{np}{\lambda}(1-\frac{1}{q})} u^{\frac{n^2}{\lambda\lambda_2}(p-1)(1-\frac{1}{q})-1} du < \infty, \tag{12}$$

$$c_2 = \frac{(\Gamma(1/\alpha))^n}{\alpha^{n-1}\Gamma(n/\alpha)} \int_0^{\infty} \{K(u)\}^{\frac{n}{\lambda}} u^{\frac{n}{\lambda_2}[1-\frac{n}{\lambda}(1-\frac{1}{q})]-1} du < \infty, \tag{13}$$

then the integral operator T *is defined by* (8): $T : L^p(E_n(\alpha)) \to L^p(\omega)$ *exists as a bounded operator and*

$$\|Tf\|_{p,\omega} \leq (\frac{c_1}{\lambda_1})^{1/p}(\frac{c_2}{\lambda_2})^{1-(1/p)}\|f\|_p,$$

This implies that

$$\|T\| = \sup_{f \neq 0} \frac{\|Tf\|_{p,\omega}}{\|f\|_p} \leq (\frac{c_1}{\lambda_1})^{1/p}(\frac{c_2}{\lambda_2})^{1-(1/p)}. \tag{14}$$

(ii) If $0 < \lambda_1 \leq \lambda_2$, and

$$c_3 = \frac{(\Gamma(1/\alpha))^n}{\alpha^{n-1}\Gamma(n/\alpha)} \int_0^\infty \{K(u)\} u^{\frac{n}{\lambda_2}(1-\frac{1}{p})-1} du < \infty, \tag{15}$$

then

$$\|T\| \geq \frac{c_3}{\lambda_2}. \tag{16}$$

In the conjugate case: $\lambda = n$, $\frac{1}{p} + \frac{1}{q} = 1$, then by (12), (13) and (15), we get

$$c_1 = c_2 = c_3 = c_0 = \frac{(\Gamma(1/\alpha))^n}{\alpha^{n-1}\Gamma(n/\alpha)} \int_0^\infty K(u) u^{\frac{n}{q\lambda_2}-1} du. \tag{17}$$

Thus, by Theorem 3, we get

Corollary 1 *Let $1 < p < \infty$, $\frac{1}{p} + \frac{1}{q} = 1$, $\omega(x) = \|x\|_\alpha^{n(\frac{\lambda_1}{\lambda_2}(p-1)-1)}$, $x \in E_n(\alpha)$, and the radial kernel $K(\|x\|_\alpha^{\lambda_1} \cdot \|y\|_\alpha^{\lambda_2})$ be a nonnegative measurable function defined on $E_n(\alpha) \times E_n(\alpha)$, $\lambda_1, \lambda_2 > 0$. If*

$$c_0 = \frac{(\Gamma(1/\alpha))^n}{\alpha^{n-1}\Gamma(n/\alpha)} \int_0^\infty K(u) u^{\frac{n}{q\lambda_2}-1} du < \infty, \tag{18}$$

then the integral operator T is defined by (8): $T : L^p(E_n(\alpha)) \to L^p(\omega)$ exists as a bounded operator and

$$\|Tf\|_{p,\omega} \leq \frac{c_0}{\lambda_1^{1/p}\lambda_2^{1/q}} \|f\|_p.$$

This implies that

$$\|T\| = \sup_{f \neq 0} \frac{\|Tf\|_{p,\omega}}{\|f\|_p} \leq \frac{c_0}{\lambda_1^{1/p}\lambda_2^{1/q}}.$$

If $0 < \lambda_1 \leq \lambda_2$, then

$$\|T\| \geq \frac{c_0}{\lambda_2}.$$

Corollary 2 *Under the same conditions as those of Corollary 1, if $\lambda_1 = \lambda_2 = \lambda_0$, then*

$$\|Tf\|_{p,\omega} \leq c\|f\|_p$$

where

$$c = \|T\| = \frac{c_0}{\lambda_0} \tag{19}$$

is the best possible.

In particular, for $\lambda_0 = n = 1$, we get the following

Corollary 3 *Under the same conditions as those of Corollary 2, if $\lambda_0 = n = 1$, then the integral operator T_0 is defined by (4):$T_0 : L^p(0, \infty) \rightarrow L^p(\omega)$ exists as a bounded operator and*

$$\|T_0 f\|_{p,\omega} \leq c\|f\|_p,$$

where $\omega(x) = x^{p-2}$ and

$$c = \|T_0\| = \int_0^\infty K(u)u^{(1/q)-1}du = \int_0^\infty K(u)u^{-(1/p)}du$$

is the best possible.

Hence, Corollary 3 reduces to Theorem 2.

3   Proof of Theorem 3

We require the following Lemmas to prove our results:

Lemma 1 ([7]) *If $a_k, b_k, p_k > 0, 1 \leq k \leq n, f$ is a measurable function on $(0, \infty)$, then*

$$\int_{B(r_1,r_2)} f\left(\sum_{k=1}^n (\frac{x_k}{a_k})^{b_k}\right)x_1^{p_1-1}\cdots x_n^{p_n-1}dx_1\cdots dx_n$$

$$= \frac{\prod_{k=1}^n a_k^{p_k}}{\prod_{k=1}^n b_k} \times \frac{\prod_{k=1}^n \Gamma(\frac{p_k}{b_k})}{\Gamma(\sum_{k=1}^n \frac{p_k}{b_k})} \int_{r_1}^{r_2} f(t)t^{(\sum_{k=1}^n \frac{p_k}{b_k}-1)}dt,$$

where $B(r_1, r_2) = \{x \in E_n(\alpha) : 0 \leq r_1 \leq \|x\|_\alpha < r_2\}$.

We get the following Lemma 2 by taking $a_k = 1, b_k = \alpha > 0, p_k = 1, 1 \le k \le n$, $r_1 = 0, r_2 = \infty$, in Lemma 1.

Lemma 2 *Let f be a measurable function on $(0, \infty)$, then*

$$\int_{E_n(\alpha)} f(\|x\|_\alpha^\alpha)dx = \frac{(\Gamma(1/\alpha))^n}{\alpha^n \Gamma(n/\alpha)} \int_0^\infty f(t)t^{(n/\alpha)-1}dt. \tag{20}$$

Lemma 3 *Let $f \in L^p(\omega), g \in L^q(E), 1 < p < \infty, \frac{1}{p}+\frac{1}{q} = 1, \omega$ be a nonnegative measurable function on E, then*

$$\|f\|_{p,\omega} = \sup\{|\int_E fg\omega^{1/p}d\mu| : \|g\|_q \le 1\}. \tag{21}$$

Proof of Lemma 3 This is an immediate consequence of the Hölder inequality with weight (see [8]).

Proof of Theorem 3 Let $p_1 = \frac{p}{p-1}, q_1 = \frac{q}{q-1}$, it follows that

$$\frac{1}{p_1} + \frac{1}{q_1} + (1 - \frac{\lambda}{n}) = 1, \frac{p}{q_1} + p(1 - \frac{\lambda}{n}) = 1.$$

By Hölder's inequality, we get

$$T(f, x) = \int_{E_n(\alpha)} K(\|x\|_\alpha^{\lambda_1} \cdot \|y\|_\alpha^{\lambda_2})f(y)dy$$

$$= \int_{E_n(\alpha)} \{\|y\|_\alpha^{(\frac{n^2}{p_1\lambda})}[K(\|x\|_\alpha^{\lambda_1} \cdot \|y\|_\alpha^{\lambda_2})]^{\frac{n}{\lambda}} f^p(y)\}^{1/q_1}$$

$$\times \{\|y\|_\alpha^{-(\frac{n^2}{\lambda q_1})}[K(\|x\|_\alpha^{\lambda_1} \cdot \|y\|_\alpha)^{\lambda_2}]^{\frac{n}{\lambda}}\}^{1/p_1}\{f(y)\}^{p(1-\frac{\lambda}{n})}dy$$

$$\le \{\int_{E_n(\alpha)} \|y\|_\alpha^{(\frac{n^2}{p_1\lambda})}[K(\|x\|_\alpha^{\lambda_1} \cdot \|y\|_\alpha^{\lambda_2})]^{\frac{n}{\lambda}}|f(y)|^pdy\}^{1/q_1}$$

$$\times \{\int_{E_n(\alpha)} \|y\|_\alpha^{-(\frac{n^2}{\lambda q_1})}[K(\|x\|_\alpha^{\lambda_1} \cdot \|y\|_\alpha^{\lambda_2})]^{\frac{n}{\lambda}}dy\}^{1/p_1}\|f\|_p^{p(1-\frac{\lambda}{n})}$$

$$= I_1^{1/q_1} \times I_2^{1/p_1} \times \|f\|_p^{p(1-\frac{\lambda}{n})}. \tag{22}$$

In I_2, by using Lemma 2 and letting $u = \|x\|_\alpha^{\lambda_1} t^{\lambda_2/\alpha}$, and using (13) we get

$$I_2 = \int_{E_n(\alpha)} \|y\|_\alpha^{-\left(\frac{n^2}{\lambda q_1}\right)} [K(\|x\|_\alpha^{\lambda_1} \cdot \|y\|_\alpha^{\lambda_2})]^{\frac{n}{\lambda}} dy$$

$$= \frac{(\Gamma(1/\alpha))^n}{\alpha^n \Gamma(n/\alpha)} \int_0^\infty t^{-\left(\frac{n^2}{q_1\lambda\alpha}\right)} [K(\|x\|_\alpha^{\lambda_1} t^{\frac{\lambda_2}{\alpha}}]^{\frac{n}{\lambda}} \times t^{\left(\frac{n}{\alpha}\right)-1} dt$$

$$= \frac{(\Gamma(1/\alpha))^n}{\lambda_2 \alpha^{n-1} \Gamma(n/\alpha)} \|x\|_\alpha^{\frac{\lambda_1 n}{\lambda_2}\left(\frac{n}{q_1\lambda}-1\right)} \int_0^\infty \{K(u)\}^{\frac{n}{\lambda}} u^{\frac{n}{\lambda_2}[1-\frac{n}{\lambda}(1-\frac{1}{q})]-1} du$$

$$= \frac{c_2}{\lambda_2} \|x\|_\alpha^{\frac{\lambda_1 n}{\lambda_2}\left(\frac{n}{q_1\lambda}-1\right)}. \tag{23}$$

Note that $\frac{1}{p} + \frac{1}{q} \geq 1$ implies that $\frac{p}{q_1} \geq 1$, thus, by (22), (23) and the Minkowski's inequality for integrals:

$$\{\int_X (\int_Y |f(x,y)| dy)^p \omega(x) dx\}^{1/p} \leq \int_Y \{\int_X |f(x,y)|^p \omega(x) dx\}^{1/p} dy, \quad 1 \leq p < \infty,$$

and letting $v = \|y\|_\alpha^{\lambda_2} t^{\left(\frac{\lambda_1}{\alpha}\right)}$, we conclude that

$$\|Tf\|_{p,\omega} = \left(\int_{E_n(\alpha)} |T(f,x)|^p \omega(x) dx\right)^{1/p} \leq \left(\int_{E_n(\alpha)} I_1^{p/q_1} I_2^{p/p_1} \|f\|_p^{p^2(1-\frac{\lambda}{n})} \omega(x) dx\right)^{1/p}$$

$$= \left(\frac{c_2}{\lambda_2}\right)^{1/p_1} \|f\|_p^{p(1-\frac{\lambda}{n})} \left\{\int_{E_n(\alpha)} \|x\|_\alpha^{\frac{\lambda_1 np}{\lambda_2 p_1}\left(\frac{n}{q_1\lambda}-1\right)+n\left(\frac{\lambda_1}{\lambda_2}(p-1)-1\right)}\right.$$

$$\times \left(\int_{E_n(\alpha)} \|y\|_\alpha^{\frac{n^2}{p_1\lambda}} [K(\|x\|_\alpha^{\lambda_1} \cdot \|y\|_\alpha^{\lambda_2})]^{\frac{n}{\lambda}} |f(y)|^p dy\right)^{\frac{p}{q_1}} dx\right\}^{1/p}$$

$$\leq \left(\frac{c_2}{\lambda_2}\right)^{1/p_1} \|f\|_p^{p(1-\frac{\lambda}{n})} \left\{\int_{E_n(\alpha)} \|y\|_\alpha^{\frac{n^2}{p_1\lambda}} |f(y)|^p\right.$$

$$\times \left(\int_{E_n(\alpha)} \|x\|_\alpha^{\left(\frac{\lambda_1 np}{\lambda_2 p_1}\right)\left(\frac{n}{q_1\lambda}-1\right)+n\left(\frac{\lambda_1}{\lambda_2}(p-1)-1\right)}\right.$$

$$\times [K(\|x\|_\alpha^{\lambda_1} \cdot \|y\|_\alpha^{\lambda_2})]^{\frac{np}{q_1\lambda}} dx\right)^{\frac{q_1}{p}} dy\right\}^{1/q_1}$$

$$= \left(\frac{c_2}{\lambda_2}\right)^{1/p_1} \|f\|_p^{p(1-\frac{\lambda}{n})} \left\{\int_{E_n(\alpha)} \|y\|_\alpha^{\frac{n^2}{p_1\lambda}} |f(y)|^p\right.$$

$$\times \left(\frac{(\Gamma(1/\alpha))^n}{\alpha^n \Gamma(n/\alpha)} \int_0^\infty t^{\left(\frac{\lambda_1 np}{p_1\alpha\lambda_2}\right)\left(\frac{n}{q_1\lambda}-1\right)+\frac{n}{\alpha}\left(\frac{\lambda_1}{\lambda_2}(p-1)-1\right)}\right.$$

$$\times [K(t^{(\lambda_1/\alpha)} \|y\|_\alpha^{\lambda_2})]^{\frac{np}{q_1\lambda}} t^{\frac{n}{\alpha}-1} dt\right)^{\frac{q_1}{p}} dy\right\}^{1/q_1}$$

$$= \left(\frac{c_2}{\lambda_2}\right)^{1/p_1} \|f\|_p^{p(1-\frac{\lambda}{n})} \times \left(\frac{c_1}{\lambda_1}\right)^{1/p} \|f\|_p^{p/q_1} = \left(\frac{c_2}{\lambda_2}\right)^{1/p_1} \left(\frac{c_1}{\lambda_1}\right)^{1/p} \|f\|_p.$$

Thus,

$$\|Tf\|_{p,\omega} \le (\frac{c_1}{\lambda_1})^{1/p}(\frac{c_2}{\lambda_2})^{1-(1/p)}\|f\|_p. \tag{24}$$

This implies that

$$\|T\| = \sup_{f \ne 0} \frac{\|Tf\|_{p,\omega}}{\|f\|_p} \le (\frac{c_1}{\lambda_1})^{1/p}(\frac{c_2}{\lambda_2})^{1-(1/p)}. \tag{25}$$

To prove the reversed inequality (16), setting f_ε and g_ε as follows:

$$f_\varepsilon(x) = \|x\|_\alpha^{-(n/p)+\varepsilon}\varphi_B(x), \tag{26}$$

$$g_\varepsilon(x) = (p\varepsilon)^{1/p_1}\{\frac{\alpha^{n-1}\Gamma(n/\alpha)}{(\Gamma(1/\alpha))^n}\}^{1/p_1}\|x\|_\alpha^{-(\frac{n}{p_1})+(\frac{\lambda_1}{\lambda_2}-p)\varepsilon}\varphi_{B^c}(x), \tag{27}$$

where $0 < \varepsilon < \frac{\alpha}{p}$, $B = B(0,1) = \{x \in E_n(\alpha) : \|x\|_\alpha < 1\}$, φ_{B^c} is the characteristic function of the set $B^c = \{x \in E_n(\alpha) : \|x\|_\alpha \ge 1\}$, that is,

$$\varphi_{B^c}(x) = \begin{cases} 1, x \in B^c \\ 0, x \in B. \end{cases}$$

Thus, we get

$$\|f_\varepsilon\|_p = (\frac{(\Gamma(1/\alpha))^n}{p\varepsilon\alpha^{n-1}\Gamma(n/\alpha)})^{1/p}, \tag{28}$$

$$\|g_\varepsilon\|_{p_1}^{p_1} = \frac{p-1}{p-\frac{\lambda_1}{\lambda_2}} \le 1. \tag{29}$$

Using Lemma 3 and the Fubini Theorem, we get

$$\|Tf_\varepsilon\|_{p,\omega} \ge \int_{E_n(\alpha)} T(f_\varepsilon, x)g_\varepsilon(x)\{\omega(x)\}^{1/p}dx$$

$$= \int_{E_n(\alpha)} \int_{E_n(\alpha)} K(\|x\|_\alpha^{\lambda_1} \cdot \|y\|_\alpha^{\lambda_2})f_\varepsilon(y)g_\varepsilon(x)(\|x\|_\alpha^{n(\frac{\lambda_1}{\lambda_2}(p-1)-1)})^{1/p}dydx$$

$$= (p\varepsilon)^{1/p_1}\{\frac{\alpha^{n-1}\Gamma(n/\alpha)}{(\Gamma(1/\alpha))^n}\}^{1/p_1}$$

$$\times \int_{B^c}\{\int_B K(\|x\|_\alpha^{\lambda_1} \cdot \|y\|_\alpha^{\lambda_2})\|y\|_\alpha^{-(n/p)+\varepsilon}dy\}\|x\|_\alpha^{\frac{\lambda_1}{\lambda_2}(\frac{n}{p_1}+\varepsilon)-p\varepsilon-n}dx. \tag{30}$$

Letting $u = t^{\frac{\lambda_2}{\alpha}} \|x\|_\alpha^{\lambda_1}$, and using (20), we have

$$\int_B K(\|x\|_\alpha^{\lambda_1} \cdot \|y\|_\alpha^{\lambda_2}) \|y\|_\alpha^{-(n/p)+\varepsilon} dy$$

$$= \frac{(\Gamma(1/\alpha))^n}{\alpha^n \Gamma(n/\alpha)} \int_0^1 K(t^{\lambda_2/\alpha} \|x\|_\alpha^{\lambda_1}) t^{-(\frac{n}{p\alpha})+\frac{\varepsilon}{\alpha}+\frac{n}{\alpha}-1} dt$$

$$= \frac{(\Gamma(1/\alpha))^n}{\alpha^{n-1} \Gamma(n/\alpha)\lambda_2} \|x\|_\alpha^{-\frac{\lambda_1}{\lambda_2}(\frac{n}{p_1}+\varepsilon)} \int_0^{\|x\|_\alpha^{\lambda_1}} K(u) u^{\frac{1}{\lambda_2}(\frac{n}{p_1}+\varepsilon)-1} du. \quad (31)$$

We insert (31) into (30) and use Fubini's theorem to obtain

$$\|Tf_\varepsilon\|_{p,\omega} \geq (p\varepsilon)^{1/p_1} \left\{ \frac{(\Gamma(1/\alpha))^n}{\alpha^{n-1} \Gamma(n/\alpha)} \right\}^{1/p} \times \frac{1}{\lambda_2}$$

$$\times \int_{B^c} \|x\|_\alpha^{-p\varepsilon-n} \left(\int_0^{\|x\|_\alpha^{\lambda_1}} K(u) u^{\frac{1}{\lambda_2}(\frac{n}{p_1}+\varepsilon)-1} du \right) dx$$

$$\geq (p\varepsilon)^{1/p_1} \left\{ \frac{(\Gamma(1/\alpha))^n}{\alpha^{n-1} \Gamma(n/\alpha)} \right\}^{1/p} \times \frac{1}{\lambda_2}$$

$$\times \int_0^\infty K(u) u^{\frac{1}{\lambda_2}(\frac{n}{p_1}+\varepsilon)-1} \left(\int_{\beta(u)}^\infty \|x\|_\alpha^{-p\varepsilon-n} dx \right) du$$

$$= (p\varepsilon)^{1/p_1} \left\{ \frac{(\Gamma(1/\alpha))^n}{\alpha^{n-1} \Gamma(n/\alpha)} \right\}^{(1/p)+1} \times \frac{1}{\alpha\lambda_2}$$

$$\times \int_0^\infty K(u) u^{\frac{1}{\lambda_2}(\frac{n}{p_1}+\varepsilon)-1} \left(\int_{\beta(u)}^\infty t^{-(p\varepsilon)/\alpha-1} dt \right) du$$

$$= (p\varepsilon)^{-(1/p)} \left\{ \frac{(\Gamma(1/\alpha))^n}{\alpha^{n-1} \Gamma(n/\alpha)} \right\}^{(1/p)+1} \times \frac{1}{\lambda_2}$$

$$\times \int_0^\infty K(u) u^{\frac{1}{\lambda_2}(\frac{n}{p_1}+\varepsilon)-1} (\beta(u))^{-(p\varepsilon)/\alpha} du, \quad (32)$$

where $\beta(u) = \max\{1, u^{1/\lambda_1}\}$. Thus, we get

$$\|T\| = \sup_{f \neq 0} \frac{\|Tf\|_{p,\omega}}{\|f\|_p} \geq \frac{\|Tf_\varepsilon\|_{p,\omega}}{\|f_\varepsilon\|_p}$$

$$\geq \frac{(\Gamma(1/\alpha))^n}{\alpha^{n-1} \Gamma(n/\alpha)\lambda_2} \int_0^\infty K(u) u^{\frac{1}{\lambda_2}(\frac{n}{p_1}+\varepsilon)-1} (\beta(u))^{-(p\varepsilon)/\alpha} du. \quad (33)$$

By letting $\varepsilon \to 0^+$ in (33) and using the Fatou lemma, we get

$$\|T\| \geq \frac{(\Gamma(1/\alpha))^n}{\alpha^{n-1}\Gamma(n/\alpha)\lambda_2} \int_0^\infty K(u)u^{\frac{n}{p_1\lambda_2}-1}du = \frac{c_3}{\lambda_2}. \tag{34}$$

The proof is complete.

4 The Discrete Versions of the Main Results

Let $a = \{a_m\}$ be a sequence of real numbers, we define

$$\|a\|_{p,\omega} = \{\sum_{m=1}^\infty |a_m|^p \omega(m)\}^{1/p}, l^p(\omega) = \{a = \{a_m\} : \|a\|_{p,\omega} < \infty\}.$$

For $m - 1 \leq x < m, n - 1 \leq y < n$, let

$$f(x) = a_m, K(x^{\lambda_1} y^{\lambda_2}) = K(m^{\lambda_1} n^{\lambda_2}),$$

then the corresponding series form of (9) is

$$T_5(a, m) = \sum_{n=1}^\infty K(m^{\lambda_1} n^{\lambda_2})a_n. \tag{35}$$

By Theorem 3, we get

Theorem 4 *Let* $1 < p, q < \infty, \frac{1}{p} + \frac{1}{q} \geq 1, 0 < \lambda = 2 - \frac{1}{p} - \frac{1}{q}$, $\omega(m) = m^{\frac{\lambda_1}{\lambda_2}(p-1)-1}, K(u)$ *be a nonnegative measurable function defined on* $(0, \infty), \lambda_1, \lambda_2 > 0$. *If*

$$c_1 = \int_0^\infty \{K(u)\}^{\frac{p}{\lambda}(1-\frac{1}{q})} u^{\frac{1}{\lambda\lambda_2}(p-1)(1-\frac{1}{q})-1} du < \infty, \tag{36}$$

$$c_2 = \int_0^\infty \{K(u)\}^{\frac{1}{\lambda}} u^{\frac{1}{\lambda_2}[1-\frac{1}{\lambda}(1-\frac{1}{q})]-1} du < \infty, \tag{37}$$

then the integral operator T_5 *is defined by (35):* $T_5 : l^p \to l^p(\omega)$ *exists as a bounded operator and*

$$\|T_5a\|_{p,\omega} \leq (\frac{c_1}{\lambda_1})^{1/p}(\frac{c_2}{\lambda_2})^{1-(1/p)}\|a\|_p.$$

This implies that

$$\|T_5\| = \sup_{a \neq 0} \frac{\|T_5a\|_{p,\omega}}{\|a\|_p} \leq (\frac{c_1}{\lambda_1})^{1/p}(\frac{c_2}{\lambda_2})^{1-(1/p)}. \tag{38}$$

If $0 < \lambda_1 \leq \lambda_2$, *and*

$$c_3 = \int_0^\infty K(u) u^{\frac{1}{\lambda_2}(1-\frac{1}{p})-1} du < \infty, \tag{39}$$

then

$$\|T_5\| \geq \frac{c_3}{\lambda_2}. \tag{40}$$

Corollary 4 *Under the same conditions as those of Theorem 4, if* $\lambda = 1$, $\frac{1}{p}+\frac{1}{q} = 1$, *and* $c_0 = \int_0^\infty K(u) u^{\frac{1}{q\lambda_2}-1} du < \infty$, *then*

$$\|T_5\| \leq \frac{c_0}{\lambda_1^{1/p}\lambda_2^{1/q}}.$$

If $0 < \lambda_1 \leq \lambda_2$, *then*

$$\|T_5\| \geq \frac{c_0}{\lambda_2}.$$

Corollary 5 *Under the same conditions as those of Corollary 4, if* $\lambda_1 = \lambda_2 = \lambda_0$, *then*

$$\|T_5 a\|_{p,\omega} \leq c_3 \|a\|_p,$$

where $\omega(m) = m^{(p-2)}$, *and*

$$c_3 = \|T_5\| = \frac{c_0}{\lambda_0} \tag{41}$$

is the best possible.

In particular, for $\lambda_0 = 1$, the corresponding series form of (35) is

$$T_6(a, m) = \sum_{n=1}^\infty K(mn)a_n. \tag{42}$$

We get the following

Corollary 6 *Under the same conditions as those of Corollary 5, if* $\lambda_0 = 1$, *then the integral operator* T_6 *is defined by* (42): $T_6 : l^p \rightarrow l^p(\omega)$ *exists as a bounded operator and*

$$\|T_6 a\|_{p,\omega} \leq c \|a\|_p, \tag{43}$$

where $\omega(m) = m^{p-2}$ and

$$c = \|T_6\| = \int_0^\infty K(u)u^{(1/q)-1}du$$

is the best possible.

5 Some Applications

As applications, a large number of known and new results have been obtained by proper choice of kernel K. In this section we present some model and interesting applications which display the importance of our results. Also these examples are of fundamental importance in analysis. In what follows, without loss of generality, we may assume $0 < \lambda_1 \le \lambda_2$, thus under the same conditions as those of Theorem 3, we have

$$\frac{c_3}{\lambda_2} \le \|T\| \le (\frac{c_1}{\lambda_1})^{1/p}(\frac{c_2}{\lambda_2})^{1-(1/p)}. \tag{44}$$

In the conjugate case $(\lambda = n)$, we get

$$\frac{c_0}{\lambda_2} \le \|T\| \le \frac{c_0}{\lambda_1^{1/p}\lambda_2^{1/q}}, \tag{45}$$

If $\lambda_1 = \lambda_2 = \lambda_0$, then

$$\|T\| = \frac{c_0}{\lambda_0},$$

where the constants c_1, c_2, c_3 and c_0 are defined by (12), (13), (15) and (18), respectively.

Example 1 If $K(\|x\|_\alpha^{\lambda_1} \cdot \|y\|_\alpha^{\lambda_2}) = \exp\{-(\|x\|_\alpha^{\lambda_1} \cdot \|y\|_\alpha^{\lambda_2})^\beta\}$, $\beta > 0$ in Theorem 3, then the operator T_7 is defined by

$$T_7(f, x) = \int_{E_n(\alpha)} \exp\{-(\|x\|_\alpha^{\lambda_1} \cdot \|y\|_\alpha^{\lambda_2})^\beta\} f(y)dy. \tag{46}$$

Thus, in (44) and (45),

$$c_1 = \frac{(\Gamma(1/\alpha))^n}{\alpha^{n-1}\Gamma(n/\alpha)} \int_0^\infty \{\exp(-u^\beta)\}^{\frac{np}{\lambda}(1-\frac{1}{q})} u^{\frac{n^2}{\lambda\lambda_2}(p-1)(1-\frac{1}{q})-1} du$$

$$= \frac{(\Gamma(1/\alpha))^n}{\beta\alpha^{n-1}\Gamma(n/\alpha)} \times (\frac{q\lambda}{np(q-1)})^{\frac{n^2}{\beta\lambda\lambda_2}(p-1)(1-\frac{1}{q})}$$

$$\times \Gamma(\frac{n^2}{\beta\lambda\lambda_2}(p-1)(1-\frac{1}{q})). \tag{47}$$

$$c_2 = \frac{(\Gamma(1/\alpha))^n}{\alpha^{n-1}\Gamma(n/\alpha)} \int_0^\infty \{\exp(-u^\beta)\}^{\frac{n}{\lambda}} u^{\frac{n}{\lambda_2}[1-\frac{n}{\lambda}(1-\frac{1}{q})]-1} du$$

$$= \frac{(\Gamma(1/\alpha))^n}{\beta\alpha^{n-1}\Gamma(n/\alpha)} (\frac{\lambda}{n})^{\frac{n}{\beta\lambda_2}(1-\frac{n}{\lambda}(1-\frac{1}{q}))} \Gamma(\frac{n}{\beta\lambda_2}(1-\frac{n}{\lambda}(1-\frac{1}{q}))). \tag{48}$$

$$c_3 = \frac{(\Gamma(1/\alpha))^n}{\alpha^{n-1}\Gamma(n/\alpha)} \int_0^\infty \{\exp(-u^\beta)\} u^{\frac{n}{\lambda_2}(1-\frac{1}{p})-1} du$$

$$= \frac{(\Gamma(1/\alpha))^n}{\beta\alpha^{n-1}\Gamma(n/\alpha)} \Gamma(\frac{n}{\beta\lambda_2}(1-\frac{1}{p})). \tag{49}$$

$$c_0 = \frac{(\Gamma(1/\alpha))^n}{\beta\alpha^{n-1}\Gamma(n/\alpha)} \Gamma(\frac{n}{\beta\lambda_2 q}). \tag{50}$$

In particular, if $\lambda_1 = \lambda_2 = \lambda_0$, then by Corollary 2, we get

$$\|T_7\| = \frac{(\Gamma(1/\alpha))^n}{\lambda_0\beta\alpha^{n-1}\Gamma(n/\alpha)} \Gamma(\frac{n}{\lambda_0\beta q}). \tag{51}$$

If $\lambda_0 = \beta = 1, \alpha = 2$, then T_7 reduces to the $n-$ dimensional Laplace transform of f, that is,

$$T_7(f,x) = \int_{\mathbb{R}_+^n} exp\{-(\|x\| \cdot \|y\|)\} f(y) dy, \tag{52}$$

thus by (51), we get

$$\|T_7\| = \frac{\pi^{n/2}}{2^{n-1}\Gamma(n/2)} \Gamma(\frac{n}{q}). \tag{53}$$

If $n = \lambda_0 = \beta = 1$, then T_7 reduces to the one-sided Laplace transform of f, thus by (53), we have

$$\|T_7\| = \Gamma(\frac{1}{q}) = \int_0^\infty u^{\frac{1}{q}-1} e^{-u} du.$$

That is, it reduces to Theorem 2.

Example 2 If $K(\|x\|_\alpha^{\lambda_1} \cdot \|y\|_\alpha^{\lambda_2}) = \frac{|\log(\|x\|_\alpha^{\lambda_1} \cdot \|y\|_\alpha^{\lambda_2})|^{\beta_1}}{|(\|x\|_\alpha^{\lambda_1} \cdot \|y\|_\alpha^{\lambda_2})^{\beta_2}-1|}$ $(\beta_1 > 0, \beta_2 > \frac{n}{q\lambda_2})$ in Theorem 3, then the operator T_8 is defined by

$$T_8(f,x) = \int_{E_n(\alpha)} \frac{|\log(\|x\|_\alpha^{\lambda_1} \cdot \|y\|_\alpha^{\lambda_2})|^{\beta_1}}{|(\|x\|_\alpha^{\lambda_1} \cdot \|y\|_\alpha^{\lambda_2})^{\beta_2} - 1|} f(y) dy. \tag{54}$$

Thus, in (44) and (45),

$$c_1 = \frac{(\Gamma(1/\alpha))^n}{\alpha^{n-1}\Gamma(n/\alpha)} \int_0^\infty \{\frac{|\log u|^{\beta_1}}{|u^{\beta_2}-1|}\}^{\frac{np}{\lambda}(1-\frac{1}{q})} u^{\frac{n^2}{\lambda\lambda_2}(p-1)(1-\frac{1}{q})-1} du;$$

$$c_2 = \frac{(\Gamma(1/\alpha))^n}{\alpha^{n-1}\Gamma(n/\alpha)} \int_0^\infty \{\frac{|\log u|^{\beta_1}}{|u^{\beta_2}-1|}\}^{\frac{n}{\lambda}} u^{\frac{n}{\lambda_2}[1-\frac{n}{\lambda}(1-\frac{1}{q})]-1} du;$$

$$c_3 = \frac{(\Gamma(1/\alpha))^n}{\alpha^{n-1}\Gamma(n/\alpha)} \int_0^\infty \{\frac{|\log u|^{\beta_1}}{|u^{\beta_2}-1|}\} u^{\frac{n}{\lambda_2}(1-\frac{1}{p})-1} du.$$

$$c_0 = \frac{(\Gamma(1/\alpha))^n}{\alpha^{n-1}\Gamma(n/\alpha)} \int_0^\infty \{\frac{|\log u|^{\beta_1}}{|u^{\beta_2}-1|}\} u^{\frac{n}{q\lambda_2}-1} du$$

$$= \frac{(\Gamma(1/\alpha))^n \Gamma(\beta_1+1)}{\alpha^{n-1}\beta_2^{\beta_1+1}\Gamma(n/\alpha)} \{\zeta(\beta_1+1, \frac{n}{q\lambda_2\beta_2}) + \zeta(\beta_1+1, 1-\frac{n}{q\lambda_2\beta_2})\}.$$

In particular, if $\lambda_1 = \lambda_2 = \lambda_0$, then by Corollary 2, we get

$$\|T_8\| = \frac{(\Gamma(1/\alpha))^n \Gamma(\beta_1+1)}{\lambda_0\alpha^{n-1}\beta_2^{\beta_1+1}\Gamma(n/\alpha)} \{\zeta(\beta_1+1, \frac{n}{q\lambda_0\beta_2}) + \zeta(\beta_1+1, 1-\frac{n}{q\lambda_0\beta_2})\}. \tag{55}$$

Example 3 If $K(\|x\|_\alpha^{\lambda_1} \cdot \|y\|_\alpha^{\lambda_2}) = \frac{\log(\|x\|_\alpha^{\lambda_1} \cdot \|y\|_\alpha^{\lambda_2})}{(\|x\|_\alpha^{\lambda_1} \cdot \|y\|_\alpha^{\lambda_2})^\beta - 1}$ $(\beta > 0)$ in Theorem 3, then the operator T_9 is defined by

$$T_9(f, x) = \int_{E_n(\alpha)} \frac{\log(\|x\|_\alpha^{\lambda_1} \cdot \|y\|_\alpha^{\lambda_2})}{(\|x\|_\alpha^{\lambda_1} \cdot \|y\|_\alpha^{\lambda_2})^\beta - 1} f(y) dy. \tag{56}$$

By (18), we have

$$c_0 = \frac{(\Gamma(1/\alpha))^n}{\beta^2 \alpha^{n-1}\Gamma(n/\alpha)} (\frac{\pi}{\sin(\frac{n\pi}{\lambda_2\beta q})})^2. \tag{57}$$

In particular, if $\lambda_1 = \lambda_2 = \lambda_0$, then by Corollary 2, we get

$$\|T_9\| = \frac{(\Gamma(1/\alpha))^n}{\lambda_0\beta^2 \alpha^{n-1}\Gamma(n/\alpha)} (\frac{\pi}{\sin(\frac{n\pi}{\lambda_0\beta q})})^2. \tag{58}$$

Example 4 If $K(\|x\|_\alpha^{\lambda_1} \cdot \|y\|_\alpha^{\lambda_2}) = \frac{\cos(\beta_1\|x\|_\alpha^{\lambda_1} \cdot \|y\|_\alpha^{\lambda_2})}{(\|x\|_\alpha^{\lambda_1} \cdot \|y\|_\alpha^{\lambda_2})^2 + \beta_2^2} \times (\|x\|_\alpha^{\lambda_1} \cdot \|y\|_\alpha^{\lambda_2})^{(1-\frac{n}{q\lambda_2})}$ $\beta_1, \beta_2 >$ 0 in Theorem 3, then the operator T_{10} is defined by

$$T_{10}(f, x) = \int_{E_n(\alpha)} \frac{\cos(\beta_1 \|x\|_\alpha^{\lambda_1} \cdot \|y\|_\alpha^{\lambda_2})}{(\|x\|_\alpha^{\lambda_1} \cdot \|y\|_\alpha^{\lambda_2})^2 + \beta_2^2} \times (\|x\|_\alpha^{\lambda_1} \cdot \|y\|_\alpha^{\lambda_2})^{(1-\frac{n}{q\lambda_2})} f(y) dy.$$

(59)

By (18), we have

$$c_0 = \frac{(\Gamma(1/\alpha))^n}{\alpha^{n-1}\Gamma(n/\alpha)} \int_0^\infty \frac{\cos(\beta_1 u)}{u^2 + \beta_2^2} du = \frac{\pi\{\Gamma(1/\alpha)\}^n}{2\beta_2 \alpha^{n-1}\Gamma(n/\alpha)} e^{-(\beta_1\beta_2)}.$$

(60)

In particular, if $\lambda_1 = \lambda_2 = \lambda_0$, then by Corollary 2, we get

$$\|T_{10}\| = \frac{\pi(\Gamma(1/\alpha))^n}{2\lambda_0\beta_2\alpha^{n-1}\Gamma(n/\alpha)} e^{-(\beta_1\beta_2)}.$$

(61)

Example 5 If $K(\|x\|_\alpha^{\lambda_1} \cdot \|y\|_\alpha^{\lambda_2}) = (2/\pi)^{n/\alpha} \sin(\beta \|x\|_\alpha^{\lambda_1} \cdot \|y\|_\alpha^{\lambda_2})$ $(\beta > 0, \lambda_1 > 0, \lambda_2 > n/q)$ in Theorem 3, then the operator T_{11} is defined by

$$T_{11}(f, x) = (2/\pi)^{n/\alpha} \int_{E_n(\alpha)} \sin(\beta \|x\|_\alpha^{\lambda_1} \cdot \|y\|_\alpha^{\lambda_2}) f(y) dy.$$

(62)

By (18), we have

$$
\begin{aligned}
c_0 &= \frac{(\Gamma(1/\alpha))^n}{\alpha^{n-1}\Gamma(n/\alpha)} \left(\frac{2}{\pi}\right)^{n/\alpha} \int_0^\infty \frac{\sin(\beta u)}{u^{1-\frac{n}{q\lambda_2}}} du \\
&= \frac{(2/\pi)^{(n/\alpha)-1}\beta^{-(\frac{n}{q\lambda_2})}(\Gamma(1/\alpha))^n}{\alpha^{n-1}\Gamma(n/\alpha)\Gamma(1 - \frac{n}{q\lambda_2})\cos(\frac{n\pi}{2q\lambda_2})}.
\end{aligned}
$$

(63)

In particular, if $\lambda_1 = \lambda_2 = \lambda_0$, then by Corollary 2, we get

$$\|T_{11}\| = \frac{\beta^{-(\frac{n}{q\lambda_0})}(2/\pi)^{(n/\alpha)-1}(\Gamma(1/\alpha))^n}{\lambda_0\alpha^{n-1}\Gamma(n/\alpha)\Gamma(1 - \frac{n}{q\lambda_0})\cos(\frac{n\pi}{2q\lambda_0})}.$$

(64)

When $\alpha = 2, \beta = \lambda_0 = 1$, T_{11} is called the $n-$ dimensional Fourier sine transform of f, then by (64), we get

$$\|T_{11}\| = \frac{\pi}{2^{n/2}\Gamma(n/2)\Gamma(1 - \frac{n}{q})\cos(\frac{n\pi}{2q})}.$$

If $n = 1$, then T_{11} reduces to the Fourier sine transform of f in [3]:

$$T_{11}(f, x) = (2/\pi)^{1/2} \int_0^\infty \sin(xy) f(y) dy.$$

Thus, we have

$$\|T_{11}\| = \frac{(\pi/2)^{1/2}}{\Gamma(\frac{1}{p})\cos(\frac{\pi}{2q})}.$$

Example 6 If $K(\|x\|_\alpha^{\lambda_1} \cdot \|y\|_\alpha^{\lambda_2}) = (\frac{2}{\pi})^{n/\alpha}\cos(\beta\|x\|_\alpha^{\lambda_1} \cdot \|y\|_\alpha^{\lambda_2})$ $(\beta > 0, \lambda_1 > 0, \lambda_2 > n/q)$ in Theorem 3, then the operator T_{12} is defined by

$$T_{12}(f, x) = (\frac{2}{\pi})^{n/\alpha}\int_{E_n(\alpha)}\cos(\beta\|x\|_\alpha^{\lambda_1} \cdot \|y\|_\alpha^{\lambda_2})f(y)dy. \tag{65}$$

By (18), we have

$$c_0 = \frac{(\Gamma(1/\alpha))^n}{\alpha^{n-1}\Gamma(n/\alpha)}(\frac{2}{\pi})^{n/\alpha}\int_0^\infty \frac{\cos(\beta u)}{u^{1-\frac{n}{q\lambda_2}}}du$$

$$= \frac{(2/\pi)^{(n/\alpha)-1}\beta^{-(\frac{n}{q\lambda_2})}(\Gamma(1/\alpha))^n}{\alpha^{n-1}\Gamma(n/\alpha)\Gamma(1-\frac{n}{q\lambda_2})\sin(\frac{n\pi}{2q\lambda_2})}. \tag{66}$$

In particular, if $\lambda_1 = \lambda_2 = \lambda_0$, then by Corollary 2, we get

$$\|T_{12}\| = \frac{(2/\pi)^{(n/\alpha)-1}\beta^{-(\frac{n}{q\lambda_0})}(\Gamma(1/\alpha))^n}{\lambda_0\alpha^{n-1}\Gamma(n/\alpha)\Gamma(1-\frac{n}{q\lambda_0})\sin(\frac{n\pi}{2q\lambda_0})}. \tag{67}$$

When $\alpha = 2, \beta = \lambda_0 = 1$, T_{12} is called the $n-$ dimensional Fourier cosine transform of f, then by (67), we get

$$\|T_{12}\| = \frac{\pi}{2^{n/2}\Gamma(n/2)\Gamma(1-\frac{n}{q})\sin(\frac{n\pi}{2q})}.$$

If $n = 1$, then T_{12} reduces to the Fourier cosine transform of f in [3]:

$$T_{12}(f, x) = (\frac{2}{\pi})^{1/2}\int_0^\infty \cos(xy)f(y)dy.$$

Thus, we have

$$\|T_{12}\| = \frac{(\pi/2)^{1/2}}{\Gamma(\frac{1}{p})\sin(\frac{\pi}{2q})}.$$

Remark 2 Because the Fourier transform of f can be decomposed into the Fourier sine and cosine transforms of f, thus, the corresponding operator norm can be derived from Examples 5 and 6.

Example 7 If $K(\|x\|_\alpha^{\lambda_1} \cdot \|y\|_\alpha^{\lambda_2}) = \dfrac{1}{1+(\|x\|_\alpha^{\lambda_1} \cdot \|y\|_\alpha^{\lambda_2})^\beta}$ in Theorem 3, then the operator T_{13} is defined by

$$T_{13}(f, x) = \int_{E_n(\alpha)} \frac{1}{1 + (\|x\|_\alpha^{\lambda_1} \cdot \|y\|_\alpha^{\lambda_2})^\beta} f(y)dy. \tag{68}$$

If $(\frac{\lambda}{\beta\lambda_2} - 1)\frac{q}{q-1} < \frac{n}{\beta\lambda_2} < \frac{p}{p-1}$, and $\lambda > n(1 - \frac{1}{q})$, then in (44) and (45),

$$c_1 = \frac{(\Gamma(1/\alpha))^n}{\alpha^{n-1}\Gamma(n/\alpha)} \int_0^\infty \{\frac{1}{1+u^\beta}\}^{\frac{np}{\lambda}(1-\frac{1}{q})} u^{\frac{n^2}{\lambda\lambda_2}(p-1)(1-\frac{1}{q})-1} du$$

$$= \frac{(\Gamma(1/\alpha))^n}{\beta\alpha^{n-1}\Gamma(n/\alpha)}$$

$$\times B\left(\frac{n^2}{\beta\lambda\lambda_2}(p-1)(1-\frac{1}{q}), \frac{n}{\lambda}(1-\frac{1}{q})(p - \frac{n}{\beta\lambda_2}(p-1))\right).$$

$$c_2 = \frac{(\Gamma(1/\alpha))^n}{\alpha^{n-1}\Gamma(n/\alpha)} \int_0^\infty \{\frac{1}{1+u^\beta}\}^{\frac{n}{\lambda}} u^{\frac{n}{\lambda_2}[1-\frac{n}{\lambda}(1-\frac{1}{q})]-1} du$$

$$= \frac{(\Gamma(1/\alpha))^n}{\beta\alpha^{n-1}\Gamma(n/\alpha)} \times B\left(\frac{n}{\beta\lambda_2}(1 - \frac{n}{\lambda}(1 - \frac{1}{q})), \frac{n}{\lambda} - \frac{n}{\beta\lambda_2}(1 - \frac{n}{\lambda}(1 - \frac{1}{q}))\right),$$

$$c_3 = \frac{(\Gamma(1/\alpha))^n}{\alpha^{n-1}\Gamma(n/\alpha)} \int_0^\infty \frac{1}{1+u^\beta} u^{\frac{n}{\lambda_2}(1-\frac{1}{p})-1} du$$

$$= \frac{(\Gamma(1/\alpha))^n}{\beta\alpha^{n-1}\Gamma(n/\alpha)} \times \frac{\pi}{\sin(\frac{n\pi}{\beta\lambda_2}(1 - \frac{1}{p}))}.$$

In the conjugate case ($\lambda = n$), if $\beta > \frac{n}{q\lambda_2}$, then

$$c_0 = \frac{(\Gamma(1/\alpha))^n}{\beta\alpha^{n-1}\Gamma(n/\alpha)} \times \frac{\pi}{\sin(\frac{n\pi}{\lambda_2\beta q})}.$$

In particular, if $\lambda_1 = \lambda_2 = \lambda_0$, then by Corollary 2, we get

$$\|T_{13}\| = \frac{(\Gamma(1/\alpha))^n}{\lambda_0\beta\alpha^{n-1}\Gamma(n/\alpha)} \times \frac{\pi}{\sin(\frac{n\pi}{\lambda_0\beta q})}. \tag{69}$$

Example 8 If $K(\|x\|_\alpha^{\lambda_1} \cdot \|y\|_\alpha^{\lambda_2}) = \dfrac{1}{\{1+(\|x\|_\alpha^{\lambda_1} \cdot \|y\|_\alpha^{\lambda_2})\}^\beta}$ in Theorem 3, then the operator T_{14} is defined by

$$T_{14}(f, x) = \int_{E_n(\alpha)} \frac{1}{\{1 + (\|x\|_\alpha^{\lambda_1} \cdot \|y\|_\alpha^{\lambda_2})\}^\beta} f(y)dy. \tag{70}$$

If $(\frac{\lambda}{\beta\lambda_2} - 1)\frac{q}{q-1} < \frac{n}{\beta\lambda_2} < \frac{p}{p-1}$ and $\lambda > n(1 - \frac{1}{q})$, then in (44) and (45),

$$c_1 = \frac{(\Gamma(1/\alpha))^n}{\alpha^{n-1}\Gamma(n/\alpha)} \int_0^\infty \{\frac{1}{(1+u)^\beta}\}^{\frac{np}{\lambda}(1-\frac{1}{q})} u^{\frac{n^2}{\lambda\lambda_2}(p-1)(1-\frac{1}{q})-1} du$$

$$= \frac{(\Gamma(1/\alpha))^n}{\alpha^{n-1}\Gamma(n/\alpha)} \times B\left(\frac{n^2}{\lambda\lambda_2}(p-1)(1-\frac{1}{q}), \frac{n}{\lambda}(1-\frac{1}{q})(p\beta - \frac{n}{\lambda_2}(p-1))\right).$$

$$c_2 = \frac{(\Gamma(1/\alpha))^n}{\alpha^{n-1}\Gamma(n/\alpha)} \int_0^\infty \{\frac{1}{(1+u)^\beta}\}^{\frac{n}{\lambda}} u^{\frac{n}{\lambda_2}[1-\frac{n}{\lambda}(1-\frac{1}{q})]-1} du$$

$$= \frac{(\Gamma(1/\alpha))^n}{\alpha^{n-1}\Gamma(n/\alpha)} \times B\left(\frac{n}{\lambda_2}(1 - \frac{n}{\lambda}(1-\frac{1}{q})), \frac{\beta n}{\lambda} - \frac{n}{\lambda_2}(1 - \frac{n}{\lambda}(1-\frac{1}{q}))\right).$$

$$c_3 = \frac{(\Gamma(1/\alpha))^n}{\alpha^{n-1}\Gamma(n/\alpha)} \int_0^\infty \{\frac{1}{(1+u)^\beta}\} u^{\frac{n}{\lambda_2}(1-\frac{1}{p})-1} du$$

$$= \frac{(\Gamma(1/\alpha))^n}{\alpha^{n-1}\Gamma(n/\alpha)} B\left(\frac{n}{\lambda_2}(1 - \frac{1}{p}), \beta - \frac{n}{\lambda_2}(1 - \frac{1}{p})\right)$$

In the conjugate case ($\lambda = n$), if $\beta > \frac{n}{q\lambda_2}$, then

$$c_0 = \frac{(\Gamma(1/\alpha))^n}{\alpha^{n-1}\Gamma(n/\alpha)} B\left(\frac{n}{\lambda_2 q}, \beta - \frac{n}{\lambda_2 q}\right).$$

In particular, if $\lambda_1 = \lambda_2 = \lambda_0$, then by Corollary 2, we get

$$\|T_{14}\| = \frac{(\Gamma(1/\alpha))^n}{\lambda_0 \alpha^{n-1}\Gamma(n/\alpha)} B\left(\frac{n}{\lambda_0 q}, \beta - \frac{n}{\lambda_0 q}\right). \tag{71}$$

If $n = 1, \lambda_1 = \lambda_2 = 1$, then by the generalized Stieltjes transform (10), we have

$$T_3(f, x) = x^\beta \int_0^\infty \frac{1}{(1+xy)^\beta} f(y)dy = x^\beta T_{14}(f, x).$$

If $\beta > 1/q$, then by (71), we get

$$\|T_3\| = B\left(\frac{1}{q}, \beta - \frac{1}{q}\right),$$

where $\omega(x) = x^{p(\beta+1)-2}$.

Example 9 If $K(\|x\|_\alpha^{\lambda_1} \cdot \|y\|_\alpha^{\lambda_2}) = \dfrac{1}{|1-\|x\|_\alpha^{\lambda_1} \cdot \|y\|_\alpha^{\lambda_2}|^\beta}$ in Theorem 3, then the operator T_{15} is defined by

$$T_{15}(f,x) = \int_{E_n(\alpha)} \frac{1}{|1 - \|x\|_\alpha^{\lambda_1} \cdot \|y\|_\alpha^{\lambda_2}|^\beta} f(y)dy. \tag{72}$$

If $(\frac{\lambda}{\beta\lambda_2} - 1)\frac{q}{q-1} < \frac{n}{\beta\lambda_2} < \frac{p}{p-1}$, and $\lambda > p\beta n(1 - \frac{1}{q})$, then in (44) and (45),

$$c_1 = \frac{(\Gamma(1/\alpha))^n}{\alpha^{n-1}\Gamma(n/\alpha)} \int_0^\infty \{\frac{1}{|1-u|^\beta}\}^{\frac{pn}{\lambda}(1-\frac{1}{q})} u^{\frac{n^2}{\lambda\lambda_2}(p-1)(1-\frac{1}{q})-1} du$$

$$= \frac{(\Gamma(1/\alpha))^n}{\alpha^{n-1}\Gamma(n/\alpha)} \{B\big(\frac{n^2}{\lambda\lambda_2}(p-1)(1-\frac{1}{q}), 1 - \frac{p\beta n}{\lambda}(1 - \frac{1}{q})\big)$$

$$+ B\big(\frac{n}{\lambda}(1 - \frac{1}{q})[p\beta - \frac{n}{\lambda_2}(p-1)], 1 - \frac{p\beta n}{\lambda}(1 - \frac{1}{q})\big)\}$$

$$c_2 = \frac{(\Gamma(1/\alpha))^n}{\alpha^{n-1}\Gamma(n/\alpha)} \int_0^\infty \{\frac{1}{|1-u|^\beta}\}^{\frac{n}{\lambda}} u^{\frac{n}{\lambda_2}[1-\frac{n}{\lambda}(1-\frac{1}{q})]-1} du$$

$$= \frac{(\Gamma(1/\alpha))^n}{\alpha^{n-1}\Gamma(n/\alpha)} \{B\big(\frac{n}{\lambda_2}[1 - \frac{n}{\lambda}(1 - \frac{1}{q})], 1 - \frac{\beta n}{\lambda}\big)$$

$$+ B\big(\frac{\beta n}{\lambda} - \frac{n}{\lambda_2}[1 - \frac{n}{\lambda}(1 - \frac{1}{q})], 1 - \frac{\beta n}{\lambda}\big)\}.$$

$$c_3 = \frac{(\Gamma(1/\alpha))^n}{\alpha^{n-1}\Gamma(n/\alpha)} \int_0^\infty \{\frac{1}{|1-u|^\beta}\} u^{\frac{n}{\lambda_2}(1-\frac{1}{p})-1} du$$

$$= \frac{(\Gamma(1/\alpha))^n}{\alpha^{n-1}\Gamma(n/\alpha)} \{B\big(\frac{n}{\lambda_2}(1 - \frac{1}{p}), 1 - \beta\big) + B\big(\beta - \frac{n}{\lambda_2}(1 - \frac{1}{p}), 1 - \beta\big)\}.$$

In the conjugate case $(\lambda = n)$, if $\frac{n}{q\lambda_2} < \beta < 1$, then

$$c_0 = \frac{(\Gamma(1/\alpha))^n}{\alpha^{n-1}\Gamma(n/\alpha)} \times \{B\big(\frac{n}{q\lambda_2}, 1 - \beta\big) + B\big(\beta - \frac{n}{q\lambda_2}, 1 - \beta\big)\}$$

In particular, if $\lambda_1 = \lambda_2 = \lambda_0$, then by Corollary 2, we get

$$\|T_{15}\| = \frac{(\Gamma(1/\alpha))^n}{\lambda_0 \alpha^{n-1}\Gamma(n/\alpha)} \times \{B(\frac{n}{q\lambda_0}, 1 - \beta) + B(\beta - \frac{n}{q\lambda_0}, 1 - \beta)\}. \tag{73}$$

Remark 3 Defining other forms of the kernel K, we can obtain new results of interest.

References

1. D.V. Widder, *The Laplace Transform* (Princeton University Press, Princeton, 1972)
2. K.B. Wolff, *Integral Transforms in Science and Engineering* (Plenum, New York, 1979)
3. E. Zauderer, *Partial Differential Equations of Applied Mathematics* (A Wiley-Interscience Publication, Wiley, 1983)
4. N. Bleistein, R.A. Handelsman, *Asymptotic Expansions of Integrals* (Dover Publications, Inc., New York, 1986)
5. G.H. Hardy, The constants of certain inequalities. J. Lond. Math. Soc. **8**, 114–119 (1933)
6. J.C. Kuang, Generalized Laplace transform inequalities in multiple Orlicz spaces, chapter 13, in *Computation, Cryptography, and Network Security* ed. by N.J. Daras, M.T. Rassias (Springer, Berlin, 2015)
7. J.C. Kuang, *Real and Functional Analysis* (coutinuation), vol. 2 (Higher Education Press, Beijing, 2015) (in Chinese)
8. J.C. Kuang, *Applied Inequalities*, 4th ed. (Shandong Science Technology Press, Jinan, 2010) (in Chinese)

On Marcinkiewicz-Zygmund Inequalities at Hermite Zeros and Their Airy Function Cousins

D. S. Lubinsky

Abstract We establish forward and converse Marcinkiewicz-Zygmund Inequalities at the zeros $\{a_j\}_{j\geq 1}$ of the Airy function $Ai(x)$, such as

$$A\frac{\pi^2}{6}\sum_{k=1}^{\infty}\frac{|f(a_k)|^p}{Ai'(a_k)^2} \leq \int_{-\infty}^{\infty}|f(t)|^p\,dt \leq B\frac{\pi^2}{6}\sum_{k=1}^{\infty}\frac{|f(a_k)|^p}{Ai'(a_k)^2}$$

under appropriate conditions on the entire function f and p. The constants A and B are those appearing in Marcinkiewicz-Zygmund inequalities at zeros of Hermite polynomials. Scaling limits are used to pass from the latter to the former.

Keywords Marcinkiewicz-Zygmund inequalities · Quadrature sums · Airy functions · Hermite polynomials

1 Introduction

There is a close relationship between the Plancherel-Polya and Marcinkiewicz-Zygmund inequalities. The former [9, p. 152] assert that for $1 < p < \infty$, and entire functions f of exponential type at most π,

$$A_p\sum_{k=-\infty}^{\infty}|f(k)|^p \leq \int_{-\infty}^{\infty}|f|^p \leq B_p\sum_{j=-\infty}^{\infty}|f(k)|^p, \tag{1}$$

Research supported by NSF grant DMS1800251.

D. S. Lubinsky (✉)
School of Mathematics, Georgia Institute of Technology, Atlanta, GA, USA
e-mail: lubinsky@math.gatech.edu

© Springer Nature Switzerland AG 2020
A. Raigorodskii, M. T. Rassias (eds.), *Trigonometric Sums and Their Applications*,
https://doi.org/10.1007/978-3-030-37904-9_6

provided either the series or integral is finite. For $0 < p \leq 1$, the left-hand inequality is still true, but the right-hand inequality requires additional restrictions [2]. We assume that B_p is taken as small as possible, and A_p as large as possible. The Marcinkiewicz-Zygmund inequalities assert [35, Vol. II, p. 30] that for $p > 1$, $n \geq 1$, and polynomials P of degree $\leq n - 1$,

$$\frac{A_p'}{n} \sum_{k=1}^{n} \left| P\left(e^{2\pi i k/n}\right) \right|^p \leq \int_0^1 \left| P\left(e^{2\pi i t}\right) \right|^p dt \leq \frac{B_p'}{n} \sum_{k=1}^{n} \left| P\left(e^{2\pi i k/n}\right) \right|^p. \qquad (2)$$

Here too, A_p' and B_p' are independent of n and P, and the left-hand inequality is also true for $0 < p \leq 1$ [15]. The author [16] proved that the inequalities (1) and (2) are equivalent, in the sense that each implies the other. Moreover, the sharp constants are the same:

Theorem A *For $0 < p < \infty$, $A_p = A_p'$ and for $1 < p < \infty$, $B_p = B_p'$.*

These inequalities are useful in studying convergence of Fourier series, Lagrange interpolation, in number theory, and weighted approximation. They have been extended to many settings, and there are a great many methods to prove them [5, 8, 13, 15, 19, 20, 22–26, 30, 31, 33, 34]. The sharp constants in (1) and (2) are unknown, except for the case $p = 2$, where of course we have equality rather than inequality, so that $A_2 = B_2 = A_2' = B_2' = 1$ [9, p. 150]. It is certainly of interest to say more about these constants.

In a recent paper, we explored the connections between Marcinkiewicz-Zygmund inequalities at zeros of Jacobi polynomials, and Polya-Plancherel type inequalities at zeros of Bessel functions. Let $\alpha, \beta > -1$ and

$$w^{\alpha,\beta}(x) = (1-x)^\alpha (1+x)^\beta, \quad x \in (-1, 1).$$

For $n \geq 1$, let $P_n^{\alpha,\beta}$ denote the standard Jacobi polynomial of degree n, so that it has degree n, satisfies the orthogonality conditions

$$\int_{-1}^1 P_n^{\alpha,\beta}(x) x^k w^{\alpha,\beta}(x) dx = 0, \quad 0 \leq k < n,$$

and is normalized by $P_n^{\alpha,\beta}(1) = \binom{n+\alpha}{n}$. Let

$$x_{nn} < x_{n-1,n} < \cdots < x_{1n}$$

denote the zeros of $P_n^{\alpha,\beta}$. Let $\{\lambda_{kn}\}$ denote the weights in the Gauss quadrature for $w^{\alpha,\beta}$, so that for all polynomials P of degree $\leq 2n - 1$,

$$\int_{-1}^1 P w^{\alpha,\beta} = \sum_{k=1}^n \lambda_{kn} P(x_{kn}).$$

There is a classical analogue of (2), established for special α, β by Richard Askey, and for all $\alpha, \beta > -1$ (and for more general "generalized Jacobi weights") by P. Nevai, and his collaborators [15, 20, 27, 29], with later work by König and Nielsen [8], and for doubling weights by Mastroianni and Totik [23]. The following special case follows from Theorem 5 in [20, eqn. (1.19), p. 534]:

Theorem B *Let $\alpha, \beta, \tau, \sigma$ satisfy $\alpha, \beta, \alpha + \sigma, \beta + \tau > -1$. Let $p > 0$. For $n \geq 1$, let $\{x_{kn}\}$ denote the zeros of the Jacobi polynomial $P_n^{\alpha, \beta}$ and $\{\lambda_{kn}\}$ denote the corresponding Gauss quadrature weights. There exists $A > 0$ such that for $n \geq 1$, and polynomials P of degree $\leq n - 1$,*

$$A \sum_{k=1}^{n} \lambda_{kn} |P(x_{kn})|^p (1 - x_{kn})^\sigma (1 + x_{kn})^\tau$$

$$\leq \int_{-1}^{1} |P(x)|^p (1 - x)^{\alpha + \sigma} (1 + x)^{\beta + \tau} dx. \tag{3}$$

The converse inequality is much more delicate, and in particular holds only for $p > 1$, and even then only for special cases of the parameters. It too was investigated by P. Nevai, with later work by Yuan Xu [33, 34], König and Nielsen [8]. König and Nielsen gave the exact range of p for which

$$\int_{-1}^{1} |P(x)|^p (1 - x)^\alpha (1 + x)^\beta dx \leq B \sum_{k=1}^{n} \lambda_{kn} |P(x_{kn})|^p, \tag{4}$$

holds with B independent of n and P. Let

$$\mu(\alpha, \beta) = \max \left\{ 1, 4\frac{\alpha + 1}{2\alpha + 5}, 4\frac{\beta + 1}{2\beta + 5} \right\};$$

$$m(\alpha, \beta) = \max \left\{ 1, 4\frac{\alpha + 1}{2\alpha + 3}, 4\frac{\beta + 1}{2\beta + 3} \right\};$$

$$M(\alpha, \beta) = \frac{m(\alpha, \beta)}{m(\alpha, \beta) - 1}. \tag{5}$$

Then (4) holds for all n and P iff

$$\mu(\alpha, \beta) < p < M(\alpha, \beta). \tag{6}$$

The most general sufficient condition for a converse quadrature inequality is due to Yuan Xu [33, pp. 881–882]. When we restrict to Jacobi weights, with the same weight on both sides, the inequality takes the following form:

Theorem C *Let $\alpha, \beta, \tau, \sigma$ satisfy $\alpha, \beta, \alpha + \sigma, \beta + \tau > -1$. Let $p > 1$, $q = \frac{p}{p-1}$, and assume that*

$$\frac{p}{2}\left(\alpha+\frac{1}{2}\right)-(\alpha+1)<\sigma<(p-1)(\alpha+1)-\max\left\{0,\frac{p}{2}\left(\alpha+\frac{1}{2}\right)\right\}. \quad (7)$$

$$\frac{p}{2}\left(\beta+\frac{1}{2}\right)-(\beta+1)<\tau<(p-1)(\beta+1)-\max\left\{0,\frac{p}{2}\left(\beta+\frac{1}{2}\right)\right\}. \quad (8)$$

Then there exists $B>0$ such that for $n \geq 1$, and polynomials P of degree $\leq n-1$,

$$\int_{-1}^{1}|P(x)|^{p}(1-x)^{\alpha+\sigma}(1+x)^{\beta+\tau}\,dx$$

$$\leq B\sum_{k=1}^{n}\lambda_{kn}|P(x_{kn})|^{p}(1-x_{kn})^{\sigma}(1+x_{kn})^{\tau}. \quad (9)$$

Inequalities of the type (9) for doubling weights have been established by Mastroianni and Totik [23] under the additional condition that one needs to restrict the degree of P in (9) further, such as $\deg(P) \leq \eta n$ for some $\eta \in (0,1)$ depending on the particular doubling weight.

Now let $\alpha > -1$ and define the Bessel function of order α,

$$J_{\alpha}(z)=\left(\frac{z}{2}\right)^{\alpha}\sum_{k=0}^{\infty}(-1)^{k}\frac{\left(\frac{z}{2}\right)^{2k}}{k!\,\Gamma(k+\alpha+1)} \quad (10)$$

and

$$J_{\alpha}^{*}(z)=J_{\alpha}(z)/z^{\alpha}, \quad (11)$$

which has the advantage of being an entire function for all $\alpha > -1$. J_{α}^{*} has real simple zeros, and we denote the positive zeros by

$$0<j_{1}<j_{2}<\cdots$$

while for $k \geq 1$,

$$j_{-k}=-j_{k}.$$

The connection between Jacobi polynomials and Bessel functions is given by the classical Mehler-Heine asymptotic, which holds uniformly for z in compact subsets of $\mathbb{C}$ [32, p. 192]:

$$\lim_{n\to\infty}n^{-\alpha}P_{n}^{\alpha,\beta}\left(1-\frac{1}{2}\left(\frac{z}{n}\right)^{2}\right)$$

$$=\lim_{n\to\infty}n^{-\alpha}P_{n}^{\alpha,\beta}\left(\cos\frac{z}{n}\right)=\left(\frac{z}{2}\right)^{-\alpha}J_{\alpha}(z)=2^{\alpha}J_{\alpha}^{*}(z). \quad (12)$$

There is an extensive literature dealing with quadrature sums and Lagrange interpolation at the $\{j_k\}$. In particular, there is the quadrature formula [6, p. 49]

$$\int_{-\infty}^{\infty} |x|^{2\alpha+1} f(x)\,dx = \frac{2}{\tau^{2\alpha+2}} \sum_{k=-\infty,k\neq 0}^{\infty} \frac{1}{\left|J_{\alpha}^{*'}(j_k)\right|^2} f\left(\frac{j_k}{\tau}\right),$$

valid for all entire functions f of exponential type at most 2τ, for which the integral on the left-hand side is finite. That same paper contains the following converse Marcinkiewicz-Zygmund type inequality: let $\alpha \geq -\frac{1}{2}$ and $p > 1$; or $-1 < \alpha < -\frac{1}{2}$ and $1 < p < \frac{2}{|1+2\alpha|}$. Then for entire functions f of exponential type $\leq \tau$ for which $|x|^{\alpha+\frac{1}{2}} f(x) \in L_p(\mathbb{R}\setminus(-\delta,\delta))$, for some $\delta > 0$, [6, Lemma 14, p. 58; Lemma 13, p. 57]

$$\int_{-\infty}^{\infty} \left||x|^{\alpha+\frac{1}{2}} f(x)\right|^p dx \leq \frac{B^*}{\tau} \sum_{k=-\infty,k\neq 0}^{\infty} \left|\frac{1}{\tau^{\alpha+\frac{1}{2}} J_{\alpha}^{*'}(j_k)} f\left(\frac{j_k}{\tau}\right)\right|^p. \qquad (13)$$

Here B^* depends on α and p. In the converse direction, since $j_{k+1} - j_k$ is bounded below by a positive constant for all k, classical inequalities from the theory of entire functions [9, p. 150] show that

$$\sum_{k=-\infty,k\neq 0}^{\infty} |f(j_k)|^p \leq C \int_{-\infty}^{\infty} |f(x)|^p dx$$

for entire functions of finite exponential type for which the right-hand side is finite.

While Grozev and Rahman note the analogous nature of Lagrange interpolation at zeros of Jacobi polynomials and Bessel functions, and also the Mehler-Heine formula, their proofs proceed purely from properties of Bessel functions. In [17, Thms. 1.1, 1.3, pp. 227–228], the author used inequalities like (3) to pass to analogues for Bessel functions using scaling limits of the form (12), keeping the same constants, much as was done in [16]: Let $L_1^p\left((0,\infty), t^{2\alpha+2\sigma+1}\right)$ denote the space of all even entire functions f of exponential type ≤ 1 with

$$\int_0^{\infty} |f(t)|^p\, t^{2\alpha+2\sigma+1} dt < \infty.$$

Theorem D *Assume that $p > 0$, α, β, $\alpha + \sigma$, $\beta + \tau > -1$, and*

$$-p\left(\frac{\alpha}{2} + \frac{5}{4}\right) + \alpha + \sigma + 1 < 0.$$

Let A be as in Theorem B. Then

$$2A \sum_{k=1}^{\infty} j_k^{2\sigma} J_{\alpha}^{*'}(j_k)^{-2} |f(j_k)|^p \leq \int_0^{\infty} |f(t)|^p\, t^{2\alpha+2\sigma+1} dt,$$

for all $f \in L_1^p\left((0,\infty), t^{2\alpha+2\sigma+1}\right)$.

Theorem E *Assume that $p > 1$, α, β, $\alpha + \sigma$, $\beta + \tau > -1$, and that (7) and (8) hold. Let B be as in Theorem C. Then for $f \in L_1^p ((0, \infty), t^{2\alpha+2\sigma+1})$, we have*

$$\int_0^\infty |f(t)|^p \, t^{2\alpha+2\sigma+1} dt \leq 2B \sum_{k=1}^\infty j_k^{2\sigma} J_\alpha^{*\prime}(j_k)^{-2} |f(j_k)|^p . \tag{14}$$

In particular this holds for $\sigma = \tau = 0$ if p satisfies (6) with $\beta = \alpha$. Moreover, for any α, β, p, it is possible to choose σ and τ satisfying (7), (8) so that this last inequality also holds.

A very recent paper of Littmann [13] provides far reaching extensions of the inequalities of Grozev and Rahman to Hermite-Biehler weights, so that $t^{2\alpha+2\sigma+1}$ is replaced by $1/|E|^p$, where E is a Hermite-Biehler function, that is, an entire function $E(z)$ satisfying $|E(z)| > |E(\bar{z})|$ for Re $z > 0$. Moreover, the zeros of Bessel functions are replaced by the zeros of $B(z) = \frac{i}{2}(E(z) - \overline{E(\bar{z})})$. Littmann then uses these to establish weighted mean convergence of certain interpolation operators for classes of entire functions.

In this paper, we shall use Marcinkiewicz-Zygmund inequalities at zeros of Hermite polynomials, to derive Plancherel-Polya type inequalities at zeros of Airy functions. We begin with our notation. Throughout,

$$W(x) = \exp\left(-\frac{1}{2}x^2\right), \quad x \in \mathbb{R}, \tag{15}$$

is the Hermite weight, and $\{p_n\}$ are the orthonormal Hermite polynomials, so that

$$\int_{-\infty}^\infty p_n p_m W^2 = \delta_{mn}. \tag{16}$$

The classical Hermite polynomial is of course denoted by H_n. The relationship between p_n and H_n is given by [32, p. 105, (5.5.1)]

$$p_n = \pi^{-1/4} 2^{-n/2} (n!)^{-1/2} H_n. \tag{17}$$

The leading coefficient of p_n is [32, p. 106, (5.5.6)]

$$\gamma_n = \pi^{-1/4} 2^{n/2} (n!)^{-1/2} . \tag{18}$$

In the sequel, $\{x_{jn}\}$ denote the zeros of the Hermite polynomials in decreasing order:

$$-\infty < x_{nn} < x_{n-1,n} < \cdots < x_{2n} < x_{1n} < \infty,$$

while $\{\lambda_{jn}\}$ denote the weights in the Gauss quadrature formula: for polynomials P of degree $\leq 2n - 1$,

$$\int_{-\infty}^{\infty} P W^2 = \sum_{j=1}^{n} \lambda_{jn} P\left(x_{jn}\right).$$

There is an extensive literature on Marcinkiewicz-Zygmund inequalities at zeros of Hermite polynomials, as well as for orthonormal polynomials for more general exponential weights [3, 4, 7, 14, 21, 28, 29]. We shall use the following forward and converse inequalities [14, p. 529], [21, p. 287]:

Theorem F *Let $1 \leq p < \infty$. Let $r, R \in \mathbb{R}$ and $S > 0$.*

(a) *Then there exists $A > 0$ such that for $n \geq 1$, and polynomials P of degree at most $n + Sn^{1/3}$,*

$$\sum_{j=1}^{n} \lambda_{jn} \left|P\left(x_{jn}\right)\right|^p W^{p-2}\left(x_{jn}\right) \left(1 + \left|x_{jn}\right|\right)^{Rp}$$

$$\leq A \int_{-\infty}^{\infty} \left|(PW)(x)(1 + |x|)^R\right|^p dx. \tag{19}$$

(b) *Assume that*

$$r < 1 - \frac{1}{p}; \quad r \leq R; \quad R > -\frac{1}{p}. \tag{20}$$

In addition if $p = 4$, we assume that $r < R$, while if $p > 4$, we assume that

$$r - \min\left\{R, 1 - \frac{1}{p}\right\} + \frac{1}{3}\left(1 - \frac{4}{p}\right) \begin{cases} \leq 0, & \text{if } R \neq 1 - \frac{1}{p} \\ < 0, & \text{if } R = 1 - \frac{1}{p} \end{cases}. \tag{21}$$

Then there exists $B > 0$ such that for $n \geq 1$, and polynomials P of degree $\leq n - 1$,

$$\int_{-\infty}^{\infty} \left|(PW)(x)(1 + |x|)^r\right|^p dx$$

$$\leq B \sum_{j=1}^{n} \lambda_{jn} \left|P\left(x_{jn}\right)\right|^p W^{p-2}\left(x_{jn}\right) \left(1 + \left|x_{jn}\right|\right)^{Rp}. \tag{22}$$

Recall that the Airy function Ai is given on the real line by [1, 10.4.32, p. 447]

$$Ai(x) = \frac{1}{\pi} \int_0^{\infty} \cos\left(\frac{1}{3}t^3 + xt\right) dt.$$

The Airy function Ai is an entire function of order $\frac{3}{2}$, with only real negative zeros $\{a_j\}$, where

$$0 > a_1 > a_2 > a_3 > \cdots .$$

These are often denoted by $\{i_j\}$ rather than $\{a_j\}$. Ai satisfies the differential equation

$$Ai''(z) - zAi(z) = 0.$$

The Airy kernel $\mathbb{A}i(\cdot, \cdot)$, much used in random matrix theory, is defined [12] by

$$\mathbb{A}i(a, b) = \begin{cases} \frac{Ai(a)Ai'(b)-Ai'(a)Ai(b)}{a-b}, & a \neq b, \\ Ai'(a)^2 - aAi(a)^2, & a = b. \end{cases}$$

Observe that

$$\mathcal{L}_j(z) = \frac{\mathbb{A}i(z, a_j)}{\mathbb{A}i(a_j, a_j)} = \frac{Ai(z)}{Ai'(a_j)(z - a_j)},$$

is the Airy analogue of a fundamental of Lagrange interpolation, satisfying

$$\mathcal{L}_j(a_k) = \delta_{jk}.$$

There is an analogue of sampling series and Lagrange interpolation series involving $\{\mathcal{L}_j\}$:

Definition 1.1 Let $\mathcal{G}$ be the class of all functions $g : \mathbb{C} \to \mathbb{C}$ with the following properties:

(a) g is an entire function of order at most $\frac{3}{2}$;
(b) There exists $L > 0$ such that for $\delta \in (0, \pi)$, some $C_\delta > 0$, and all $z \in \mathbb{C}$ with $|\arg z| \leq \pi - \delta$,

$$|g(z)| \leq C_\delta (1 + |z|)^L \left| \exp\left(-\frac{2}{3}z^{\frac{3}{2}}\right) \right|;$$

(c)

$$\sum_{j=1}^{\infty} \frac{|g(a_j)|^2}{|a_j|^{1/2}} < \infty. \tag{23}$$

In [12, Corollary 1.3, p. 429], it was shown that each $g \in \mathcal{G}$ admits the locally uniformly convergent expansion

$$g(z) = \sum_{j=1}^{\infty} g(a_j) \frac{\mathbb{A}i(z, a_j)}{\mathbb{A}i(a_j, a_j)} = \sum_{j=1}^{\infty} g(a_j) \mathcal{L}_j(z).$$

We let

$$S_M[g] = \sum_{j=1}^{M} g(a_j) \mathcal{L}_j, \quad M \geq 1, \tag{24}$$

denote the Mth partial sum of this expansion. Moreover, for $f, g \in \mathcal{G}$, there is the quadrature formula [12, Corollary 1.4, p. 429]

$$\int_{-\infty}^{\infty} f(x) g(x) \, dx = \sum_{j=1}^{\infty} \frac{(fg)(a_j)}{\mathbb{A}i(a_j, a_j)}.$$

In particular,

$$\int_{-\infty}^{\infty} g^2(x) \, dx = \sum_{j=1}^{\infty} \frac{|g(a_j)|^2}{\mathbb{A}i(a_j, a_j)},$$

and the series on the right converges because of (23), and the fact that $\mathbb{A}i(a_j, a_j) = Ai'(a_j)^2$ grows like $j^{1/3}$ – see Lemma 2.2.

Lagrange interpolation at zeros of Airy functions was considered in [18]. We shall need a class of functions that are limits in L_p of the partial sums of the Airy series expansion:

Definition 1.2 Let $0 < p < \infty$ and $f \in L_p(\mathbb{R})$. We write $f \in \mathcal{G}_p$ if

$$\lim_{M \to \infty} \|f - S_M[f]\|_{L_p(\mathbb{R})} = 0.$$

The relationship between Hermite polynomials and Airy functions lies in the asymptotic [32, p. 201],

$$e^{-x^2/2} H_n(x) = 3^{1/3} \pi^{-3/4} 2^{n/2+1/4} (n!)^{1/2} n^{-1/12} \{Ai(-t) + o(1)\} \tag{25}$$

as $n \to \infty$, uniformly for

$$x = \sqrt{2n}(1 - 6^{-1/3} (2n)^{-2/3} t), \tag{26}$$

and t in compact subsets of $\mathbb{C}$. This follows from the formulation in [32] because of the uniformity. Using this and part (a) of Theorem F with $R = r = 0$, we shall prove:

Theorem 1.3 *Let $p \geq 1$. Let A be the constant in (19) with $R = r = 0$ there.*

(a) *Then for $f \in \mathcal{G}_p$, we have*

$$\sum_{k=1}^{\infty} \frac{|f(a_k)|^p}{Ai'(a_k)^2} \leq A \frac{6}{\pi^2} \int_{-\infty}^{\infty} |f(t)|^p \, dt. \tag{27}$$

(b) *In particular, if $p \geq 2$, $f \in \mathcal{G}$ and for some $C > 0$, $\beta > \frac{1}{4}$, we have*

$$|f(x)| \leq C(1 + |x|)^{-\beta}, \quad x \in \mathbb{R}, \tag{28}$$

then (27) is true.

Remark We expect that (27) also holds for $0 < p < 1$, but this would require (19) for such p, and that does not seem to appear in the literature.

Using part (b) of Theorem F, we shall prove:

Theorem 1.4 *Let $1 < p < 4$. Let B be the constant in (22) with $R = r = 0$ there.*

(a) *For $f \in \mathcal{G}_p$, we have*

$$\frac{6}{\pi^2} \int_{-\infty}^{\infty} |f(t)|^p \, dt \leq B \sum_{k=1}^{\infty} \frac{|f(a_k)|^p}{Ai'(a_k)^2}. \tag{29}$$

(b) *In particular, if $f \in \mathcal{G}$ and*

$$\sum_{k=1}^{\infty} \frac{|f(a_k)|^p}{k^{1/3}} < \infty, \tag{30}$$

then (29) is true.

In the sequel, $C, C_1, C_2, \ldots$ denote constants independent of n, z, x, t, and polynomials of degree $\leq n$. The same symbol does not necessarily denote the same constant in different occurrences. $[x]$ denotes the greatest integer $\leq x$. Given two sequences $\{x_n\}, \{y_n\}$ of non-zeros real numbers, we write

$$x_n \sim y_n$$

if there exist constants C_1 and C_2 such that

$$C_1 \leq x_n / y_n \leq C_2$$

for $n \geq 1$. Similar notation is used for functions and sequences of functions. We establish some basic estimates and then prove Theorems 1.3 and 1.4 in Section 2.

2   Proof of Theorems 1.3 and 1.4

We start with properties of Hermite polynomials. Throughout $\{p_n\}$ denote the orthonormal Hermite polynomials satisfying (16), with leading coefficient γ_n, and with zeros $\{x_{jn}\}$. In the sequel, we let

$$\psi_n(x) = \left| 1 - \frac{|x|}{\sqrt{2n}} \right| + n^{-2/3}.$$

We also let

$$K_n(x, y) = \sum_{j=0}^{n-1} p_j(x) p_j(y)$$

denote the nth reproducing kernel, and

$$\lambda_n(x) = 1/K_n(x, x)$$

denote the nth Christoffel function. In particular, $\lambda_{jn} = \lambda_n(x_{jn})$. The jth fundamental polynomial at the zeros of $p_n(x)$ is

$$\ell_{jn}(x) = \frac{p_n(x)}{p_n'(x_{jn})(x - x_{jn})}.$$

It also admits the identity

$$\ell_{jn}(x) = \lambda_{jn} K_n(x, x_{jn}). \tag{31}$$

Lemma 2.1

(a)

$$\frac{\gamma_{n-1}}{\gamma_n} = \sqrt{\frac{n}{2}}. \tag{32}$$

(b) *For each fixed j, as $n \to \infty$,*

$$x_{jn} = \sqrt{2n}(1 - 6^{-1/3}(2n)^{2/3}\{|a_j| + o(1)\}). \tag{33}$$

(c) *Uniformly for t in compact subsets of $\mathbb{C}$, and for*

$$x = \sqrt{2n}\left(1 - 6^{-1/3}(2n)^{-2/3}t\right), \tag{34}$$

we have

$$(p_n W)(x) = 3^{1/3} \pi^{-1} 2^{1/4} n^{-1/12} \{Ai(-t) + o(1)\}. \tag{35}$$

(d) *For each fixed j, as* $n \to \infty$,

$$(p_n' W)(x_{jn}) = 3^{2/3} \pi^{-1} 2^{3/4} n^{1/12} \{Ai'(a_j) + o(1)\}. \tag{36}$$

(e) *For each fixed j, and uniformly for t in compact subsets of* $\mathbb{C}$*, and x of the form* (34)

$$\lim_{n \to \infty} (\ell_{jn} W)(x) W^{-1}(x_{jn}) = \mathcal{L}_j(-t). \tag{37}$$

(f) *For all* $1 \le j \le n$ *and all* $x \in \mathbb{R}$,

$$\left| \ell_{jn} W \right|(x) W^{-1}(x_{jn}) \le C \left(\frac{\psi_n(x)}{\psi_n(x_{jn})} \right)^{1/4} \frac{1}{1 + n^{1/2} \psi_n(x)^{1/2} \left| x - x_{jn} \right|}. \tag{38}$$

(g) *In particular for fixed j, and* $n \ge n_0(j)$ *and all* $x \in \mathbb{R}$,

$$\left| \ell_{jn} W \right|(x) W^{-1}(x_{jn}) \le C \frac{n^{1/6} \psi_n(x)^{1/4}}{1 + n^{1/2} \psi_n(x)^{1/2} \left| x - \sqrt{2n} \right|}. \tag{39}$$

(h) *For each fixed j,*

$$\lambda_{jn}^{-1} W^2(x_{jn}) = 3^{4/3} \pi^{-2} 2^{3/2} n^{1/6} Ai'(a_j)^2 (1 + o(1)). \tag{40}$$

Proof

(a) This follows from (18).
(b) See [32, p. 132, (6.32.5)]. We note that Szego uses $Ai(-x)$ as the Airy function, so there zeros are positive there. Moreover there the symbol i_j is used for $|a_j|$.
(c) This follows from (25) and (17).
(d) Because of the uniform convergence, we can differentiate the relation (35): uniformly for t in compact sets,

$$W(x)\{-xp_n(x) + p_n'(x)\} \frac{dx}{dt} = 3^{1/3} \pi^{-1} 2^{1/4} n^{-1/12} \{-Ai'(-t) + o(1)\}$$

so setting $x = x_{jn}$ and using (34), we obtain (36).
(e) From (33–36),

$$(\ell_{jn} W)(x) W^{-1}(x_{jn}) = \frac{(p_n W)(x)}{(p_n' W)(x_{jn})(x - x_{jn})}$$

$$= \frac{3^{1/3}\pi^{-1}2^{1/4}n^{-1/12}\{Ai(-t)+o(1)\}}{3^{2/3}\pi^{-1}2^{3/4}n^{1/12}\{Ai'(a_j)+o(1)\}\left(-6^{-1/3}(2n+1)^{-1/6}\left(t-|a_j|+o(1)\right)\right)}$$

$$= \frac{Ai(-t)}{Ai'(a_j)(-t-a_j)}(1+o(1)) = \mathcal{L}_j(-t)+o(1).$$

(f) We note the following estimates [10, p. 465–467]: uniformly for $n \geq 1$ and $x \in \mathbb{R}$,

$$n^{1/4}|p_n(x)|W(x) \leq C\psi_n(x)^{-1/4}. \tag{41}$$

Note that for the Hermite weight, the Mhaskar-Rakhmanov number is $a_n = \sqrt{2n}$. We have uniformly for $n \geq 1$ and $x \in \left[-\sqrt{2n}, \sqrt{2n}\right]$,

$$\lambda_n(x) \sim \frac{W^2(x)}{\sqrt{n}}\psi_n(x)^{-1/2}, \tag{42}$$

while for all $x \in (-\infty, \infty)$,

$$\lambda_n(x) \geq C\frac{W^2(x)}{\sqrt{n}}\psi_n(x)^{1/2}. \tag{43}$$

Also uniformly for $1 \leq k \leq n$,

$$|p_{n-1}W|(x_{kn}) \sim n^{-1/4}\psi_n(x_{kn})^{-1/4} \tag{44}$$

and

$$\left|p_n'W\right|(x_{kn}) \sim n^{1/4}\psi_n(x_{kn})^{1/4}. \tag{45}$$

Hence

$$\left|\ell_{jn}W\right|(x)W^{-1}(x_{jn}) = \frac{|p_nW|(x)}{|p_n'W|(x_{jn})|x-x_{jn}|}$$

$$\leq C\frac{n^{-1/4}\psi_n(x)^{-1/4}}{n^{1/4}\psi_n(x_{jn})^{1/4}|x-x_{jn}|}.$$

Next by Cauchy-Schwarz, and then (42), (43),

$$\left|\ell_{jn}W\right|(x)W^{-1}(x_{jn}) = \lambda_{jn}W^{-1}(x_{jn})W(x)\left|K_n(x,x_{jn})\right|$$

$$\leq \lambda_{jn}W^{-1}(x_{jn})W(x)\left(K_n(x,x)K_n(x_{jn},x_{jn})\right)^{1/2}$$

$$= \left(\lambda_{jn} W^{-2}\left(x_{jn}\right)\right)^{1/2} \left(\lambda_n\left(x\right) W^{-2}\left(x\right)\right)^{-1/2}$$

$$\leq C \psi_n\left(x_{jn}\right)^{-1/4} \psi_n\left(x\right)^{1/4} .$$

Thus combining the two estimates,

$$\left|\ell_{jn} W\right|\left(x\right) W^{-1}\left(x_{jn}\right) \leq C \left(\frac{\psi_n\left(x\right)}{\psi_n\left(x_{jn}\right)}\right)^{1/4} \min\left\{1, \frac{1}{n^{1/2}\psi_n\left(x\right)^{1/2}\left|x - x_{jn}\right|}\right\},$$

which can be recast as (38).

(g) First note that as $\left|1 - \frac{x_{jn}}{\sqrt{2n}}\right| \leq Cn^{-2/3}$, we have $\psi_n\left(x_{jn}\right) \sim n^{-2/3}$. We have to show that uniformly in n and for $x \in \mathbb{R}$,

$$1 + n^{1/2}\psi_n\left(x\right)^{1/2}\left|x - x_{jn}\right| \sim 1 + n^{1/2}\psi_n\left(x\right)^{1/2}\left|x - \sqrt{2n}\right|. \tag{46}$$

Let L be some large positive number. If firstly $\left|x - \sqrt{2n}\right| \geq L\sqrt{2n}n^{-2/3}$, then from (33),

$$\left|\frac{x - x_{jn}}{x - \sqrt{2n}} - 1\right| = \frac{\left|x_{jn} - \sqrt{2n}\right|}{\left|x - \sqrt{2n}\right|} \leq \frac{C\sqrt{2n}n^{-2/3}}{L\sqrt{2n}n^{-2/3}}$$

so that

$$\left|\frac{x - x_{jn}}{x - \sqrt{2n}}\right| \leq 1 + C/L,$$

so that

$$1 + n^{1/2}\psi_n\left(x\right)^{1/2}\left|x - x_{jn}\right| \leq C \left(1 + n^{1/2}\psi_n\left(x\right)^{1/2}\left|x - \sqrt{2n}\right|\right).$$

Also, for some C_1 independent of L,

$$1 + n^{1/2}\psi_n\left(x\right)^{1/2}\left|x - \sqrt{2n}\right|$$

$$\leq 1 + n^{1/2}\psi_n\left(x\right)^{1/2}\left(\left|x - x_{jn}\right| + C_1\sqrt{2n}n^{-2/3}\right)$$

$$\leq \left(1 + n^{1/2}\psi_n\left(x\right)^{1/2}\left|x - x_{jn}\right|\right) + n^{1/2}\psi_n\left(x\right)^{1/2}\frac{C_1}{L}\left|x - \sqrt{2n}\right|$$

$$\leq \left(1 + n^{1/2}\psi_n\left(x\right)^{1/2}\left|x - x_{jn}\right|\right) + \frac{C_1}{L}\left(1 + n^{1/2}\psi_n\left(x\right)^{1/2}\left|x - \sqrt{2n}\right|\right)$$

so that

$$\left(1 + n^{1/2}\psi_n(x)^{1/2}\left|x - \sqrt{2n}\right|\right)\left(1 - \frac{C_1}{L}\right) \le \left(1 + n^{1/2}\psi_n(x)^{1/2}\left|x - x_{jn}\right|\right).$$

Then we have (46) if L is large enough. Next, if $\left|x - \sqrt{2n}\right| < L\sqrt{2n}n^{-2/3}$, $\psi_n(x) \sim n^{-2/3}$ and then

$$1 \le 1 + n^{1/2}\psi_n(x)^{1/2}\left|x - \sqrt{2n}\right|$$

$$\le 1 + Cn^{1/2}n^{-1/3}\sqrt{2n}n^{-2/3}$$

$$\le C_2 \le C_2\left(1 + n^{1/2}\psi_n(x)^{1/2}\left|x - x_{jn}\right|\right).$$

Again we have (46).

(h) We use the confluent form of the Christoffel-Darboux formula:

$$\lambda_{jn}^{-1} = \frac{\gamma_{n-1}}{\gamma_n}p_n'\left(x_{jn}\right)p_{n-1}\left(x_{jn}\right).$$

Here since [32, p. 106, (5.5.10)], $H_n'(x) = 2nH_{n-1}(x)$ so from (17),

$$p_n'(x) = \sqrt{2n}\,p_{n-1}(x).$$

Together with (32) this gives

$$\lambda_{jn}^{-1} = p_n'\left(x_{jn}\right)^2.$$

Then (40) follows from (36).

∎

Next, we record some estimates involving the Airy function:

Lemma 2.2

(a) *For $x \in [0, \infty)$,*

$$|Ai(x)| \le C(1+x)^{-1/4}\exp\left(-\frac{2}{3}x^{\frac{3}{2}}\right); \tag{47}$$

$$|Ai(-x)| \le C(1+x)^{-1/4}. \tag{48}$$

(b) *As* $x \to \infty$,

$$Ai'(-x) = -\pi^{-1/2}x^{1/4}\left[\cos\left(\frac{2}{3}x^{\frac{3}{2}} + \frac{\pi}{4}\right) + O\left(x^{-3/2}\right)\right]. \qquad (49)$$

$$Ai'(a_j) = (-1)^{j-1}\pi^{-1/2}\left(\frac{3\pi}{8}(4j-1)\right)^{1/6}\left(1 + O\left(j^{-2}\right)\right)$$

$$= (-1)^{j-1}\pi^{-1/2}|a_j|^{1/4}(1 + o(1)). \qquad (50)$$

(c)

$$a_j = -[3\pi(4j-1)/8]^{2/3}\left(1 + O\left(\frac{1}{j^2}\right)\right)$$

$$= -\left(\frac{3\pi j}{2}\right)^{2/3}(1 + o(1)). \qquad (51)$$

(d)

$$|a_j| - |a_{j-1}| = \pi|a_j|^{-1/2}(1 + o(1)). \qquad (52)$$

(e) *For* $j \geq 1$ *and* $t \in [0, \infty)$,

$$|\mathcal{L}_j(t)| \leq Cj^{-5/6}(1 + t)^{-1/4}\exp\left(-\frac{2}{3}t^{\frac{3}{2}}\right) \qquad (53)$$

and

$$|\mathcal{L}_j(-t)| \leq \frac{C}{1 + (1+t)^{1/4}|a_j|^{1/4}|t - |a_j||}. \qquad (54)$$

Proof (a) The following asymptotics and estimates for Airy functions are listed on pages 448–449 of [1]: see (10.4.59–61) there.

$$Ai(x) = \frac{1}{2\pi^{1/2}}x^{-1/4}\exp\left(-\frac{2}{3}x^{\frac{3}{2}}\right)(1 + o(1)), \quad x \to \infty;$$

$$Ai(-x) = \pi^{-1/2}x^{-1/4}\left[\sin\left(\frac{2}{3}x^{\frac{3}{2}} + \frac{\pi}{4}\right) + O\left(x^{-\frac{3}{2}}\right)\right], \quad x \to \infty.$$

Then (47) and (48) follow as Ai is entire.
(b), (c), (d) The zeros $\{a_j\}$ of Ai satisfy [1, p. 450, (10.4.94,96)]

$$a_j = -[3\pi(4j-1)/8]^{2/3}\left(1 + O\left(\frac{1}{j^2}\right)\right) = -\left(\frac{3\pi j}{2}\right)^{2/3}(1 + o(1)).$$

$$Ai'\left(a_j\right) = (-1)^{j-1}\,\pi^{-1/2}\left(\frac{3\pi}{8}\left(4j-1\right)\right)^{1/6}\left(1+O\left(j^{-2}\right)\right)$$

$$= (-1)^{j-1}\,\pi^{-1/2}\left|a_j\right|^{1/4}\left(1+o\left(1\right)\right).$$

Then (52) also follows, as was shown in [12, p. 431, eqn. (2.7)].

(e) We first prove (54). For $t \in [0, \infty)$,

$$\left|\mathcal{L}_j\left(-t\right)\right| = \left|\frac{Ai\left(-t\right)}{Ai'\left(a_j\right)\left(-t-a_j\right)}\right|$$

$$\leq \frac{C\left(1+t\right)^{-1/4}}{j^{1/6}\left|t-\left|a_j\right|\right|}$$

by (48), (50). If $(1+t)^{1/4}\,j^{1/6}\left|t-\left|a_j\right|\right| \geq \frac{1}{2}\left|a_1\right|$, we then obtain (54). In the contrary case,

$$(1+t)^{1/4}\,j^{1/6}\left|t-\left|a_j\right|\right| < \frac{1}{2}\left|a_1\right|$$

$$\Rightarrow \left|t-\left|a_j\right|\right| < \frac{1}{2}\left|a_1\right| \leq \frac{1}{2}\left|a_j\right|.$$

We then have for some ξ between $-t$ and a_j, from (49),

$$\left|\mathcal{L}_j\left(t\right)\right| = \left|\frac{Ai'\left(\xi\right)}{Ai'\left(a_j\right)}\right| \leq C\left(\frac{\left|\xi\right|}{\left|a_j\right|}\right)^{1/4} \leq C.$$

We again obtain (54). Next, for $t \in (0, \infty)$, we have from (48), (50),

$$\left|\mathcal{L}_j\left(t\right)\right| = \left|\frac{Ai\left(t\right)}{Ai'\left(a_j\right)\left(t-a_j\right)}\right|$$

$$\leq \frac{C\left(1+t\right)^{-1/4}}{j^{1/6}\left|a_j\right|}\exp\left(-\frac{2}{3}t^{\frac{3}{2}}\right)$$

$$\leq Cj^{-5/6}\left(1+t\right)^{-1/4}\exp\left(-\frac{2}{3}t^{\frac{3}{2}}\right).$$

∎

Next, we record a restricted range inequality:

Lemma 2.3 *Let* $\eta \in (0, 1)$, $0 < p < \infty$. *There exists* B, n_0 *such that for* $n \geq n_0$ *and polynomials* P *of degree* $\leq n + n^{1/3}$,

$$\|PW\|_{L_p(\mathbb{R})} \leq (1+\eta)\,\|PW\|_{L_p[-D_n, D_n]}, \tag{55}$$

where

$$D_n = \sqrt{2n}\left(1 + Bn^{-2/3}\right).$$

Proof It suffices to prove that

$$\|PW\|_{L_p(\mathbb{R}\setminus[-D_n, D_n])} \le \eta \, \|PW\|_{L_p[-D_n, D_n]}. \tag{56}$$

For $p \ge 1$, the triangle inequality then yields (55). For $p < 1$, we can use the triangle inequality on the integral inside the norm and then just reduce the size of η appropriately. Let $m = m(n) = n + n^{1/3}$. It follows from Theorem 4.2(b) in [11, p. 96] that for $B \ge 0$, P of degree $\le m$,

$$\|PW\|_{L_p\left(\mathbb{R}\setminus\left[-\sqrt{2m}\left(1 + \frac{1}{2}Bm^{-2/3}\right), \sqrt{2m}\left(1 + \frac{1}{2}Bm^{-2/3}\right)\right]\right)}$$

$$\le C_1 \exp\left(-C_2 B^{3/2}\right) \|PW\|_{L_p\left[-\sqrt{2m}, \sqrt{2m}\right]}. \tag{57}$$

Here C_1 and C_2 are independent of m, P, B. Choose $B \ge 2$ so large that

$$C_1 \exp\left(-C_2 B^{3/2}\right) \le \eta. \tag{58}$$

Now

$$\sqrt{2m}\left(1 + \frac{1}{2}Bm^{-2/3}\right)/D_n$$

$$= \sqrt{\frac{m}{n}} \frac{1 + \frac{1}{2}Bm^{-2/3}}{1 + Bn^{-2/3}}$$

$$\le \sqrt{1 + n^{-2/3}} \frac{1 + \frac{1}{2}Bn^{-2/3}}{1 + Bn^{-2/3}} \le 1$$

for $n \ge n_0(B)$ as $B \ge 2$. Then also $\sqrt{2m}/D_n \le 1$, and

$$\mathbb{R}\setminus\left[-\sqrt{2m}\left(1 + \frac{1}{2}Bm^{-2/3}\right), \sqrt{2m}\left(1 + \frac{1}{2}Bm^{-2/3}\right)\right] \supseteq \mathbb{R}\setminus[-D_n, D_n]$$

and (56) follows from (57) and (58). ∎

Following is the main part of the proof of Theorem 1.3:

Lemma 2.4 *Fix $M \ge 1$ and let*

$$P(x) = \sum_{k=1}^{M} c_k \mathcal{L}_k(x). \tag{59}$$

Then

$$\sum_{k=1}^{M} \frac{|P(a_k)|^p}{Ai'(a_k)^2} \le A \frac{6}{\pi^2} \int_{-\infty}^{\infty} |P(t)|^p \, dt. \tag{60}$$

Here A is the constant in (19) with $R = r = 0$.

Proof Choose $\eta \in (0, 1)$ and D_n, B as in the above lemma. Let

$$R_n(x) = U_n(x) \sum_{k=1}^{M} c_k \ell_{kn}(x) W^{-1}(x_{kn}). \tag{61}$$

Here we set

$$U_n(x) = \left(\frac{T_m\left(\frac{x}{D_n}\right) - T_m(1)}{m^2\left(\frac{x}{D_n} - 1\right)} \right)^L, \tag{62}$$

where T_m is the usual Chebyshev polynomial, L is some large enough even positive integer, and $m = \left[\frac{\varepsilon}{L} n^{1/3}\right]$, while $\varepsilon \in (0, 1)$. Since R_n has degree $\le n + n^{1/3}$, we have by Lemma 2.3, at least for large enough n, that

$$\|R_n W\|_{L_p(\mathbb{R})} \le (1 + \eta) \|R_n W\|_{L_p[-D_n, D_n]}. \tag{63}$$

We first estimate the norm on the right by splitting the integral inside the norm into ranges near 1 and away from 1. First let us deal with the range

$$\mathcal{I}_1 = \left[\sqrt{2n}\left(1 - 6^{-1/3}(2n)^{-2/3} R\right), D_n\right],$$

where R is some fixed (large) number. For $x \in \mathcal{I}_1$, write for $t \in [-R, 6^{1/3}2^{2/3} B]$,

$$x = \sqrt{2n}\left(1 + 6^{-1/3}(2n)^{-2/3} t\right). \tag{64}$$

To find the asymptotics for U_n, also write

$$\frac{x}{D_n} = \cos\frac{s}{m}$$

$$\Rightarrow 1 - \frac{x}{D_n} = 2\sin^2\frac{s}{2m} = \frac{1}{2}\left(\frac{s}{m}\right)^2 (1 + o(1))$$

$$\Rightarrow s = \sqrt{2m^2\left(1 - \frac{x}{D_n}\right)} + o(1)$$

$$\Rightarrow s = \frac{\varepsilon}{L}\sqrt{2\left(B - 6^{-1/3}2^{-2/3}t\right)} + o(1).$$

Then if $\mathbb{S}(u) = \frac{\sin u}{u}$ is the sinc kernel,

$$\frac{T_m\left(\frac{x}{D_n}\right) - T_m(1)}{m^2\left(\frac{x}{D_n} - 1\right)} = \frac{\cos s - 1}{m^2\left(\frac{x}{D_n} - 1\right)} = \frac{-2\sin^2\frac{s}{2}}{-\frac{1}{2}s^2} + o(1)$$

$$= \left(\mathbb{S}\left(\frac{s}{2}\right)\right)^2 + o(1)$$

$$= \mathbb{S}\left(\frac{\varepsilon}{L}\sqrt{\frac{B - 6^{-1/3}2^{-2/3}t}{2}}\right) + o(1),$$

and uniformly in such x,

$$U_n(x) = \mathbb{S}\left(\frac{\varepsilon}{L}\sqrt{\frac{B - 6^{-1/3}2^{-2/3}t}{2}}\right)^L + o(1).$$

In particular, for each fixed k, as $n \to \infty$, recalling (33), and that $a_k < 0$,

$$U_n(x_{kn}) = \mathbb{S}\left(\frac{\varepsilon}{L}\sqrt{\frac{B + 6^{-1/3}2^{-2/3}|a_k|}{2}}\right)^L + o(1). \qquad (65)$$

Then uniformly for x in this range, from Lemma 2.1(e) and recalling (59),

$$|R_n W|(x) = \left|U_n(x)\sum_{k=1}^{M} c_k(\ell_{kn}W)(x)W^{-1}(x_{kn})\right|$$

$$= \left|\mathbb{S}\left(\frac{\varepsilon}{L}\sqrt{\frac{B - 6^{-1/3}2^{-2/3}t}{2}}\right)^L P(-t)\right| + o(1). \qquad (66)$$

Then as $|\mathbb{S}(u)| \leq 1$,

$$\int_{\mathcal{I}_1} |R_n W|^P(x)\,dx$$

$$\leq 6^{-1/3}(2n)^{-1/6}\left(\int_{-R}^{6^{1/3}2^{2/3}B} |P(-t)|^P\,dt + o(1)\right). \qquad (67)$$

Next, for $x \in [-D_n, D_n]$,

$$
|U_n(x)| \leq \left(\min \left\{ 1, \frac{2}{\left| m^2 \left(\frac{x}{D_n} - 1 \right) \right|} \right\} \right)^L
$$

$$
\leq \frac{C}{\left(1 + m^2 \left| \frac{x}{D_n} - 1 \right| \right)^L}
$$

$$
\leq C n^{-2L/3} \frac{1}{\left(n^{-2/3} + \left| \frac{x}{a_n} - 1 \right| \right)^L}
$$

by straightforward estimation. Here C depends on ε. Then from Lemma 2.1(g),

$$
|R_n(x) W(x)| \leq C n^{-2L/3} \frac{1}{\left(n^{-2/3} + \left| \frac{x}{a_n} - 1 \right| \right)^L} \frac{n^{1/6} \psi_n(x)^{1/4}}{1 + n^{1/2} \psi_n(x)^{1/2} |x - a_n|}.
$$

$$(68)$$

Of course here C depends on the particular P and ε, but not on n nor R nor x. Then

$$
\int_{[-D_n, D_n] \setminus \mathcal{I}_1} |R_n W|(x)^p \, dx
$$

$$
\leq C n^{-2Lp/3 + p/6} \int_{-D_n}^{\sqrt{2n}\left(1 - 6^{-1/3}(2n)^{-2/3} R\right)}
$$

$$
\times \left[\frac{1}{\left(n^{-2/3} + \left| \frac{x}{\sqrt{2n}} - 1 \right| \right)^L} \frac{n^{1/6} \psi_n(x)^{1/4}}{1 + n^{1/2} \psi_n(x)^{1/2} \left| x - \sqrt{2n} \right|} \right]^p \, dx
$$

$$
\leq C n^{-2Lp/3 + p/6 + 1/2} \int_{-(1 + Bn^{-2/3})}^{1 - 6^{-1/3}(2n)^{-2/3} R}
$$

$$
\times \left[\frac{1}{\left(n^{-2/3} + |y - 1| \right)^L} \frac{\left(|1 - |y|| + n^{-2/3} \right)^{1/4}}{1 + n \left(|1 - |y|| + n^{-2/3} \right)^{1/2} |y - 1|} \right]^p \, dy
$$

$$
\leq C n^{-2Lp/3 + p/6 + 1/2} \left\{ \int_{-(1 + Bn^{-2/3})}^{0} \left[\frac{\left(|1 - |y|| + n^{-2/3} \right)^{1/4}}{1 + n \left(|1 - |y|| + n^{-2/3} \right)^{1/2}} \right]^p \, dy \right.
$$

$$
\left. + \int_{0}^{1 - 6^{-1/3}(2n)^{-2/3} R} \left[\frac{1}{n |y - 1|^{L + 5/4}} \right]^p \, dy \right\}
$$

$$\leq Cn^{-2Lp/3+p/6+1/2} \left\{ n^{-2/3} \int_{-B}^{n^{2/3}} \left[\frac{n^{-1/6}(|s|+1)^{1/4}}{1+n^{2/3}(|s|+1)^{1/2}} \right]^{p} ds \right.$$

$$\left. + n^{-p} \left(Rn^{-2/3} \right)^{1-(L+5/4)p} \right\}$$

$$\leq Cn^{-2Lp/3+p/6+1/2} \left\{ n^{-2/3-5p/6} \int_{-B}^{n^{2/3}} \frac{1}{(|s|+1)^{p/4}} ds \right.$$

$$\left. + n^{-p} \left(Rn^{-2/3} \right)^{1-(L+5/4)p} \right\}$$

$$\leq Cn^{-2Lp/3+p/6+1/2} \left\{ n^{-5p/6} + n^{-p} \left(Rn^{-2/3} \right)^{1-(L+5/4)p} \right\}$$

$$\leq Cn^{-2Lp/3-2p/3+1/2} + Cn^{-1/6} R^{1-(L+5/4)p}.$$

Assuming that L is large enough so that

$$-2Lp/3 - 2p/3 + 1/2 < -1/6$$

and

$$1 - (L+5/4)\, p < -1,$$

we have

$$\int_{[-D_n,D_n]\setminus \mathcal{I}_1} |R_n W|\,(x)^p\, dx \leq o\left(n^{-1/6} \right) + Cn^{-1/6} R^{-1}.$$

Then combined with (67) and (63) this gives

$$(1+\eta)^{-p} \int_{-\infty}^{\infty} |R_n W|^p$$

$$\leq 6^{-1/3}(2n)^{-1/6} \int_{-R}^{6^{1/3}2^{2/3}B} |P(t)|^p\, dt + o\left(n^{-1/6} \right) + Cn^{-1/6} R^{-1}. \qquad (69)$$

Next from (40), and (65–66), for each fixed k, as $P(a_k) = c_k$,

$$\lambda_{kn} W^{-2}(x_{kn}) |R_n W(x_{kn})|^p = \left[3^{4/3}\pi^{-2}2^{3/2}n^{1/6} Ai'(a_k)^2 \right]^{-1}$$

$$\times \left\{ \left| \mathbb{S}\left(\frac{\varepsilon}{L}\sqrt{\frac{B+6^{-1/3}2^{-2/3}|a_k|}{2}} \right) \right|^{Lp} |P(a_k)|^p + o(1) \right\}$$

so

$$\sum_{k=1}^{M} \lambda_{kn} W^{-2}(x_{kn}) |R_n W(x_{kn})|^p = \left[3^{4/3}\pi^{-2}2^{3/2}n^{1/6}\right]^{-1}$$

$$\times \left\{ \sum_{k=1}^{M} \frac{|P(a_k)|^p}{Ai'(a_k)^2} \left| \mathbb{S}\left(\frac{\varepsilon}{L}\sqrt{\frac{B+6^{-1/3}2^{-2/3}|a_k|}{2}}\right)\right|^{Lp} + o(1) \right\}. \quad (70)$$

Together with (19) and (69), this gives as $n \to \infty$,

$$(1+\eta)^{-p}\left[3^{4/3}\pi^{-2}2^{3/2}\right]^{-1}\sum_{k=1}^{M}\frac{|P(a_k)|^p}{Ai'(a_k)^2}\left|\mathbb{S}\left(\frac{\varepsilon}{L}\sqrt{\frac{B+6^{-1/3}2^{-2/3}|a_k|}{2}}\right)\right|^{Lp}$$

$$\leq 6^{-1/3}2^{-1/6}A\int_{-R}^{6^{1/3}2^{2/3}B}|P(t)|^p\,dt+CR^{-1}.$$

Here B, ε are independent of R. We let $R \to \infty$ and obtain

$$(1+\eta)^{-p}\sum_{k=1}^{M}\frac{|P(a_k)|^p}{Ai'(a_k)^2}\left|\mathbb{S}\left(\frac{\varepsilon}{L}\sqrt{\frac{B+6^{-1/3}2^{-2/3}|a_k|}{2}}\right)\right|^{Lp}$$

$$\leq 6\pi^{-2}A\int_{-\infty}^{6^{1/3}2^{1/6}B}|P(t)|^p\,dt.$$

Now let $\varepsilon \to 0+$:

$$(1+\eta)^{-p}\sum_{k=1}^{M}\frac{|P(a_k)|^p}{Ai'(a_k)^2}\leq 6\pi^{-2}A\int_{-\infty}^{\infty}|P(t)|^p\,dt.$$

Finally we can let $\eta \to 0$:

$$\sum_{k=1}^{M}\frac{|P(a_k)|^p}{Ai'(a_k)^2}\leq 6\pi^{-2}A\int_{-\infty}^{\infty}|P(t)|^p\,dt.$$

∎

Proof of Theorem 1.3(a) Recall that $S_M[f]$ is the partial sum defined in (24). As $f \in \mathcal{G}_p$,

$$\lim_{M\to\infty}\int_{-\infty}^{\infty}|f(t)-S_M[f](t)|^p\,dt=0.$$

Then for a fixed positive integer L, and by Lemma 2.4, and as $S_M[f](a_k) = f(a_k)$ for $k \leq M$,

$$
\begin{aligned}
\left(\sum_{k=1}^{L} \frac{|f(a_k)|^p}{Ai'(a_k)^2} \right)^{1/p} &= \lim_{M \to \infty} \left(\sum_{k=1}^{L} \frac{|S_M[f](a_k)|^p}{Ai'(a_k)^2} \right)^{1/p} \\
&\leq \limsup_{M \to \infty} \left(\sum_{k=1}^{M} \frac{|S_M[f](a_k)|^p}{Ai'(a_k)^2} \right)^{1/p} \\
&\leq \left(\frac{6}{\pi^2} A \right)^{1/p} \limsup_{M \to \infty} \left(\int_{-\infty}^{\infty} |S_M[f](t)|^p \, dt \right)^{1/p} \\
&\leq \left(\frac{6}{\pi^2} A \right)^{1/p} \limsup_{M \to \infty} \left\{ \left(\int_{-\infty}^{\infty} |S_M[f](t) - f(t)|^p \, dt \right)^{1/p} \right. \\
&\qquad \left. + \left(\int_{-\infty}^{\infty} |f(t)|^p \, dt \right)^{1/p} \right\} \\
&= \left(\frac{6}{\pi^2} A \right)^{1/p} \left(\int_{-\infty}^{\infty} |f(t)|^p \, dt \right)^{1/p}.
\end{aligned}
$$

Now let $L \to \infty$.   ∎

For Theorem 1.3(b), we need:

Lemma 2.5 *Assume that for some $\beta > \frac{1}{4}$, we have*

$$
|f(x)| \leq C (1 + |x|)^{-\beta}, \quad x \in (-\infty, 0). \tag{71}
$$

Then for $M \geq 1$, and all $t \in (-\infty, 0]$,

$$
|S_M[f]|(t) \leq C (1 + |t|)^{-\beta} \log(2 + |t|). \tag{72}
$$

For $t \in (0, \infty)$,

$$
|S_M[f]|(t) \leq C (1 + t)^{-1/4} \exp\left(-\frac{2}{3} t^{\frac{3}{2}} \right). \tag{73}
$$

Proof From (71) and (54), followed by (52), for $t \geq 0$,

$$
|S_M[f]|(-t) \leq C \sum_{j=1}^{M} \frac{|a_j|^{-\beta}}{1 + (1+t)^{1/4} |a_j|^{1/4} |t - |a_j||}
$$

$$\leq C \sum_{j=1}^{M} \left(|a_j| - |a_{j-1}| \right) \frac{|a_j|^{-\beta+1/2}}{1 + (1+t)^{1/4} |a_j|^{1/4} |t - |a_j||}$$

$$\leq C \int_0^\infty \frac{s^{-\beta+1/2}}{1 + (1+t)^{1/4} s^{1/4} |t - s|} ds.$$

If $0 \leq t \leq 1$, we can bound this by

$$C \int_0^2 s^{-\beta+1/2} ds + C \int_2^\infty s^{-\beta-3/4} ds \leq C,$$

recall $\beta > \frac{1}{4}$. If $t \geq 1$, we can bound this by

$$C \int_0^\infty \frac{s^{-\beta+1/2}}{1 + t^{1/4} s^{1/4} |t - s|} ds$$

$$= C t^{-\beta+3/2} \int_0^\infty \frac{u^{-\beta+1/2}}{1 + t^{3/2} u^{1/4} |u - 1|} du$$

$$\leq C t^{-\beta+3/2} \left[\begin{array}{l} t^{-3/2} \int_0^{1-1/t^{3/2}} \frac{u^{-\beta+1/4} du}{|u-1|} + \int_{1-1/t^{3/2}}^{1+1/t^{3/2}} 1 du \\ + t^{-3/2} \int_{1+1/t^{3/2}}^2 \frac{du}{|u-1|} + t^{-3/2} \int_2^\infty u^{-\beta-3/4} du \end{array} \right]$$

$$\leq C t^{-\beta} \left[\log(1 + |t|) + 1 + \log(1 + |t|) + 1 \right].$$

Thus we have the bound (72). Next, if $t \geq 0$, we obtain from (53) and (51),

$$|S_M [f]| (-t) \leq C (1+t)^{-1/4} \exp\left(-\frac{2}{3} t^{\frac{3}{2}} \right) \sum_{j=1}^{M} |a_j|^{-\beta} j^{-5/6}$$

$$\leq C (1+t)^{-1/4} \exp\left(-\frac{2}{3} t^{\frac{3}{2}} \right) \sum_{j=1}^{M} j^{-5/6-2\beta/3}$$

$$\leq C (1+t)^{-1/4} \exp\left(-\frac{2}{3} t^{\frac{3}{2}} \right),$$

as $5/6 + 2\beta/3 > 5/6 + 1/6 > 1$. ∎

Proof of Theorem 1.3(b). Recall that we are assuming $p \geq 2$. If $N > M$, we have in view of the lemma and our bound on f

$$\int_{-\infty}^{\infty} |S_N [f] - S_M [f]|^p (t) dt$$

$$\leq C \int_{-\infty}^{\infty} |S_N [f] - S_M [f]|^2 (t) \, dt$$

$$\to 0 \text{ as } M, N \to \infty,$$

as $f \in \mathcal{G}$ implies that $S_M [f] \to f$ in $L_2 (\mathbb{R})$ as $M \to \infty$. It follows that $\{S_M [f]\}$ is Cauchy in $L_p (\mathbb{R})$, so has a limit there. This limit must be f, as $f \in \mathcal{G}$. Then also $f \in \mathcal{G}_p$ and the result follows. ∎

Lemma 2.6 *Assume that* (22) *holds with* $R = r = 0$. *Let* $P = \sum_{k=1}^{M} P(a_k) \cdot \mathcal{L}_k$ *and* $1 < p < 4$. *Then*

$$\int_{-\infty}^{\infty} |P(t)|^p \, dt \leq B \frac{\pi^2}{6} \sum_{j=1}^{M} \frac{|P(a_k)|^p}{Ai'(a_k)^2}. \tag{74}$$

Proof We use (22) with $R = r = 0$. If R_n is a polynomial of degree $\leq n - 1$,

$$\int_{-\infty}^{\infty} |(R_n W)(x)|^p \, dx \leq B \sum_{j=1}^{n} \lambda_{jn} |R_n(x_{jn})|^p W^{p-2}(x_{jn}). \tag{75}$$

Let

$$R_n(x) = \sum_{k=1}^{M} P(a_k) \ell_{kn}(x) W^{-1}(x_{kn}).$$

Let $R > 0$ and

$$\mathcal{I}_1 = \left[\sqrt{2n} \left(1 - 6^{-1/3} (2n)^{-2/3} R \right), \sqrt{2n}(1 + 6^{-1/3} (2n)^{-2/3} R) \right].$$

From (37) with x of the form (34), we have

$$|R_n W|(x) = |P(-t)| + o(1),$$

so

$$\int_{\mathcal{I}_1} |R_n W|(x)^p \, dx = 6^{-1/3} (2n)^{-1/6} \left(\int_{-R}^{R} |P(t)|^p \, dt + o(1) \right).$$

Also, as at (70),

$$\sum_{j=1}^{n} \lambda_{jn} |R_n(x_{jn})|^p W^{p-2}(x_{jn})$$

$$= \sum_{j=1}^{M} \lambda_{jn} \left| R_n \left(x_{jn} \right) \right|^p W^{p-2} \left(x_{jn} \right)$$

$$= (1 + o(1)) \left[3^{4/3} \pi^{-2} 2^{3/2} n^{1/6} \right]^{-1} \sum_{k=1}^{M} \frac{\left| P \left(a_k \right) \right|^p}{Ai' \left(a_k \right)^2}.$$

Then (75) gives

$$6^{-1/3} (2n)^{-1/6} \left(\int_{-R}^{R} \left| P(t) \right|^p dt + o(1) \right)$$

$$\leq B (1 + o(1)) \left[3^{4/3} \pi^{-2} 2^{3/2} n^{1/6} \right]^{-1} \sum_{k=1}^{M} \frac{\left| P \left(a_k \right) \right|^p}{Ai' \left(a_k \right)^2}$$

or

$$\left(\int_{-R}^{R} \left| P(t) \right|^p dt + o(1) \right) \leq B (1 + o(1)) \frac{\pi^2}{6} \sum_{k=1}^{M} \frac{\left| P \left(a_k \right) \right|^p}{Ai' \left(a_k \right)^2}.$$

Letting $R \to \infty$ gives (74). ∎

Proof of Theorem 1.4. (a) Lemma 2.6 gives

$$\| f \|_{L_p(\mathbb{R})} \leq \| f - S_M [f] \|_{L_p(\mathbb{R})} + \| S_M [f] \|_{L_p(\mathbb{R})}$$

$$\leq \| f - S_M [f] \|_{L_p(\mathbb{R})} + \left(B \frac{\pi^2}{6} \sum_{k=1}^{M} \frac{\left| f \left(a_k \right) \right|^p}{Ai' \left(a_k \right)^2} \right)^{1/p}$$

$$\to 0 + \left(B \frac{\pi^2}{6} \sum_{k=1}^{\infty} \frac{\left| f \left(a_k \right) \right|^p}{Ai' \left(a_k \right)^2} \right)^{1/p},$$

as $M \to \infty$.

(b) Our assumption that $f \in \mathcal{G}$ ensures that $f = \lim_{M \to \infty} S_M [f]$ uniformly in compact sets. Next, given $N > M$, we have from Lemma 2.6,

$$\int_{-\infty}^{\infty} \left| S_N [f] - S_M [f] \right|^p (t) \, dt \leq B \frac{\pi^2}{6} \sum_{k=M+1}^{N} \frac{\left| f \left(a_k \right) \right|^p}{Ai' \left(a_k \right)^2}$$

$$\leq C \sum_{k=M+1}^{\infty} \frac{\left| f \left(a_k \right) \right|^p}{k^{1/3}} \to 0,$$

as $k \to \infty$ – recall (50) and our hypothesis (30). So $\{ S_M [f] \}$ is Cauchy in complete $L_p (\mathbb{R})$ and as above, its limit in $L_p (\mathbb{R})$ must be f, so that (a) is applicable. ∎

References

1. M. Abramowitz, I.A. Stegun, *Handbook of Mathematical Functions* (Dover, New York, 1965)
2. R.P. Boas, *Entire Functions* (Academic Press, New York, 1954)
3. S.B. Damelin, D.S. Lubinsky, Necessary and sufficient conditions for mean convergence of Lagrange interpolation for Erdös weights. Can. Math. J. **40**, 710–736 (1996)
4. S.B. Damelin, D.S. Lubinsky, Necessary and sufficient conditions for mean convergence of Lagrange interpolation for Erdös weights II. Can. Math. J. **40**, 737–757 (1996)
5. F. Filbir, H.N. Mhaskar, Marcinkiewicz-Zygmund measures on manifolds. J. Complex. **27**, 568–596 (2011)
6. G.R. Grozev, Q.I. Rahman, Lagrange interpolation in the zeros of Bessel functions by entire functions of exponential type and mean convergence. Methods Appl. Anal. **3**, 46–79 (1996)
7. H. König, Vector-valued Lagrange interpolation and mean convergence of Hermite series, in *Functional Analysis* (Essen, 1991). Lecture notes in pure and applied mathematics, vol. 150 (Dekker, New York, 1994), pp. 227–247
8. H. König, N.J. Nielsen, Vector-valued L_p convergence of orthogonal series and Lagrange interpolation. Forum Math. **6**, 183–207 (1994)
9. B. Ja Levin, *Lectures on Entire Functions*, Translations of Mathematical Monographs (American Mathematical Society, Providence, 1996)
10. E. Levin, D.S. Lubinsky, Christoffel functions, orthogonal polynomials, and Nevai's conjecture for Freud weights.Constr. Approx. **8**, 463–535 (1992)
11. E. Levin, D.S. Lubinsky, *Orthogonal Polynomials for Exponential Weights* (Springer, New York, 2001)
12. E. Levin, D.S. Lubinsky, On the Airy reproducing kernel, sampling series, and quadrature formula. Integr. Equ. Oper. Theory **63**, 427–438 (2009)
13. F. Littmann, Marcinkiewicz inequalities for entire functions in spaces with Hermite-Biehler weights, manuscript
14. D.S. Lubinsky, Converse quadrature sum inequalities for polynomials with Freud weights. Acta Sci. Math. **60**, 527–557 (1995)
15. D.S. Lubinsky, Marcinkiewicz-Zygmund inequalities: methods and results, in *Recent Progress in Inequalities*, ed. by G.V. Milovanovic et al. (Kluwer Academic Publishers, Dordrecht, 1998), pp. 213–240
16. D.S. Lubinsky, On sharp constants in Marcinkiewicz-Zygmund and Plancherel-Polya inequalities. Proc. Am. Math. Soc. **142**, 3575–3584 (2014)
17. D.S. Lubinsky, On Marcinkiewicz-Zygmund inequalities at Jacobi zeros and their Bessel function cousins. Contemp. Math. **669**, 223–245 (2017)
18. D.S. Lubinsky, Mean convergence of interpolation at zeros of airy functions, in *Contemporary Computational Mathematics – A Celebration of the 80th Birthday of Ian Sloan*, ed. by J. Dick, F.Y. Kuo, H. Wozniakowski, vol. 2 (Springer International, Switzerland, 2018), pp. 889–909
19. D.S. Lubinsky, G. Mastroianni, Converse quadrature sum inequalities for Freud Weights. II Acta Math. Hungar. **96**, 161–182 (2002)
20. D.S. Lubinsky, A. Maté, P. Nevai, Quadrature sums involving pth powers of polynomials. SIAM J. Math. Anal. **18**, 531–544 (1987)
21. D.S. Lubinsky, D. Matjila, Full quadrature sums for pth powers of polynomials with Freud weights. J. Comp. Appl. Math. **60**, 285–296 (1995)
22. G. Mastorianni, G.V. Milovanovic, *Interpolation Processes: Basic Theory and Applications* (Springer, Berlin, 2008)
23. G. Mastroianni, V. Totik, Weighted polynomial inequalities with doubling and A_∞ Weights. Constr. Approx. **16**, 37–71 (2000)
24. H.N. Mhaskar, F.J. Narcowich, J.D. Ward, Spherical Marcinkiewicz-Zygmund inequalities and positive quadrature. Math. Comp. **70**(235), 1113–1130 (2001)

25. H.N. Mhaskar, J. Prestin, On Marcinkiewicz-Zygmund-Type Inequalities, in *Approximation Theory: In Memory of A.K. Varma*, ed. by N.K. Govil et al. (Marcel Dekker, New York, 1998), pp. 389–404
26. P. Nevai, Lagrange interpolation at zeros of orthogonal polynomials, in *Approximation Theory II*, ed. by G.G. Lorentz et al. (Academic Press, New York, 1976), pp. 163–201
27. P. Nevai, Orthogonal polynomials. Mem. Am. Math. Soc. **18**(213), 185 (1979)
28. P. Nevai, Mean convergence of Lagrange interpolation II. J. Approx. Theory **30**, 263–276 (1980)
29. P. Nevai, Geza Freud, orthogonal polynomials and Christoffel functions. A case study. J. Approx. Theory **40**, 3–167 (1986)
30. J. Ortega-Cerda, J. Saludes, Marcinkiewicz-Zygmund inequalities. J. Approx. Theory **147**, 237–252 (2007)
31. J. Szabados, Weighted Lagrange and Hermite-Fejer interpolation on the real line. J. Inequal. Appl. **1**, 99–123 (1997)
32. G. Szegő, *Orthogonal Polynomials* (American Mathematical Society, Providence, 1939)
33. Y. Xu, On the Marcinkiewicz-Zygmund inequality, in *Progress in Approximation Theory*, ed. by P. Nevai, A. Pinkus (Academic Press, Boston, 1991), pp. 879–891
34. Y. Xu, Mean convergence of generalized Jacobi series and interpolating polynomials II. J. Approx. Theory **76**, 77–92 (1994)
35. A. Zygmund, *Trigonometric Series*, vols. 1, 2, Second Paperback edition (Cambridge University Press, Cambridge, 1988)

The Maximum of Cotangent Sums Related to the Nyman-Beurling Criterion for the Riemann Hypothesis

Helmut Maier, Michael Th. Rassias, and Andrei Raigorodskii

Abstract In a previous paper (see H. Maier, M. Th. Rassias, The maximum of cotangent sums related to Estermann's zeta function in rational numbers in short intervals. Appl. Anal. Discrete Math. 11, 166–176 (2017)) we investigate the maximum of certain cotangent sums. These cotangent sums can be associated to the study of the Riemann Hypothesis through its relation with the so-called Vasyunin sum. Here we continue this research by restricting the rational numbers in short intervals to rational numbers of special type.

Keywords Cotangent sums · Estermann zeta function · Nyman-Beurling criterion · Riemann zeta function · Riemann Hypothesis

2000 Mathematics Subject Classification: 26A12; 11L03.

H. Maier
Department of Mathematics, University of Ulm, Ulm, Germany
e-mail: helmut.maier@uni-ulm.de

M. Th. Rassias (✉)
Institute of Mathematics, University of Zurich, Zurich, Switzerland

Moscow Institute of Physics and Technology, Dolgoprudny, Russia

Institute for Advanced Study, Program in Interdisciplinary Studies, Princeton, NJ, USA
e-mail: michail.rassias@math.uzh.ch

A. Raigorodskii
Moscow Institute of Physics and Technology, Dolgoprudny, Russia

Moscow State University, Moscow, Russia

Buryat State University, Ulan-Ude, Russia

Caucasus Mathematical Center, Adyghe State University, Maykop, Russia
e-mail: raigorodsky@yandex-team.ru

1   Introduction

The authors in various papers [8–12] and the second author in his thesis [13], studied
the distribution of cotangent sums

$$c_0\left(\frac{r}{b}\right) = -\sum_{m=1}^{b-1} \frac{m}{b} \cot\left(\frac{\pi m r}{b}\right)$$

as r ranges over the set

$$\{r \,:\, (r, b) = 1, \; A_0 b \le r \le A_1 b\}\,,$$

where A_0, A_1 are fixed with $1/2 < A_0 < A_1 < 1$ and b tends to infinity.

Bettin [2] succeeded in replacing the inequality $1/2 < A_0 < A_1 < 1$ by
$0 < A_0 < A_1 \le 1$.

These sums are related to the values of the Estermann zeta function $E(s, r/b, \alpha)$,
which are defined by the Dirichlet series

$$E\left(s, \frac{r}{b}, \alpha\right) = \sum_{n \ge 1} \frac{\sigma_\alpha(n) \exp(2\pi i n r/b)}{n^s}\,,$$

where $Re\,s > Re\,\alpha + 1, b \ge 1, (r, b) = 1$ and

$$\sigma_\alpha(n) = \sum_{d \mid n} d^\alpha\,.$$

Estermann (see [4]) introduced the above function in the special case when $\alpha = 0$
and Kiuchi (see [7]) for $\alpha \in (-1, 0]$. Ishibashi (see [6]) proved the following
relation regarding the value of $E\left(s, \frac{r}{b}, \alpha\right)$ at $s = 0$:

Let $b \ge 2, 1 \le r \le b, (r, b) = 1, \alpha \in \mathbb{N} \cup \{0\}$. Then, for even α, it holds that

$$E\left(0, \frac{r}{b}, \alpha\right) = \left(-\frac{i}{2}\right)^{\alpha+1} \sum_{m=1}^{b-1} \frac{m}{b} \cot^{(\alpha)}\left(\frac{\pi m r}{b}\right) + \frac{1}{4}\delta_{\alpha,0}\,,$$

where $\delta_{\alpha,0}$ is the Kronecker delta function.

For $\alpha = 0$, the sum on the right reduces to $c_0(r/b)$.

The cotangent sum $c_0(r/b)$ can be associated to the study of the Riemann
Hypothesis, also through its relation with the so-called Vasyunin sum. The Vasyunin
sum is defined as follows:

$$V\left(\frac{r}{b}\right) := \sum_{m=1}^{b-1} \left\{\frac{mr}{b}\right\} \cot\left(\frac{\pi m r}{b}\right)\,,$$

where $\{u\} = u - \lfloor u \rfloor, u \in \mathbb{R}$.

It can be shown that

$$V\left(\frac{r}{b}\right) = -c_0\left(\frac{\bar{r}}{b}\right),$$

where $\bar{r}$ is such that $\bar{r}r \equiv 1 \pmod{b}$.

The Vasyunin sum is itself associated to the study of the Riemann hypothesis through the following identity (see [1, 3]):

$$\frac{1}{2\pi(rb)^{1/2}} \int_{-\infty}^{+\infty} \left|\zeta\left(\frac{1}{2}+it\right)\right|^2 \left(\frac{r}{b}\right)^{it} \frac{dt}{\frac{1}{4}+t^2} \tag{1}$$

$$= \frac{\log 2\pi - \gamma}{2}\left(\frac{1}{r}+\frac{1}{b}\right) + \frac{b-r}{2rb}\log\frac{r}{b} - \frac{\pi}{2rb}\left(V\left(\frac{r}{b}\right)+V\left(\frac{b}{r}\right)\right).$$

Note that the only non-explicit function in the right hand side of (1) is the Vasyunin sum. According to this approach, the Riemann Hypothesis is true if and only if

$$\lim_{N\to+\infty} d_N = 0,$$

where

$$d_N^2 = \inf_{D_N} \frac{1}{2\pi} \int_{-\infty}^{+\infty} \left|1 - \zeta\left(\frac{1}{2}+it\right)D_N\left(\frac{1}{2}+it\right)\right|^2 \frac{dt}{\frac{1}{4}+t^2}$$

and the infimum is taken over all Dirichlet polynomials

$$D_N(s) = \sum_{n=1}^{N} \frac{a_n}{n^s}.$$

The authors in several papers [8–12] and the second author in his thesis [13] investigated moments of the form

$$\frac{1}{\phi(b)} \sum_{\substack{(r,b)=1 \\ A_0b<r\leq A_1b}} c_0\left(\frac{r}{b}\right)^{2k}, \quad \frac{1}{2} < A_0 < A_1 < 1$$

and could show that

$$\frac{1}{\phi(b)} \sum_{\substack{(r,b)=1 \\ A_0b<r\leq A_1b}} c_0\left(\frac{r}{b}\right)^{2k} = H_k b^{2k}(1+o(1)), \quad (b\to+\infty),$$

where

$$H_k := \int_0^1 \left(\frac{g(x)}{\pi} \right)^{2k} dx \, ,$$

$$g(x) := \sum_{l \geq 1} \frac{1 - 2\{lx\}}{l} \, .$$

The range $1/2 < A_0 < A_1 < 1$ was later extended to $0 < A_0 < A_1 < 1$ by Bettin in [2].

In the paper [12] the authors investigated the maximum of $\left| c_0 \left(\frac{r}{b} \right) \right|$ for the values r/b in a short interval. They gave the following definition:

Definition 1.1 (of [12]) Let $0 < A_0 < 1, 0 < C < 1/2$. For $b \in \mathbb{N}$ we set

$$\Delta := \Delta(b, C) = b^{-C}.$$

We set

$$M(b, C, A_0) := \max_{A_0 b \leq r < (A_0 + \Delta)b} \left| c_0 \left(\frac{r}{b} \right) \right| \, .$$

They proved the following results:

Theorem 1.2 (of [12]) *With Definition* 1.1 *of [12] let D satisfy $0 < D < \frac{1}{2} - C$. Then we have for sufficiently large b:*

$$M(b, C, A_0) \geq \frac{D}{\pi} b \log b \, .$$

Theorem 1.3 (of [12]) *Let C be as in Theorem* 1.2 *of [12] and let D satisfy $D > 2 - C - E$, where $E \geq 0$ is a fixed constant. Let B be sufficiently large. Then we have:*

$$M(b, C, A_0) \leq \frac{D}{\pi} b \log b \, ,$$

for all b with $B \leq b < 2B$ with at most B^E exceptions.

In this paper we replace b by a prime number q and the numerators r of the rational numbers r/b by prime numbers. For the statement and proof of the next theorem we need the following definition.

Definition 1.4 The letter p always denotes a prime number. Let $0 < A_0 < 1$, $0 < C < 1/2$. For q a prime number we set:

$$\Delta := \Delta(q, C) = q^{-C} \, .$$

We set

$$L(q, C, A_0) := \max_{A_0 q \le p \le (A_0 + \Delta) q} \left| c_0 \left(\frac{p}{q} \right) \right| .$$

We shall prove the following results:

Theorem 1.5 *With Definition* 1.4, *let D satisfy*

$$0 < D < \frac{1}{32} - C .$$

Then we have for sufficiently large q:

$$L(q, C, A_0) \ge \frac{D}{\pi} q \log q .$$

As in [12] a crucial role is played by the simultaneous localisation of a residue-class and its multiplicative inverse. In [12] the inverse $\bar{r}$ (mod b) of r (mod b) is defined by $\bar{r} r \equiv 1$ (mod b). The quantities r and $\bar{r}$ are simultaneously located by the application of Kloosterman sums. Here we additionally have the primality condition. We have to localise simultaneously the prime p and its inverse $\bar{p}$ (mod q).

This will be achieved by the application of a result of Fouvry and Michel [5], Lemma 2.1, on exponential sums in finite fields over primes. The analogue of [5] follows immediately, since the upper bound for the cardinality of a set is always also an upper bound for all of its subsets. We have the following:

Corollary 1.6 (of Theorem 1.3 of [12]) *Let C be as in Theorem* 1.5 *and let D satisfy* $D > 2 - C - E$, *where* $E \ge D$ *is a fixed constant. Let Q be sufficiently large. Then we have*

$$L(q, C, A_0) \le \frac{D}{\pi} q \log q$$

for all q with $Q \le q \le 2Q$, *q prime, with at most* Q^E *exceptions.*

We now give the proof of Theorem 1.5.

2   Exponential Sums over Primes in Finite Fields

Lemma 2.1 *Let* $\mathbb{F}_r$ *be a finite field with r elements and* ψ *be a non-trivial additive character over* $\mathbb{F}_r$, *f a rational function of the form*

$$f(x) = \frac{P(x)}{Q(x)} ,$$

P and Q relatively prime monic polynomials

$$S(f; r, x) := \sum_{p \leq x} \psi(f(p))$$

(p denotes the p-fold sum of the element 1 in $\mathbb{F}_r$).
   Then we have

$$S(f; r, x) \ll r^{3/16+\epsilon} x^{25/32}.$$

The implied constant depends only on ϵ and the degrees of P and Q.

Proof This is due to Fouvry and Michel [5].   □

3   Other Preliminary Lemmas

Definition 3.1 For $x \in \mathbb{R}$, $\mathrm{Re}(s) > 1$ we set

$$D_{\sin}(s, x) := \sum_{n \geq 1} \frac{d(n) \sin(2\pi n x)}{n^s}. \tag{2}$$

Lemma 3.2 *Let $(a_0; a_1, a_2, \ldots)$ be the continued fraction expansion of $x \in \mathbb{R}$. Moreover, let u_r/v_r be the r-th partial quotient of x. Then*

$$D_{sin}(1, x) = -\frac{\pi^2}{2} \sum_{l \geq 1} \frac{(-1)^l}{v_l} \left(\left(\frac{1}{\pi v_l} \right) + \psi \left(\frac{v_{l-1}}{v_l} \right) \right), \tag{3}$$

whenever either of the two series (2) and (3) is convergent.
   If $x = (a_0; a_1, a_2, \ldots, a_r)$ is a rational number, then the range of summation of the series on the right is to be interpreted to be $1 \leq l \leq r$. Here ψ is an analytic function satisfying

$$\psi(x) = -\frac{\log(2\pi x) - \gamma}{\pi x} + O(\log x), \quad (x \to 0).$$

Proof This is Proposition 1 of Bettin [2].   □

Lemma 3.3

$$c_0 \left(\frac{r}{b} \right) = \frac{1}{2} D_{sin} \left(0, \frac{r}{b} \right) = 2b \, \pi^{-2} D_{sin} \left(1, \frac{\bar{r}}{b} \right),$$

where $r\bar{r} \equiv 1 \pmod{b}$.

Proof This is due to Ishibashi [6]. □

Definition 3.2 Let Δ be as in Definition 1.4 and $\Omega > 0$. We set

$$N(q, \Delta, \Omega) := \#\{p \,:\, A_0 q \leq p < (A_0 + \Delta)q \,,\, |\bar{p}| \leq \Omega q\}.$$

Let the functions χ_1, χ_2 be defined by

$$\chi_1(u, v) := \begin{cases} 1, & \text{if } A_0 + v < u \leq A_0 + \Delta - v \\ 0, & \text{otherwise}. \end{cases}$$

and

$$\chi_2(u, v) := \Delta^{-1} \int_0^\Delta \chi_1(u, v) \, dv.$$

Lemma 3.5 *We have*

$$\chi_2(u) = \sum_{n=-\infty}^\infty a(n) e(nu),$$

where $a(0) = \Delta/2$ and

$$a(n) = \begin{cases} O(\Delta), & \text{if } |n| \leq \Delta^{-1} \\ O(\Delta^{-1} n^{-2}), & \text{if } |n| > \Delta^{-1}. \end{cases}$$

Proof This is Lemma 2.10 of [12]. □

Definition 3.6 Let

$$\chi_3(u, v) := \begin{cases} 1, & \text{if } -\Omega + v < u < \Omega \\ 0, & \text{otherwise}. \end{cases}$$

and

$$\chi_4(u) := \Omega^{-1} \int_0^\Omega \chi_3(u, v) dv.$$

Lemma 3.7 *We have*

$$\chi_4(u) = \sum_{n=-\infty}^{+\infty} c(n) \, e(nu),$$

where $c(0) = \Omega$ and

$$c(n) = \begin{cases} O(\Omega), & \text{if } |n| \leq \Omega^{-1} \\ O(\Omega^{-1} n^{-2}), & \text{if } |n| > \Omega^{-1}. \end{cases}$$

Proof This is Lemma 2.12 of [12]. □

Lemma 3.8 *Let $\epsilon > 0$ be such that*

$$D + \epsilon < \frac{1}{32} - C.$$

Set

$$\Omega := q^{-(D+\epsilon)}.$$

Then

$$N(q, \Delta, \Omega) > 0$$

for q sufficiently large.

Proof By Definitions 3.2 and 3.6 we have

$$N(q, \Delta, \Omega) \geq \phi(q)a(0)c(0) + \sum_{\substack{m,n=-\infty \\ (m,n)\neq(0,0)}}^{+\infty} a(m)c(n)E(m, n, q),$$

where

$$E(m, n, q) := \sum_{1 \leq p \leq q-1} e\left(\frac{mp + n\bar{p}}{q}\right).$$

For $(m, n) \neq (0, 0)$ we estimate $E(m, n, q)$ by application of Lemma 2.1, where we set

$$r = q, \quad P(x) = x^2 + 1, \quad Q(x) = x, \quad \psi(l \bmod q) = e\left(\frac{ml}{q}\right)$$

and obtain

$$E(m, n, q) \ll q^{31/32}.$$

Lemma 3.8 follows from Lemmas 3.5 and 3.7. □

Lemma 3.9 *Let $\epsilon > 0$, $q \geq q(\epsilon)$. For $1 \leq p < q$, let*

$$\frac{p}{q} = \langle 0; w_1, \ldots, w_s \rangle$$

be the continued fraction expansion of p/q with partial fractions u_i/v_i. Then there are at most 3 values of l for which

$$\frac{1}{v_l}\psi\left(\frac{v_{l-1}}{v_l}\right) \geq \log\log q$$

and at most one value of l, for which

$$\frac{1}{v_l}\psi\left(\frac{v_{l-1}}{v_l}\right) \geq \epsilon\log q.$$

Proof This is Lemma 2.14 of [12].   □

4 Proof of Theorem 1.5

By Lemma 3.8 there is at least one

$$p \in [A_0\, q,\ (A_0 + \Delta)q],$$

such that $\frac{\bar{p}}{q} \in (0, \Omega)$. By Lemmas 3.2 and 3.3 we have:

$$c_0\left(\frac{p}{q}\right) = -q\sum_{l\geq 1}\frac{(-1)^l}{v_l}\left(\left(\frac{1}{\pi v_l}\right) + \psi\left(\frac{v_{l-1}}{v_l}\right)\right).$$

Let $(u_i/v_i)_{i=1}^s$ be the sequence of partial fractions of $\bar{p}/q$. From

$$\Omega \geq \frac{\bar{p}}{q} \geq \frac{1}{v_1 + 1}\,,$$

we obtain

$$v_1 + 1 \geq \Omega^{-1}.$$

By Lemma 3.9 we have

$$\sum_{l>1}\left(\frac{1}{\pi v_l} + \psi\left(\frac{v_{l-1}}{v_l}\right)\right) < 2\epsilon\log q,\ \text{for } q \geq q_0(\epsilon).$$

Therefore,

$$\left|D_{sin}\left(0, \frac{p}{q}\right)\right| \geq \frac{1}{\pi}\log(\Omega^{-1})(1 + o(1))\ (q \to +\infty).$$

This proves Theorem 1.5.   □

Acknowledgements M. Th. Rassias: I would like to express my gratitude to the John S. Latsis Public Benefit Foundation for their financial support provided under the auspices of my current "Latsis Foundation Senior Fellowship" position.

References

1. S. Bettin, A generalization of Rademacher's reciprocity law. Acta Arith. **159**(4), 363–374 (2013)
2. S. Bettin, On the distribution of a cotangent sum. Int. Math. Res. Not. (2015). https://doi.org/10.1093/imrn/rnv036
3. S. Bettin, B. Conrey, Period functions and cotangent sums. Algebra Number Theory **7**(1), 215–242 (2013)
4. T. Estermann, On the representation of a number as the sum of two products. Proc. Lond. Math. Soc. **31**(2), 123–133 (1930)
5. E. Fouvry, P. Michel, Sur certaines sommes d'exponentielles sur les nombres premiers. Ann. Sci. Écope Norm. Sup. (4), **31**, 93–130 (1998)
6. M. Ishibashi, The value of the Estermann zeta function at $s = 0$. Acta Arith. **73**(4), 357–361 (1995)
7. I. Kiuchi, On an exponential sum involving the arithmetic function $\sigma_\alpha(n)$. Math. J. Okayama Univ. **29**, 193–205 (1987)
8. H. Maier, M.Th. Rassias, Generalizations of a cotangent sum associated to the Estermann zeta function. Commun. Contemp. Math. **18**(1) (2016). https://doi.org/10.1142/S0219199715500789
9. H. Maier, M.Th. Rassias, The order of magnitude for moments for certain cotangent sums. J. Math. Anal. Appl. **429**(1), 576–590 (2015)
10. H. Maier, M.Th. Rassias, The rate of growth of moments of certain cotangent sums. Aequationes Math. 2015, **90**(3), 581–595 (2016)
11. H. Maier, M.Th. Rassias, *Asymptotics for moments of certain cotangent sums*. Houst. J. Math. **43**(1), 207–222 (2017)
12. H. Maier, M.Th. Rassias, The maximum of cotangent sums related to Estermann's zeta function in rational numbers in short intervals. Appl. Anal. Discrete Math. **11**, 166–176 (2017)
13. M.T. Rassias, Analytic investigation of cotangent sums related to the Riemann zeta function. Doctoral Dissertation, ETH-Zürich, Switzerland, 2014

Double-Sided Taylor's Approximations and Their Applications in Theory of Trigonometric Inequalities

Branko Malešević, Tatjana Lutovac, Marija Rašajski, and Bojan Banjac

Abstract In this paper the double-sided TAYLOR's approximations are used to obtain generalisations and improvements of some trigonometric inequalities.

Keywords Trigonometric inequalities · Double-sided Taylor's approximations

1 Introduction

Many mathematical and engineering problems cannot be solved without TAYLOR's approximations [1–3]. Particularly, their application in proving various analytic inequalities is of great importance [4–8]. Recently, numerous inequalities have been generalized and improved by the use of the so-called double-sided TAYLOR's approximations [6, 9–14] and [15]. Many topics regarding these approximations are presented in [15]. Some of the basic concepts and results about the double-sided TAYLOR's approximations presented in [15], which will be used in this paper, are given in the next section.

In this paper, using the double-sided TAYLOR's approximations, we obtain generalizations and improvements of some trigonometric inequalities proved by Sandor [16].

Statement 1 ([16], Theorem 1)

$$\frac{3}{8} < \frac{1 - \dfrac{\cos x}{\cos \frac{x}{2}}}{x^2} < \frac{4}{\pi^2}, \tag{1}$$

for any $x \in (0, \pi/2)$.

B. Malešević (✉) · T. Lutovac · M. Rašajski
School of Electrical Engineering, University of Belgrade, Belgrade, Serbia
e-mail: branko.malesevic@etf.bg.ac.rs; tatjana.lutovac@etf.bg.ac.rs; marija.rasajski@etf.bg.ac.rs

B. Banjac
Faculty of Technical Sciences, University of Novi Sad, Novi Sad, Serbia
e-mail: bojan.banjac@uns.ac.rs

© Springer Nature Switzerland AG 2020
A. Raigorodskii, M. T. Rassias (eds.), *Trigonometric Sums and Their Applications*,
https://doi.org/10.1007/978-3-030-37904-9_8

Note that D'Aurizio [17] used the infinite products as well as some inequalities connected with the RIEMANN zeta function ζ to prove the right-hand side inequality (1).

Statement 2 ([16], Theorem 2)

$$\frac{4}{\pi^2}\left(2 - \sqrt{2}\right) < \frac{2 - \frac{\sin x}{\sin \frac{x}{2}}}{x^2} < \frac{1}{4}, \tag{2}$$

for any $x \in (0, \pi/2)$.

Inequalities (1) and (2) are reducible to mixed trigonometric-polynomial inequalities and can be proved by methods and algorithms that have been developed and shown in papers [5, 7] and dissertation [18].

In this paper, we propose and prove generalizations of inequality (1) by determining the sequence of the polynomial approximations. Also, an improvement of inequality (2) is given for some intervals. The proposed generalizations and improvements are based on the double-sided TAYLOR's approximations and the corresponding results presented in [15].

2 An Overview of the Results Related to Double-Sided TAYLOR's Approximations

Let us consider a real function $f : (a, b) \longrightarrow \mathbb{R}$, such that there exist finite limits $f^{(k)}(a+) = \lim_{x \to a+} f^{(k)}(x)$, for $k = 0, 1, \ldots, n$.
TAYLOR's polynomial

$$T_n^{f,\,a+}(x) = \sum_{k=0}^{n} \frac{f^{(k)}(a+)}{k!}(x - a)^k, \quad n \in \mathbb{N}_0,$$

and the polynomial

$$\mathbb{T}_n^{f;\,a+,\,b-}(x) = \begin{cases} T_{n-1}^{f,\,a+}(x) + \dfrac{1}{(b-a)^n} R_n^{f,\,a+}(b-)(x - a)^n \, , \, n \geq 1 \\ \\ f(b-) \hspace{4.5cm} , \, n = 0, \end{cases}$$

are called the *first TAYLOR's approximation for the function f in the right neighborhood of a*, and the *second TAYLOR's approximation for the function f in the right neighborhood of a*, respectively.

Also, the following functions:

$$R_n^{f,\,a+}(x) = f(x) - T_{n-1}^{f,\,a+}(x), \quad n \in \mathbb{N}$$

and

$$\mathbb{R}_n^{f;\,a+,\,b-}(x) = f(x) - \mathbb{T}_{n-1}^{f;\,a+,\,b-}(x), \quad n \in \mathbb{N}$$

are called the *remainder of the first* TAYLOR*'s approximation in the right neighborhood of* a, and the *remainder of the second* TAYLOR*'s approximation in the right neighborhood of* a, respectively.

The following Theorem, which has been proved in [19] and whose variants are considered in [20, 21] and [22], provides an important result regarding TAYLOR's approximations.

Theorem 1 ([19], Theorem 2) *Suppose that* $f(x)$ *is a real function on* (a, b), *and that* n *is a positive integer such that* $f^{(k)}(a+)$, *for* $k \in \{0, 1, 2, \ldots, n\}$, *exist. Supposing that* $f^{(n)}(x)$ *is increasing on* (a, b), *then for all* $x \in (a, b)$ *the following inequality also holds*:

$$T_n^{f,\,a+}(x) < f(x) < \mathbb{T}_n^{f;\,a+,\,b-}(x). \tag{3}$$

Furthermore, if $f^{(n)}(x)$ *is decreasing on* (a, b), *then the reversed inequality of* (3) *holds.*

The above theorem is called *Theorem on double-sided* TAYLOR*'s approximations* in [15], i.e. *Theorem WD* in [10–14].

The proof of the following proposition is given in [15].

Proposition 1 ([15], Proposition 1) *Consider a real function* $f : (a, b) \longrightarrow \mathbb{R}$ *such that there exist its first and second* TAYLOR*'s approximations, for some* $n \in N_0$. *Then,*

$$\mathrm{sgn}\Big(\mathbb{T}_n^{f,\,a+,\,b-}(x) - \mathbb{T}_{n+1}^{f,\,a+,\,b-}(x)\Big) = \mathrm{sgn}\Big(f(b-) - T_n^{f,\,a+}(b)\Big),$$

for all $x \in (a, b)$.

From the above proposition, as shown in [15], the following theorem directly follows:

Theorem 2 ([15], Theorem 4) *Consider the real analytic functions* $f : (a, b) \longrightarrow \mathbb{R}$:

$$f(x) = \sum_{k=0}^{\infty} c_k (x - a)^k,$$

where $c_k \in \mathbb{R}$ and $c_k \geq 0$ for all $k \in \mathbb{N}_0$. Then,

$$T_0^{f,a+}(x) \leq \ldots \leq T_n^{f,a+}(x) \leq T_{n+1}^{f,a+}(x) \leq \ldots$$

$$\ldots \leq f(x) \leq \ldots$$

$$\ldots \leq \mathbb{T}_{n+1}^{f;a+,b-}(x) \leq \mathbb{T}_n^{f;a+,b-}(x) \leq \ldots \leq \mathbb{T}_0^{f;a+,b-}(x),$$

for all $x \in (a, b)$.

3   Main Results

3.1   *Generalization of Statement 1*

Consider the function:

$$f(x) = \begin{cases} \dfrac{3}{8} & , \quad x = 0, \\[2mm] \dfrac{1 - \dfrac{\cos x}{\cos \frac{x}{2}}}{x^2} & , \quad x \in (0, \pi). \end{cases}$$

First, we prove that f is a real analytic function on $[0, \pi)$. Based on the elementary equality:

$$1 - \frac{\cos x}{\cos \frac{x}{2}} = 1 + \sec \frac{x}{2} - 2\cos \frac{x}{2},$$

and well known power series expansions [23] (formula 1.411):

$$\cos t = \sum_{k=0}^{\infty} \frac{(-1)^k}{(2k)!} t^{2k} \qquad t \in R,$$

$$\sec t = \sum_{k=0}^{\infty} \frac{|E_{2k}|}{(2k)!} t^{2k} \quad t \in \left(-\frac{\pi}{2}, \frac{\pi}{2}\right);$$

where E_k are EULER's numbers [23], for $t = \dfrac{x}{2} \in \left[0, \dfrac{\pi}{2}\right)$, i.e. for $x \in [0, \pi)$, we have:

$$f(x) = \sum_{k=1}^{\infty} \frac{|E_{2k}| - 2(-1)^k}{2^{2k}(2k)!} x^{2k-2}$$

i.e.

$$f(x) = \frac{3}{8} + \frac{1}{128}x^2 + \frac{7}{5120}x^4 + \frac{461}{3440640}x^6 + \frac{16841}{1238630400}x^8 + \dots$$

where the power series converges for $x \in [0, \pi)$.

Further, based on the elementary well-known features of EULER's numbers E_k, we have:

$$c_{2k-2} = \frac{|E_{2k}| - 2(-1)^k}{2^{2k}(2k)!} > 0 \quad \text{and} \quad c_{2k-1} = 0,$$

for $k = 1, 2, \dots$.

Finally, from Theorem 2 the following result directly follows.

Theorem 3 *For the function*

$$f(x) = \begin{cases} \dfrac{3}{8} & , \ x = 0, \\[3mm] \dfrac{1 - \dfrac{\cos x}{\cos \frac{x}{2}}}{x^2} & , \ x \in (0, \pi) \end{cases}$$

and any $c \in (0, \pi)$ the following inequalities hold true:

$$\frac{3}{8} = T_0^{f,\,0+}(x) \le T_2^{f,\,0+}(x) \le \dots \le T_{2n}^{f,\,0+}(x) \le \dots$$

$$\dots \le f(x) \le \dots$$

$$\le \mathbb{T}_{2m}^{f;\,0+,\,c-}(x) \le \dots \le \mathbb{T}_2^{f;\,0+,\,c-}(x) \le \mathbb{T}_0^{f;\,0+,\,c-}(x) = \left(1 - \frac{\cos c}{\cos \frac{c}{2}}\right)\Big/c^2. \tag{4}$$

for every $x \in (0, c)$, where $m, n \in \mathbb{N}_0$.

Note that inequalities from Statement 1 can be directly obtained from (4), for $c = \dfrac{\pi}{2}$

$$\frac{3}{8} < \frac{1 - \dfrac{\cos x}{\cos \frac{x}{2}}}{x^2} < \frac{4}{\pi^2}.$$

Also, Theorem 3 gives a generalization and a sequence of improvements of results from Statement 1. For example, for $c = \pi/2$ i.e. for $x \in \left(0, \dfrac{\pi}{2}\right)$ we have:

$$\frac{3}{8} \le T_2^{f,\,0+}(x) = \frac{3}{8} + \frac{1}{128}x^2 \le f(x) \le \mathbb{T}_2^{f;\,0+,\,\pi/2-}(x) = \frac{3}{8} + \left(\frac{16}{\pi^4} - \frac{3}{2\pi^2}\right)x^2 \le \frac{4}{\pi^2}.$$

Using standard numerical methods it is easy to verify:

$$\max_{x\in[0,\pi/2]} |R_3^{f,0+}(x)| = f(\pi/2) = 0.01100\ldots$$

and

$$\max_{x\in[0,\pi/2]} |R_3^{f;\,0+,\,\pi/2-}(x)| = f(1.14909\ldots) = 0.00315\ldots.$$

3.2 An Improvement of Statement 2

Let $\beta \in (0, \pi)$ be a fixed real number. Consider the function:

$$g(x) = \begin{cases} \dfrac{1}{4} & , \ x = 0, \\[2ex] \dfrac{1 - \dfrac{\sin x}{\sin \frac{x}{2}}}{x^2} & , \ x \in (0, \beta]. \end{cases}$$

We prove that g is a real analytic function on $[0, \beta]$.
Notice that

$$g(x) = g_1(x) - g_2(x) \tag{5}$$

for $x \in [0, \beta]$, where

$$g_1(x) = \begin{cases} \dfrac{1}{4} & , \ x = 0, \\[2ex] \dfrac{\cosh\frac{x}{2} - \cos\frac{x}{2}}{x^2} & , \ x \in (0, \beta] \end{cases}$$

and

$$g_2(x) = \begin{cases} 0 & , \ x = 0, \\[2ex] \dfrac{\cosh\frac{x}{2} + \cos\frac{x}{2} - 2}{x^2} & , \ x \in (0, \beta]. \end{cases}$$

Since the functions g_1 and g_2 are real analytic functions on $[0, \beta]$, with the following power series expansions:

$$g_1(x) = \sum_{k=0}^{\infty} \frac{1}{2^{4k}(4k+2)!} x^{4k}$$

and

$$g_2(x) = \sum_{k=0}^{\infty} \frac{1}{2^{4k+2}(4k+4)!} x^{4k+2},$$

the function g must also be a real analytic function on $[0, \beta]$.

Also, from Theorem 2 the following results directly follow.

Theorem 4 *For all $c \in (0, \pi)$ the following inequalities hold true*:

$$\frac{1}{4} = T_0^{g_1,0+}(x) \leq \ldots \leq T_{4n}^{g_1,0+}(x) \leq T_{4n+4}^{g_1,0+}(x) \leq \ldots$$

$$\ldots \leq g_1(x) \leq \ldots$$

$$\ldots \leq \mathbb{T}_{4m+4}^{g_1;0+,c-}(x) \leq \mathbb{T}_{4m}^{g_1;0+,c-}(x) \leq \ldots \leq \mathbb{T}_0^{g_1;0+,c-}(x) = g_1(c).$$

for all $x \in (0, c)$, where $m, n \in \mathbb{N}_0$.

Theorem 5 *For all $c \in (0, \pi)$ the following inequalities hold true*:

$$\frac{1}{192}x^2 = T_2^{g_2,0+}(x) \leq \ldots \leq T_{4n+2}^{g_2,0+}(x) \leq T_{4n+6}^{g_2,0+}(x) \leq \ldots$$

$$\ldots \leq g_2(x) \leq \ldots$$

$$\ldots \leq \mathbb{T}_{4m+6}^{g_2;0+,c-}(x) \leq \mathbb{T}_{4m+2}^{g_2;0+,c-}(x) \leq \ldots \leq \mathbb{T}_2^{g_2;0+,c-}(x) = \frac{g_2(c)}{c^2}x^2$$

for all $x \in (0, c)$, where $m, n \in \mathbb{N}_0$.

Thus, from (5), Theorems 4 and 5, for $c = \frac{\pi}{2}$, an improvement of inequalities from Statement 2 are obtained, as shown bellow.

First, for all $x \in (0, \pi/2)$ the following inequalities hold true:

$$\frac{1}{4} - \frac{4}{\pi^2} g_2\left(\frac{\pi}{2}\right) x^2 \leq g(x) \leq g_1\left(\frac{\pi}{2}\right) - \frac{1}{192}x^2$$

i.e.

$$\frac{1}{4} - \frac{16}{\pi^4}\left(\cosh\frac{\pi}{4} + \frac{\sqrt{2}}{2} - 2\right)x^2 \leq g(x) \leq \frac{4}{\pi^2}\left(\cosh\frac{\pi}{4} - \frac{\sqrt{2}}{2}\right) - \frac{1}{192}x^2.$$

It is easy to check

$$\frac{4}{\pi^2}\left(\cosh\frac{\pi}{4} - \frac{\sqrt{2}}{2}\right) - \frac{1}{192}x^2 \leq \frac{1}{4}$$

for all $x \in [\delta_2, \pi/2]$, where $\delta_2 = \dfrac{4\sqrt{3}e^{-\frac{\pi}{8}}}{\pi}\sqrt{8 + 8e^{\frac{\pi}{2}} - (\pi^2 + 8\sqrt{2})e^{\frac{\pi}{4}}} = 0.22525 \ldots$.

Also,

$$\frac{4}{\pi^2}(2 - \sqrt{2}) \leq \frac{1}{4} - \frac{16}{\pi^4}\left(\cosh\frac{\pi}{4} + \frac{\sqrt{2}}{2} - 2\right)x^2$$

for all $x \in [0, \delta_1]$, where $\delta_1 = \dfrac{\sqrt{2}\pi e^{\frac{\pi}{8}}\sqrt{\pi^2 + 16\sqrt{2} - 32}}{8\sqrt{(\sqrt{2} - 4)e^{\frac{\pi}{4}} + e^{\frac{\pi}{2}} + 1}} = 1.55456\ldots$

4 Conclusion

In this paper, we showed a way to prove some trigonometric inequalities using the double-sided TAYLOR's approximations. The presented approach enabled generalizations of inequalities (1) i.e. produced sequences of polynomial approximations of the given trigonometric function f.

Note that Theorem 2 cannot be applied directly to inequality (2) because the function g has an alternating series expansion. We overcame this obstacle by representing this function by a linear combination of two functions whose power series expansions have nonnegative coefficients.

Our approach makes a good basis for the systematic proving of trigonometric inequalities. Developing general, automated-oriented methods for proving of trigonometric inequalities is an area our continuing interest [5–7, 10–15, 24–30] and [18].

Acknowledgements Research of the first and second and third author was supported in part by the Serbian Ministry of Education, Science and Technological Development, under Projects ON 174032 & III 44006, ON 174033 and TR 32023, respectively.

References

1. D.S. Mitrinović, *Analytic Inequalities* (Springer, Berlin, 1970)
2. G. Milovanović, M. Rassias (eds.), Topics in special functions III, in *Analytic Number Theory, Approximation Theory and Special Functions*, ed. by G.D. Anderson, M. Vuorinen, X. Zhang (Springer, New York, 2014), pp. 297–345
3. M.J. Cloud, B.C. Drachman, L.P. Lebedev, *Inequalities with Applications to Engineering* (Springer, Cham, 2014)
4. C. Mortici, The natural approach of Wilker-Cusa-Huygens inequalities. Math. Inequal. Appl. **14**(3), 535–541 (2011)
5. B. Malešević, M. Makragić, A method for proving some inequalities on mixed trigonometric polynomial functions. J. Math. Inequal. **10**(3), 849–876 (2016)
6. M. Makragić, A method for proving some inequalities on mixed hyperbolic-trigonometric polynomial functions. J. Math. Inequal. **11**(3), 817–829 (2017)
7. T. Lutovac, B. Malešević, C. Mortici, The natural algorithmic approach of mixed trigonometric-polynomial problems. J. Inequal. Appl. **2017**(116), 1–16 (2017)

8. B. Malešević, M. Rašajski, T. Lutovac, Refinements and generalizations of some inequalities of Shafer-Fink's type for the inverse sine function. J. Inequal. Appl. **2017**(275), 1–9 (2017)
9. H. Alzer, M.K. Kwong, On Jordan's inequality. Period. Math. Hung. **77**(2), 191–200 (2018)
10. B. Malešević, T. Lutovac, M. Rašajski, C. Mortici, Extensions of the natural approach to refinements and generalizations of some trigonometric inequalities. Adv. Differ. Equ. **2018**(90), 1–15 (2018)
11. M. Rašajski, T. Lutovac, B. Malešević, Sharpening and generalizations of Shafer-Fink and Wilker type inequalities: a new approach. J. Nonlinear Sci. Appl. **11**(7), 885–893 (2018)
12. M. Rašajski, T. Lutovac, B. Malešević, About some exponential inequalities related to the sinc function. J. Inequal. Appl. **2018**(150), 1–10 (2018)
13. T. Lutovac, B. Malešević, M. Rašajski, A new method for proving some inequalities related to several special functions. Results Math. **73**(100), 1–15 (2018)
14. M. Nenezić, L. Zhu, Some improvements of Jordan-Steckin and Becker-Stark inequalities. Appl. Anal. Discrete Math. **12**, 244–256 (2018)
15. B. Malešević, M. Rasajski, T. Lutovac, Double-sided Taylor's approximations and their applications in theory of analytic inequalities, in *Differential and Integral Inequalities*, ed. by D. Andrica, T. Rassias. Springer Optimization and Its Applications, vol. 151 (Springer, 2019), pp. 569–582. https://doi.org/10.1007/978-3-030-27407-8_20
16. J. Sándor, On D'aurizio's trigonometric inequality. J. Math. Inequal. **10**(3), 885–888 (2016)
17. J. D'Aurizio, Refinements of the Shafer-Fink inequality of arbitrary uniform precision. Math. Inequal. Appl. **17**(4), 1487–1498 (2014)
18. B.D. Banjac, System for automatic proving of some classes of analytic inequalities. Doctoral dissertation (in Serbian), School of Electrical Engineering, Belgrade, May 2019. Available on: http://nardus.mpn.gov.rs/
19. S.-H. Wu, L. Debnath, A generalization of L'Hospital-type rules for monotonicity and its application. Appl. Math. Lett. **22**(2), 284–290 (2009)
20. S.-H. Wu, H.M. Srivastva, A further refinement of a Jordan type inequality and its applications. Appl. Math. Comput. **197**, 914–923 (2008)
21. S.-H. Wu, L. Debnath, Jordan-type inequalities for differentiable functions and their applications. Appl. Math. Lett. **21**(8), 803–809 (2008)
22. S.-H. Wu, H.M. Srivastava, A further refinement of Wilker's inequality. Integral Transforms Spec. Funct. **19**(9–10), 757–765 (2008)
23. I.S. Gradshteyn, I.M Ryzhik, *Table of Integrals Series and Products*, 8th edn. (Academic Press, San Diego, 2015)
24. B. Banjac, M. Nenezić, B. Malešević, Some applications of Lambda-method for obtaining approximations in filter design, in *Proceedings of 23-rd TELFOR Conference*, Beograd, 2015, pp. 404–406
25. M. Nenezić, B. Malešević, C. Mortici, New approximations of some expressions involving trigonometric functions. Appl. Math. Comput. **283**, 299–315 (2016)
26. B. Banjac, M. Makragić, B. Malešević, Some notes on a method for proving inequalities by computer. Results Math. **69**(1), 161–176 (2016)
27. B. Malešević, I. Jovović, B. Banjac, A proof of two conjectures of Chao-Ping Chen for inverse trigonometric functions. J. Math. Inequal. **11**(1), 151–162 (2017)
28. B. Malešević, T. Lutovac, B. Banjac, A proof of an open problem of Yusuke Nishizawa for a power-exponential function. J. Math. Inequal. **12**(2), 473–485 (2018)
29. B. Malešević, M. Rašajski, T. Lutovac, Refined estimates and generalizations of inequalities related to the arctangent function and Shafer's inequality. Math. Probl. Eng. **2018**, Article ID 4178629, 1–8
30. B. Malešević, T. Lutovac, B. Banjac, One method for proving some classes of exponential analytical inequalities. Filomat **32**(20), 6921–6925 (2018)

The Second Moment of the First Derivative of Hardy's Z-Function

Maxim A. Korolev and Andrei V. Shubin

Abstract We give a new estimate of the error term in the asymptotic formula for the second moment of first derivative of Hardy's function $Z(t)$. This estimate improves the previous result of R.R. Hall.

Keywords Riemann zeta-function · Hardy's Z-function · Approximate functional equation · Second moment · Third derivative test

1 Introduction

We use the following notation. For $t > 0$, let $\vartheta(t)$ be an increment of any fixed continuous branch of the function $\arg\{\pi^{-s/2}\Gamma(s/2)\}$ along the segment with endpoints $s = 0.5$ and $s = 0.5 + it$. Hardy's function $Z(t)$ is defined as

$$Z(t) = e^{i\vartheta(t)}\zeta(0.5 + it).$$

It is known that $Z(t)$ is real for real t and its real zeros coincide with the ordinates of zeros of $\zeta(s)$ lying on the critical line (see, for example, [1, Ch. III, §4]).

One of the significant branches in the theory of Riemann zeta function deals with mean values of the functions $\zeta(0.5 + it)$, $Z(t)$ and its derivatives. For example, it is known that

$$\int_0^T Z^2(t)dt = TP_1\left(\ln\frac{T}{2\pi}\right) + E(T),$$

M. A. Korolev (✉)
Steklov Mathematical Institute of Russian Academy of Sciences, Moscow, Russia
e-mail: korolevma@mi.ras.ru

A. V. Shubin
Department of Mathematics and Statistics, McGill University, Montreal, QC, Canada
e-mail: andrei.shubin@mail.mcgill.ca

© Springer Nature Switzerland AG 2020
A. Raigorodskii, M. T. Rassias (eds.), *Trigonometric Sums and Their Applications*,
https://doi.org/10.1007/978-3-030-37904-9_9

169

where $P_1(u) = u + 2\gamma_0 - 1$, γ_0 is the Euler constant and $E(T)$ is the error term which has a long history of exploration (see, for example, [2, Ch. XV]). Its best present estimate belongs to J. Bourgain and N. Watt [3] and has the form $E(T) \ll T^{\alpha + \varepsilon}$, $\alpha = \frac{1515}{4816} = 0.314576\ldots$.

Studying the value distribution of $Z(t)$ at the points of local extremum, R.R. Hall [4] obtained the following asymptotic formula for the second moment of kth derivative of Hardy's function:

$$\int_0^T \{Z^{(k)}(t)\}^2 dt = \frac{T}{4^k(2k+1)} P_{2k+1}\left(\ln \frac{T}{2\pi}\right) + O\left(T^{3/4}(\ln T)^{2k+1/2}\right), \quad T \to +\infty, \tag{1}$$

where $k \geq 1$ is any fixed integer and $P_{2k+1}(u)$ is a polynomial of degree $2k + 1$ whose leading coefficient equals to 1.

One of the main tools for studying the behavior of the Riemann zeta function $\zeta(s)$ on the critical line $\operatorname{Re} s = \frac{1}{2}$ is the following Riemann-Siegel approximate functional equation

$$Z(t) = 2 \sum_{n=1}^m \frac{\cos(\vartheta(t) - t \ln n)}{\sqrt{n}} + R(t), \quad m = \left[\sqrt{t/(2\pi)}\right]. \tag{2}$$

The error term $R(t)$ in (2) obeys the estimate $R(t) \ll t^{-1/4}$; moreover, it has the asymptotic expansion of the form

$$R(t) = (-1)^{m-1}\left(\frac{t}{2\pi}\right)^{-1/4}\left(H_0(t) + \left(\frac{t}{2\pi}\right)^{-1/2} H_1(t) + \ldots + \left(\frac{t}{2\pi}\right)^{-r/2} H_r(t)\right) + \\ + O_r\left(t^{-(r+1)/2}\right), \tag{3}$$

where $r \geq 0$ is any fixed integer. The functions $H_j(t)$ are expressed as linear combinations of the values of the function

$$\Phi(z) = \frac{\cos \pi \left(\frac{1}{2} z^2 - z - \frac{1}{8}\right)}{\cos \pi z} \tag{4}$$

and its derivatives at the point $z = 2\alpha = 2\{\sqrt{t/(2\pi)}\}$.

The expansion (2), (3) was discovered by B. Riemann. Its complete proof was reconstructed by C.L. Siegel, who used Riemann's drafts from Göttingen University's library, and published in [5] in 1932 (for the history of this question, see [6, Ch. 7]).

First analogues of the approximate functional equation (2) for the derivatives $Z^{(k)}(t)$, $k = 1, 2, \ldots$ were obtained by A.A. Karatsuba [7] and A.A. Lavrik [8] who studied the gaps between consecutive zeros of $Z^{(k)}(t)$. However, this problem does not require an explicit form of the error term $R(t)$. These authors obtained the estimate $R(t) \ll (1.5 \ln t)^k t^{-1/4}$, which is uniform over k, and this bound was sufficient for their purposes.

In [9], the full analogues of the expansion (2), (3) for the derivatives $Z^{(k)}(t)$, $k = 1, 2$ of Hardy's function, were obtained. These expansions allow one to specify significantly the well-known formulas for the sums

$$\sum_{n \leq N} Z(t_n), \quad \sum_{n \leq N} \zeta(0.5 + it_n), \quad \sum_{n \leq N} Z(t_n)Z(t_{n+1}),$$

where t_n is the sequence of Gram points (for the definition and basic properties of Gram points, see, for example, [6, Ch. 6, §5].[1]) Moreover, the following interesting effect was found in [9]: it appears that first k terms in the expression (3) for $Z^{(k)}(t)$ are equal to zero in the cases $k = 1$ and $k = 2$. It seems that the same fact is true for any $k \geq 3$.

In this paper, we give another application of the approximate formula for $Z'(t)$ which yields a new estimate of the error term in (1). Our main result is the following

Theorem *If $T \to +\infty$ then*

$$\int_0^T \{Z'(t)\}^2 dt = \frac{T}{12} P_3 \left(\ln \frac{T}{2\pi} \right) + O\left(T^{5/12} (\ln T)^3 \right),$$

where $P_3(u) = u^3 + c_2 u^2 + c_1 u + c_0$,

$$c_0 = 12\gamma_0 + 24\gamma_1 + 24\gamma_2 - 6 = -7.053561\ldots,$$

$$c_1 = 6 - 12\gamma_0 - 24\gamma_1 = 6.820992\ldots,$$

$$c_2 = 6\gamma_0 - 3 = 0.463294\ldots,$$

γ_k are Stieltjes constants defined by the expansion

$$\zeta(s) = \frac{1}{s-1} + \sum_{m=0}^{+\infty} \frac{(-1)^m}{m!} \gamma_m (s-1)^m.$$

2 Auxiliary Lemmas

In this section, we give some lemmas which are necessary for the proof of the main theorem.

---

[1] In [6], another notation for Gram points is used: g_n instead of t_n. The corresponding results of the first author concerning the above sums will appear soon.

Lemma 1 *If* $x \to +\infty$, *then*

$$\sum_{n \le \sqrt{x}} \frac{1}{n} = \frac{1}{2} \ln x + \gamma_0 + \frac{\varrho(x)}{\sqrt{x}} + O\left(\frac{1}{x}\right),$$

$$\sum_{n \le \sqrt{x}} \frac{\ln n}{n} = \frac{1}{8} \ln^2 x + \gamma_1 + \frac{\varrho(x)}{2\sqrt{x}} \ln x + O\left(\frac{\ln x}{x}\right),$$

$$\sum_{n \le \sqrt{x}} \frac{(\ln n)^2}{n} = \frac{1}{24} \ln^3 x + \gamma_2 + \frac{\varrho(x)}{4\sqrt{x}} \ln^2 x + O\left(\frac{\ln^2 x}{x}\right),$$

where $\varrho(x) = 0.5 - \{x\}$.

These expansions follow directly from Euler's summation formula (see, for example, [1, Appendix, § 1, Th. 2]).

Lemma 2 *Let* $b - a > 1$, $f \in C^3[a, b]$ *and let* $\lambda_3 \le |f^{(3)}(x)| \le h\lambda_3$ *for some* $\lambda_3 > 0$, $h \ge 1$. *Then*

$$\sum_{a < x \le b} \exp(2\pi i f(x)) \ll \sqrt{h}(b-a)\lambda_3^{1/6} + \sqrt{b-a}\,\lambda_3^{-1/6},$$

where the implied constant is absolute.

This is a particular case of van der Corput's theorem (see [1, Appendix, §11, Th. 2]).

Lemma 3 *Let* $c < b$ *and suppose that real functions* $f(t)$, $g(t)$ *satisfy the following conditions on* $[c, b]$: *(1) the derivatives* $g''(t)$ *and* $f^{(4)}(t)$ *are continuous; (2)* $f''(t) > 0$; *(3)* $f'(c) = 0$. *Then the integral*

$$I = \int_c^b g(t)e^{2\pi i f(t)}\,dt$$

is expressed as follows:

$$I = A - B + \frac{1}{2\pi i}(C + D - E + H - K),$$

where

$$A = \frac{1+i}{2} \cdot \frac{g(c)e^{2\pi i f(c)}}{\sqrt{2f''(c)}}, \quad B = \frac{g(c)e^{2\pi i f(c)}}{\sqrt{2f''(c)}} \int_\lambda^{+\infty} \frac{e^{2\pi i u}}{\sqrt{u}}\,du, \quad \lambda = f(b) - f(c),$$

$$C = \left(\frac{g(b)}{f'(b)} - \frac{g(c)}{\sqrt{2f''(c)}} \cdot \frac{1}{\sqrt{\lambda}}\right) e^{2\pi i f(b)}, \quad H = \frac{g(c)}{3} \cdot \frac{f^{(3)}(c)}{(f''(c))^2} e^{2\pi i f(c)},$$

$$K = \frac{g'(c)}{f''(c)} e^{2\pi i f(c)},$$

and D, E denote the integrals along the segment $[c, b]$ with the integrands

$$g(c)\left(\frac{f''(t)}{(f'(t))^2} - \frac{|f'(t)|}{\sqrt{8f''(c)}(f(t) - f(c))^{3/2}}\right)e^{2\pi i f(t)},$$

$$\frac{g'(t)f'(t) - (g(t) - g(c))f''(t)}{(f'(t))^2}e^{2\pi i f(t)}.$$

This is a particular case of lemma 7 from [10].

Lemma 4 *The following equality holds:* $\vartheta(t) = \theta(t) + \delta(t)$, *where*

$$\theta(t) = \frac{t}{2}\ln\frac{t}{2\pi} - \frac{t}{2} - \frac{\pi}{8}$$

and $\delta(t)$ is a smooth function such that

$$\delta(t) = \frac{1}{48t} + O\left(\frac{1}{t^3}\right), \quad \delta'(t) = -\frac{1}{48t^2} + O\left(\frac{1}{t^4}\right), \quad as \quad t \to +\infty.$$

These relations follow from the asymptotic expansion for $\vartheta(t)$ obtained by C.L. Siegel [5, formula (43)].

Lemma 5 *If $t \to +\infty$ then*

$$Z'(t) = -2\sum_{n\leq m}\frac{1}{\sqrt{n}}\left(\vartheta'(t) - \ln n\right)\sin\left(\vartheta(t) - t\ln n\right)$$

$$+(-1)^{m-1}\left(\frac{t}{2\pi}\right)^{-3/4}\frac{\Phi(2\alpha)}{2\pi} + O\left(t^{-5/4}\right), \tag{5}$$

where $m = \left[\sqrt{t/(2\pi)}\right]$, $\alpha = \left\{\sqrt{t/(2\pi)}\right\}$, and the function $\Phi(z)$ is defined by (4).

This approximate functional equation for $Z'(t)$ is a corollary of theorem 2 from [9].

3 Proof of the Main Theorem

Let $p(t)$, $q(t)$ and $r(t)$ denote the terms in the right hand side of (5). Then the integral $I(T)$ is expressed as follows:

$$I(T) = I_1 + I_2 + I_3 + 2(I_4 + I_5 + I_6),$$

Here $I_1, \ldots, I_6$ denote the integrals of the functions $p^2(t)$, $q^2(t)$, $r^2(t)$, $q(t)r(t)$, $p(t)r(t)$ and $p(t)q(t)$, respectively. Then we have

$$I_2 = \frac{1}{4\pi^2} \int_{2\pi}^{T} \Phi^2(2\alpha) \left(\frac{t}{2\pi}\right)^{-3/2} dt \ll 1,$$

$$I_3 \ll \int_{2\pi}^{T} t^{-5/2} dt \ll 1, \quad I_4 \ll \int_{2\pi}^{T} t^{-3/4} \cdot t^{-5/4} dt \ll 1.$$

Further,

$$I_5 = \int_{2\pi}^{T} p(t)r(t)dt = \int_{2\pi}^{T} \left(Z'(t) - q(t) - r(t)\right) r(t)dt =$$

$$= \int_{2\pi}^{T} Z'(t)r(t)dt - I_3 - I_4 \ll \int_{2\pi}^{T} |Z'(t)|t^{-5/2}dt + 1.$$

Splitting the interval of integration by points 2^k, $k = 1, 2, \ldots$ to the segments of the form $\tau \le t \le \tau_1$, $\tau_1 \le 2\tau$, and using (1) we get:

$$\int_{\tau}^{\tau_1} |Z'(t)|t^{-5/4}dt \ll \tau^{-5/4} \int_{\tau}^{\tau_1} |Z'(t)|dt \ll \tau^{-5/4} \left(\tau \int_{\tau}^{\tau_1} |Z'(t)|^2 dt\right)^{1/2} \ll$$

$$\ll \tau^{-5/4} \left(\tau^2 (\ln \tau)^3\right)^{1/2} \ll \tau^{-1/4} (\ln \tau)^{3/2},$$

whence

$$\int_{2\pi}^{T} |Z'(t)|t^{-5/4}dt \ll \sum_{\tau} \tau^{-1/4} (\ln \tau)^{3/2} \ll \sum_{k \ge 1} 2^{-k/4} k^{3/2} \ll 1, \quad I_5 \ll 1.$$

Similarly,

$$I_6 = \int_{2\pi}^{T} p(t)q(t)dt = \int_{2\pi}^{T} \left(Z'(t) - q(t) - r(t)\right) q(t)dt \ll$$

$$\ll \int_{2\pi}^{T} |Z'(t)|t^{-3/4}dt + 1 \ll \sum_{\tau} \tau^{1/4} (\ln \tau)^{3/2} \ll T^{1/4} (\ln T)^{3/2}.$$

Thus,

$$I(T) = I_1 + O\left(T^{1/4} (\ln T)^{3/2}\right).$$

To calculate I_1, we note that the error arising from the replacement of $\vartheta'(t)$ by $\theta'(t) = \frac{1}{2} \ln(t/(2\pi))$ in the expression

$$p(t) = -2 \sum_{n \le m} \frac{1}{\sqrt{n}} \left(\vartheta'(t) - \ln n\right) \sin\left(\vartheta(t) - t \ln n\right) \qquad (6)$$

is estimated as

$$-4 \int_{2\pi}^{T} \delta'(t) p(t) \left(\sum_{n=1}^{m} \frac{1}{\sqrt{n}} \sin\left(\vartheta(t) - t \ln n\right)\right) dt -$$

$$-4 \int_{2\pi}^{T} (\delta'(t))^2 \left(\sum_{n=1}^{m} \frac{1}{\sqrt{n}} \sin\left(\vartheta(t) - t \ln n\right)\right)^2 dt \ll$$

$$\ll \int_{2\pi}^{T} |p(t)| t^{-7/4} dt + \int_{2\pi}^{T} t^{-7/2} dt \ll \int_{2\pi}^{T} |Z'(t) - q(t) - r(t)| t^{-7/4} + 1 \ll 1.$$

Further, we replace $\vartheta(t)$ by $\theta(t)$ in (6). One can check that

$$\sin\left(\vartheta(t) - t \ln n\right) - \sin\left(\theta(t) - t \ln n\right) = 2 \sin\left(0.5\delta(t)\right) \cos\left(\theta(t) - t \ln n + 0.5\delta(t)\right) =$$

$$= 2 \sin\left(0.5\delta(t)\right) \{\cos\left(\theta(t) - t \ln n\right) \cos\left(0.5\delta(t)\right) - \sin\left(\theta(t) - t \ln n\right) \sin\left(0.5\delta(t)\right)\} =$$

$$= 2 \sin\left(0.5\delta(t)\right) \cos\left(\theta(t) - t \ln n\right) + O(\delta^2(t)) = \delta(t) \cos\left(\theta(t) - t \ln n\right) + O(\delta^2(t)) =$$

$$= \frac{1}{48t} \cos\left(\theta(t) - t \ln n\right) + O\left(t^{-2}\right).$$

Setting

$$\varphi_n = \varphi_n(t) = \theta(t) - t \ln n, \quad s(t) = -2 \sum_{n=1}^{m} \frac{\varphi_n}{\sqrt{n}} \left(\sin \varphi_n + \frac{1}{48t} \cos \varphi_n\right),$$

we get

$$I_1 = I_7 + O(1), \quad I_7 = \int_{2\pi}^{T} s^2(t) dt.$$

Next, we have

$$s^2(t) =$$

$$4 \sum_{k,n=1}^{m} \frac{\varphi'_n(t) \varphi'_k(t)}{\sqrt{kn}} \left(\sin \varphi_n \sin \varphi_k + \frac{1}{48t}(\sin \varphi_n \cos \varphi_k + \cos \varphi_n \sin \varphi_k)\right.$$

$$\left. + \frac{1}{(48t)^2} \cos \varphi_n \cos \varphi_k\right) =$$

$$2 \sum_{k,n=1}^{m} \frac{\varphi_n'(t)\varphi_k'(t)}{\sqrt{kn}} \left(\cos(\varphi_n - \varphi_k) - \cos(\varphi_n + \varphi_k) + \frac{1}{24t} \sin(\varphi_n + \varphi_k) \right)$$

$$+ O\left(t^{-3/2} \ln^2 t\right). \qquad (7)$$

Hence,

$$I_7 = J_1 + J_2 + J_3 + O(1),$$

where J_1, J_2 and J_3 denote the contributions arising from the first, second and third term in right-hand side of (7).

Let $J_{1,1}$ and $J_{1,2}$ denote the contributions to J_1 coming from the terms with $n = k$ and $n \neq k$, respectively. Then

$$J_{1,1} = 2 \int_{2\pi}^{T} \sum_{n=1}^{m} \frac{1}{n} \left(\frac{1}{2} \ln \frac{t}{2\pi} - \ln n \right)^2 dt =$$

$$= 2 \int_{2\pi}^{T} \left\{ \frac{1}{4} \left(\ln \frac{t}{2\pi} \right)^2 \sum_{n=1}^{m} \frac{1}{n} - \left(\ln \frac{t}{2\pi} \right) \sum_{n=1}^{m} \frac{\ln n}{n} + \sum_{n=1}^{m} \frac{(\ln n)^2}{n} \right\} dt.$$

Applying lemma 1 with $x = t/(2\pi)$, we conclude that the integrand equals to

$$\frac{1}{24} \left(\ln \frac{t}{2\pi} \right)^3 + \frac{\gamma_0}{4} \left(\ln \frac{t}{2\pi} \right)^2 - \gamma_1 \left(\ln \frac{t}{2\pi} \right) + \gamma_2 + O\left(\frac{\ln^2 t}{t} \right).$$

The integration over t yields

$$J_{1,1} = \frac{T}{12} P_3 \left(\ln \frac{T}{2\pi} \right) + O(\ln^3 T).$$

Next, we have

$$J_{1,2} = 4 \operatorname{Re} \int_{2\pi}^{T} \sum_{1 \le k < n \le m} \frac{\varphi_n'(t)\varphi_k'(t)}{\sqrt{nk}} e^{it \ln(n/k)} dt = 4 \operatorname{Re} \sum_{1 \le k < n \le P} \frac{j_{n,k}}{\sqrt{nk}}, \qquad (8)$$

where

$$j_{n,k} = \int_{2\pi n^2}^{T} \left(\frac{1}{2} \ln \frac{t}{2\pi} - \ln n \right) \left(\frac{1}{2} \ln \frac{t}{2\pi} - \ln k \right) e^{it \ln(n/k)} dt, \quad P = \sqrt{T/(2\pi)}.$$

By taking $t = 2\pi\tau^2$ and integrating $j_{n,k}$ by parts twice, we obtain

$$j_{n,k} = 4\pi \int_n^P \tau \left(\ln\frac{\tau}{n}\right)\left(\ln\frac{\tau}{k}\right)e^{2\pi i\tau^2\ln(n/k)}d\tau =$$

$$= -i\left(\ln\frac{n}{k}\right)^{-1}\left(\ln\frac{P}{n}\right)\left(\ln\frac{P}{k}\right)e^{2\pi iP^2\ln(n/k)}$$

$$+ \frac{1}{4\pi P^2}\left(\ln\frac{n}{k}\right)^{-2}\left(\ln\frac{P}{n} + \ln\frac{P}{k}\right)e^{2\pi iP^2\ln(n/k)} -$$

$$- \frac{1}{4\pi n^2}\left(\ln\frac{n}{k}\right)^{-1}e^{2\pi in^2\ln(n/k)} +$$

$$+ \frac{1}{2\pi}\left(\ln\frac{n}{k}\right)^{-2}\int_n^P e^{2\pi i\tau^2\ln(n/k)}\left(\ln\frac{\tau}{n} + \ln\frac{\tau}{k} - 1\right)\frac{d\tau}{\tau^3} =$$

$$= -iA_{n,k} + B_{n,k} - C_{n,k} + D_{n,k},$$

where the notation is obvious. The following inequalities hold:

$$|B_{n,k}| \le \frac{1}{4\pi n^2}\left(\ln\frac{n}{k}\right)^{-2}\left(\ln\frac{P}{n} + \ln\frac{P}{k}\right) \le \frac{1}{2\pi n^2}\left(\ln\frac{n}{k}\right)^{-2}\ln P,$$

$$|C_{n,k}| \le \frac{1}{4\pi n^2}\left(\ln\frac{n}{k}\right)^{-1} < \frac{1}{4\pi n^2}\left(\ln\frac{n}{k}\right)^{-2}\ln P,$$

$$|D_{n,k}| \le \frac{1}{2\pi}\left(\ln\frac{n}{k}\right)^{-2}\int_n^P\left(\ln\frac{\tau}{n} + \ln\frac{\tau}{k} - 1\right)\frac{d\tau}{\tau^3} < \frac{1}{2\pi n^2}\left(\ln\frac{n}{k}\right)^{-2}.$$

Consequently, the contribution to the sum on the right hand side of (8) coming from $B_{n,k}$, $C_{n,k}$ and $D_{n,k}$ does not exceed

$$(\ln P)\sum_{1<n\le P}\frac{1}{n^{5/2}}\sum_{1\le k\le n-1}\frac{1}{\sqrt{k}}\left(\ln\frac{n}{k}\right)^{-2} \ll$$

$$\ll (\ln P)\sum_{1<n\le P}\frac{1}{n^{5/2}}\left(\sum_{1\le k\le n/2}\frac{1}{\sqrt{k}} + \sum_{1\le r\le n/2}\frac{1}{\sqrt{n-r}}\left(\frac{n}{r}\right)^2\right) \ll (\ln T)^2.$$

The contribution to (8) from $A_{n,k}$ is expressed as

$$4\operatorname{Im}\sum_{1\le k<n\le P}\frac{1}{\sqrt{nk}}\left(\ln\frac{n}{k}\right)^{-1}\left(\ln\frac{P}{n}\right)\left(\ln\frac{P}{k}\right)e^{2\pi iP^2\ln(n/k)} = 4\operatorname{Im}\left(\Sigma_1 + \Sigma_2\right),$$

where Σ_1 and Σ_2 stand for terms with the conditions $1 \le k \le n/2$ and $n/2 < k \le n - 1$, respectively. Changing the order of summation in Σ_1, we have

$$\Sigma_1 = \sum_{1 \le k \le P/2} \frac{1}{\sqrt{k}} \left(\ln \frac{P}{k} \right) e^{-2\pi i f(k)} S_k, \quad S_k = \sum_{2k < n \le P} g(n) e^{2\pi i f(n)},$$

where

$$g(x) = \frac{1}{\sqrt{x}} \left(\ln \frac{P}{x} \right) \left(\ln \frac{x}{k} \right)^{-1}, \quad f(x) = P^2 \ln x.$$

Splitting the range of summation over n by points $2^{-s} P$, $s = 1, 2, \ldots$ to the segments of the form $N < n \le N_1$, where $N_1 \le 2N$, we find

$$S_k = \sum_s S_k(N), \quad S_k(N) = \sum_{N < n \le N_1} g(n) e^{2\pi i f(n)}.$$

Since $g(x) \ll N^{-1/2} \ln P$, $g'(x) \ll N^{-3/2} \ln P$ for $N \le x \le N_1$, Abel summation formula yields

$$S_k(N) \ll \frac{\ln P}{\sqrt{N}} \left| \sum_{N < n \le N_2} e^{2\pi i f(n)} \right|,$$

where $N < N_2 \le N_1$. Obviously, $|f^{(3)}(x)| \asymp P^2 N^{-3}$ for $N \le x \le N_2$. Then the application of lemma 2 gives

$$S_k(N) \ll \frac{\ln P}{\sqrt{N}} \left(\sqrt{N} P^{1/3} + N P^{-1/3} \right) \ll P^{1/3} \ln P, \quad S_k \ll P^{1/3} (\ln P)^2,$$

$$\Sigma_1 \ll P^{5/6} (\ln P)^3 \ll T^{5/12} (\ln T)^3.$$

Further, setting $k = n - r$, where $1 \le r \le n/2$, we write the sum Σ_2 in the form

$$\Sigma_2 = \sum_{1 \le r < P/2} \frac{S_r}{r}, \quad S_r = \sum_{2r < n \le P} g_r(n) e^{2\pi i f_r(n)},$$

where

$$g_r(x) = \frac{r}{\sqrt{x(x-r)}} \left(\ln \frac{P}{x} \right) \left(\ln \frac{P}{x-r} \right) \left(\ln \frac{x}{x-r} \right)^{-1}, \quad f_r(x) = P^2 \ln \frac{x}{x-r}.$$

Using the same arguments as above, we find

$$S_r \ll \sum_s S_r(N), \quad N = 2^{-s}P, \quad s = 1, 2, \ldots,$$

$$S_r(N) = \sum_{N < n \le N_1} g_r(n) e^{2\pi i f_r(n)} \ll (\ln P)^2 \left| \sum_{N < n \le N_3} e^{2\pi i f_r(n)} \right|, \quad N < N_3 \le N_1.$$

Since $|f_r^{(3)}(x)| \asymp r P^2 N^{-4}$ for $N \le x \le N_3$, lemma 2 yields

$$S_r(N) \lll (\ln P)^2 \left(N \left(\frac{P^2 r}{N^4} \right)^{1/6} + \sqrt{N} \left(\frac{N^4}{P^2 r} \right)^{1/6} \right)$$

$$\ll \left((NP)^{1/3} r^{1/6} + P^{-1/3} N^{7/6} r^{-1/6} \right) (\ln P)^2,$$

$$S_r \ll \left(P^{2/3} r^{1/6} + P^{5/6} r^{-1/6} \right) (\ln P)^2,$$

$$\Sigma_2 \ll (\ln P)^2 \sum_{1 \le r \le P/2} \left(P^{2/3} r^{-5/6} + P^{5/6} r^{-7/6} \right) \ll T^{5/12} (\ln T)^2.$$

Therefore, $J_{1,2} \ll T^{5/12} (\ln T)^2$.

Further, we express the integral

$$J_2 = 2 \int_{2\pi}^{T} \sum_{k,n=1}^{m} \frac{\varphi_n'(t)\varphi_k'(t)}{\sqrt{nk}} \cos(\varphi_n(t) + \varphi_k(t)) \, dt$$

as a sum $J_{2,1} + J_{2,2}$, where $J_{2,1}$ and $J_{2,2}$ denote the contributions coming from the pairs n, k with the conditions $n = k$ and $n \ne k$, respectively. Then

$$J_{2,1} = 2 \int_{2\pi}^{T} \left(\sum_{n=1}^{m} \frac{(\varphi_n'(t))^2}{n} \cos(2\varphi_n(t)) \right) dt = 2 \operatorname{Re} \left(e^{-\pi i/4} \sum_{n \le P} \frac{j(n)}{n} \right).$$

$$(9)$$

Taking $t = 2\pi \tau^2$, we get

$$j(n) = \int_n^P g(\tau) e^{2\pi i f(\tau)} d\tau, \quad g(x) = x \left(\ln \frac{x}{n} \right)^2, \quad f(x) = 2x^2 \left(\ln \frac{x}{n} - \frac{1}{2} \right).$$

Since $g(n) = g'(n) = 0$, the application of lemma 3 to $j(n)$ yields: $j(n) = (C(n) - E(n))/(2\pi i)$, where

$$C(n) = \frac{1}{4} \left(\ln \frac{P}{n} \right) e^{2\pi i f(P)}, \quad E(n) = \frac{1}{4} \int_n^P e^{2\pi i f(\tau)} \frac{d\tau}{\tau}.$$

Therefore,

$$|j(n)| \leq \frac{1}{4\pi} \ln \frac{P}{n}, \quad J_{2,1} \ll (\ln T)^2.$$

Using the same arguments as above, we write $J_{2,2}$ as

$$4 \operatorname{Re}\left(e^{-\pi i/4} \sum_{1<n\leq P} \frac{j_{n,k}}{\sqrt{nk}} \right),$$

$$j_{n,k} = \int_{2\pi n^2}^{T} \left(\frac{1}{2} \ln \frac{t}{2\pi} - \ln n \right)\left(\frac{1}{2} \ln \frac{t}{2\pi} - \ln k \right) e^{it(\ln(t/(2\pi))-\ln kn-1)} dt =$$

$$= 4\pi \int_{n}^{P} \tau \left(\ln \frac{\tau}{n} \right)\left(\ln \frac{\tau}{k} \right) e^{2\pi i f(\tau)} d\tau,$$

where $f(\tau) = \tau^2(2\ln\tau - \ln kn - 1)$. Integrating by parts twice, we get: $j_{n,k} = -i A_{n,k} + B_{n,k} - C_{n,k} - D_{n,k}$, where

$$A_{n,k} = \left(\ln \frac{P^2}{kn} \right)^{-1}\left(\ln \frac{P}{n} \right)\left(\ln \frac{P}{k} \right) e^{2\pi i f(P)},$$

$$B_{n,k} = \frac{1}{4\pi P^2}\left(\ln \frac{P^2}{kn} \right)^{-3}\left(\left(\ln \frac{P}{n} \right)^2 + \left(\ln \frac{P}{k} \right)^2 \right) e^{2\pi i f(P)},$$

$$C_{n,k} = \frac{1}{4\pi n^2}\left(\ln \frac{n}{k} \right)^{-1} e^{2\pi i f(n)},$$

$$D_{n,k} = \frac{1}{2\pi} \int_{n}^{P} \left(\frac{\ln^2(\tau/n) + \ln^2(\tau/k)}{(\ln(\tau/n) + \ln(\tau/k))^3} + \right.$$

$$\left. + 2 \cdot \frac{\ln^2(\tau/n) + \ln^2(\tau/k) + \ln(\tau/n)\ln(\tau/k)}{(\ln(\tau/n) + \ln(\tau/k))^4} \right) \frac{e^{2\pi i f(\tau)}}{\tau^3} d\tau.$$

Next, one can check that

$$|B_{n,k}| \leq \frac{1}{4\pi P^2}\left(\ln \frac{P}{n} + \ln \frac{P}{k} \right)^{-1} \leq \frac{1}{4\pi n^2}\left(\ln \frac{n}{k} \right)^{-1},$$

$$|C_{n,k}| \leq \frac{1}{4\pi n^2}\left(\ln \frac{n}{k} \right)^{-1}, \quad |D_{n,k}| \leq \frac{1}{4\pi n^2}\left(\left(\ln \frac{n}{k} \right)^{-1} + \left(\ln \frac{n}{k} \right)^{-2} \right).$$

Hence, the contribution arising from the terms $B_{n,k}$, $C_{n,k}$ and $D_{n,k}$ in the sum over n in right hand side of (9) does not exceed

$$\sum_{1<n\leq P}\frac{1}{n^{5/2}}\sum_{1\leq k\leq n-1}\frac{1}{\sqrt{k}}\left(\left(\ln\frac{n}{k}\right)^{-1}+\left(\ln\frac{n}{k}\right)^{-2}\right)\ll 1.$$

Further, the contribution coming from $A_{n,k}$ to (9) is expressed as follows:

$$-2\,\mathrm{Re}\left(e^{\pi i/4+2\pi i P^2(2\ln P-1)}\Sigma\right),$$

$$\Sigma=\sum_{1\leq k<n\leq P}\frac{1}{\sqrt{kn}}\frac{\ln(P/k)\ln(P/n)}{\ln(P/k)+\ln(P/n)}e^{2\pi i P^2\ln(kn)}=\Sigma_3+\Sigma_4,$$

where Σ_3 and Σ_4 are responsible for terms with the conditions $1\leq k\leq n/2$ and $n/2<k\leq n-1$, respectively. Taking $k=n-r$ in Σ_4 and changing the order of summation in both sums we get

$$\Sigma_3=\sum_{1\leq k<P/2}\frac{1}{\sqrt{k}}\left(\ln\frac{P}{k}\right)e^{2\pi if(k)}\sum_{2k<n\leq P}g(n)e^{2\pi if(n)},$$

$$\Sigma_4=\sum_{1\leq r<P/2}\sum_{2r<n\leq P}g_r(n)e^{2\pi if_r(n)},$$

where

$$g(x)=\frac{\ln(P/x)}{\ln(P/x)+\ln(P/k)}\frac{1}{\sqrt{x}},\qquad f(x)=P^2\ln x,$$

$$g_r(x)=\frac{\ln(P/x)\ln(P/(x-r))}{\ln(P/x)+\ln(P/(x-r))}\frac{1}{\sqrt{x(x-r)}},\qquad f_r(x)=P^2(\ln x+\ln(x-r)).$$

One can check that

$$g(x)\ll\frac{\ln P}{\sqrt{x}},\quad g'(x)\ll\frac{\ln P}{x^{3/2}},\quad |f^{(3)}(x)|\asymp P^2x^{-3}\quad\text{for}\quad 2k\leq x\leq P,$$

$$g_r(x)\ll\frac{\ln P}{x},\quad g_r'(x)\ll\frac{\ln P}{x^2},\quad |f_r^{(3)}(x)|\asymp P^2x^{-3}\quad\text{for}\quad 2r\leq x\leq P.$$

Estimating Σ_3,Σ_4 similarly to Σ_1,Σ_2, we finally get

$$\Sigma_3,\ \Sigma_4\ll T^{5/12}(\ln T)^3.$$

Theorem is proved.

Remark Of course, the exponent $5/12$ in our theorem is not the best one. It can be decreased by different ways. First of all, one can apply one-dimensional method of exponent pairs to the estimation of the sums $S_k(N)$, $S_r(N)$ defined above (see [11, Ch. 3]). Next, it is possible to apply two-dimensional method of exponent pairs to the sums Σ_j, $1 \leq j \leq 4$, to take into account the oscillation over both variables of summation (see [12–16]). Finally, one can try to derive an explicit formula of Atkinson type similar to well-known formula for the remainder $E(T)$ (see [17, Ch. 2]). Each of these approaches requires a long paper. Since our main purpose was only to demonstrate one more application of the functional equation for $Z'(t)$, we limited ourselves by the simplest estimates.

References

1. A.A. Karatsuba, S.M. Voronin, *The Riemann Zeta-Function* (Walter de Gruyter, Berlin/New-York, 1992)
2. A. Ivić, *The Riemann Zeta-Function. Theory and Applications*, 2nd edn. (Dover Publication, Mineola/New York, 2003)
3. J. Bourgain, N. Watt, Decoupling for perturbed cones and mean square of $|\zeta(\frac{1}{2} + it)|$. arXiv:1505.04161 [math.NT]
4. R.R. Hall, The behaviour of the Riemann zeta-function on the critical line. Mathematika **46**(2), 281–313 (1999)
5. C.L. Siegel, Über Riemanns Nachlaß zur analytischen Zahlentheorie. Quellen und Studien zur Geshichte der Mathematik. Astronomie und Physik **2**, 45–80 (1932); (see also: C.L. Siegel, Gesammelte ˙Abhandlungen, B. 1. K. Chandrasekharan, H. Maass (eds.), Springer, Berlin, 1966, 275–310)
6. H.M. Edwards, *Riemann's Zeta Function* (Dover Publication, Mineola/New York, 2001)
7. A.A. Karatsuba, On the distance between consecutive zeros of the Riemann zeta function that lie on the critical line. Trudy Mat. Inst. Steklov. **157**, 49–63 (1981), (Russian); Proc. Steklov Inst. Math. **157**, 51–66 (1983) (English)
8. A.A. Lavrik, Uniform approximations and zeros in short intervals of the derivatives of the Hardy's Z-function. Anal. Math. **17**(4), 257–279 (1991) (Russian)
9. M.A. Korolev, On Rieman-Siegel formula for the derivatives of Hardy's function. Algebra i Analiz, **29**(4), 53–81 (2017) (Russian); St. Petersburg Math. J. **29**(4), 581–601 (2018)
10. M.A. Korolev, On the integral of Hardy's function $Z(t)$. Izv. RAN. Ser. Mat. **72**(3), 19–68 (2008) (Russian); Izv. Math. **72**(3), 429–478 (2008) (English)
11. S.W. Graham, G. Kolesnik, *Van der Corput's Method of Exponential Sums*. Lecture Notes Series, vol. 126 (Cambridge University Press, Cambridge, 1991)
12. E.C. Titchmarsh, On Epstein's zeta-function. Proc. Lond. Math. Soc. **36**(2), 485–500 (1934)
13. E.C. Titchmarsh, The lattice-points in a circle. Proc. Lond. Math. Soc. **38**(2), 96–115 (1934)
14. B.R. Srinivasan, The lattice point problem of many-dimensional hyperboloids II. Acta Arith. **8**(2), 173–204 (1963)
15. B.R. Srinivasan, The lattice point problem of many-dimensional hyperboloids III. Math. Ann. **16**, 280–311 (1965)
16. G. Kolesnik, On the method of exponential pairs. Acta Arith. **45**(2), 115–143 (1985)
17. A. Ivić, *Lectures on Mean Values of the Remann Zeta Function* (Tata Institute of Fundamental Research, Bombay, 1991)

Dedekind and Hardy Type Sums and Trigonometric Sums Induced by Quadrature Formulas

Gradimir V. Milovanović and Yilmaz Simsek

Abstract The Dedekind and Hardy sums and several their generalizations, as well as the trigonometric sums obtained from the quadrature formulas with the highest (algebraic or trigonometric) degree of exactness are studied. Beside some typical trigonometric sums mentioned in the introductory section, the Lambert and Eisenstein series are introduced and some remarks and observations for Eisenstein series are given. Special attention is dedicated to Dedekind and Hardy sums, as well as to Dedekind type Daehee-Changhee (DC) sums and their trigonometric representations and connections with some special functions. Also, the reciprocity law of the previous mentioned sums is studied. Finally, the trigonometric sums obtained from Gauss-Chebyshev quadrature formulas, as well as ones obtained from the so-called trigonometric quadrature rules, are considered.

Keywords Trigonometric sums · Dedekind sums · Hardy sums · Eisenstein series · Gauss-Chebyshev quadrature sums · Degree of exactness

1 Introduction and Preliminaries

Trigonometric sums play very important role in many various branches of mathematics (number theory, approximation theory, numerical analysis, Fourier analysis, etc.), physics, as well as in other computational and applied sciences. Inequalities with trigonometric sums, in particular their positivity and monotonicity are also important in many subjects (for details see [63, Chap. 4] and [66]).

G. V. Milovanović (✉)
The Serbian Academy of Sciences and Arts, Belgrade, Serbia
Faculty of Sciences and Mathematics, University of Niš, Niš, Serbia
e-mail: gvm@mi.sanu.ac.rs

Y. Simsek
Faculty of Arts and Science, Department of Mathematics, Akdeniz University,
Antalya, Turkey
e-mail: ysimsek@akdeniz.edu.tr

© Springer Nature Switzerland AG 2020
A. Raigorodskii, M. T. Rassias (eds.), *Trigonometric Sums and Their Applications*,
https://doi.org/10.1007/978-3-030-37904-9_10

There are several trigonometric sums in the well-known books [70, 71], [42, pp. 36–40] and [47]. The famous Dedekind and Hardy sums and many generalized sums have also trigonometric representations. In this introduction we mention some typical trigonometric sums obtained lately.

In 2000 Cvijović and Klinowski [26] gave closed form of the finite cotangent sums

$$S_n(q; \xi) = \sum_{p=0}^{q-1} \cot^n \frac{(\xi + p)\pi}{q} \quad \text{and} \quad S_n^*(q) = \sum_{p=1}^{q-1} \cot^n \frac{p\pi}{q},$$

where n and q are positive integers ($q \geq 2$) and ξ is a non-integer real number. They obtained $S_n(q; \xi)$ in a determinant form, as well as the following differential recurrence relation

$$S_{n+2}(q; \xi) = -S_n(q; \xi) - \frac{q}{\pi(n+1)} \frac{d}{d\xi} S_{n+1}(q; \xi) \quad (n \geq 1),$$

where

$$S_1(q; \xi) = q \cot(\pi\xi),$$
$$S_2(q; \xi) = q^2[\cot^2(\pi\xi) + 1] - q,$$
$$S_3(q; \xi) = q^3[\cot^3(\pi\xi) + \cot(\pi\xi)] - q\cot(\pi\xi),$$

etc. Evidently, according to the properties of the cotangent function, $S_{2n+1}^*(q) = 0$, as well as

$$\sum_{p=1}^{q-1} \cot^{2n} \frac{p\pi}{2q} = \frac{1}{2} S_{2n}^*(2q) \quad \text{an} \quad \sum_{p=1}^{q} \cot^{2n} \frac{p\pi}{2q+1} = \frac{1}{2} S_{2n}^*(2q+1).$$

For example, $S_2^*(q) = (q^2 - 3q + 2)/3$, $S_4^*(q) = (q^4 - 20q^2 + 45q - 26)/45$, $S_6^*(q) = (2q^6 - 42q^4 + 483q^2 - 945q + 502)/945$, etc. In general, $S_{2n}^*(q)$ is a polynomial of degree $2n$ with rational coefficients [26] (see also [70, p. 646] for $n = 1$ and $n = 2$).

Using contour integrals and the Cauchy residue theorem, Cvijović and Srivastava [27] derived formulas for general family of secant sums

$$S_{2n}(q, r) = \sum_{\substack{p=0 \\ p \neq \frac{q}{2} \, (q \text{ is even})}}^{q-1} \cos\left(\frac{2rp\pi}{q}\right) \sec^{2n}\left(\frac{p\pi}{q}\right) \quad (r = 0, 1, \ldots, q-1),$$

when $n \in \mathbb{N}$ and $q \in \mathbb{N} \setminus \{1\}$, as well as for various special cases including ones for $r = 0$, i.e.,

$$S_{2n}(q) = \sum_{p=0}^{q-1} \sec^{2n}\left(\frac{p\pi}{q}\right).$$

They also obtained sums which were considered earlier by Chen [17] by using the method of generating functions. In the Appendix of [17], Chen gave tables of power sums of secant, cosecant, tangent and cotangent. Among various such trigonometric summation formulae, we mention only a few of them for tangent function:

$$\sum_{k=1}^{n-1} \tan^2\left(\frac{k\pi}{n}\right) = n(n-1),$$

$$\sum_{k=1}^{n-1} \tan^4\left(\frac{k\pi}{n}\right) = \frac{1}{3}n(n-1)(n^2+n-3),$$

$$\sum_{k=1}^{n-1} \tan^6\left(\frac{k\pi}{n}\right) = \frac{1}{15}n(n-1)(2n^4+2n^3-8n^2-8n+15).$$

Using their earlier method, Cvijović and Srivastava [28] obtained closed-form summation formulas for 12 general families of trigonometric sums of the form

$$\sum_{k=1}^{n-1}(\pm 1)^{k-1} f\left(\frac{2rk\pi}{n}\right) g\left(\frac{k\pi}{n}\right)^m \qquad (n \in \mathbb{N} \setminus \{1\}, \ r = 1, \ldots, n-1),$$

for different combinations of the functions $x \mapsto f(x)$ and $x \mapsto g(x)$ and different values (even and odd) of $m \in \mathbb{N}$. The first function can be $f(x) = \sin x$ or $f(x) = \cos x$, while the second one can be one of the functions $\cot x$, $\tan x$, $\sec x$, and $\csc(x)$. Such a family of cosecant sums. I.e.,

$$C_{2m}(n, r) = \sum_{k=1}^{n-1} \cos\left(\frac{2rk\pi}{n}\right) \csc^{2m}\left(\frac{k\pi}{n}\right),$$

where $m \in \mathbb{N}$, $n \in \mathbb{N} \setminus \{1\}$, and $r = 0, 1, \ldots, n-1$, was previously considered by Dowker [30]. All obtained formulas in [28] involve the higher-order Bernoulli polynomials (see also [23] and [24]).

In [32] da Fonseca, Glasser, and Kowalenko have considered the trigonometric sums of the form

$$C_{2m}(n) = \sum_{k=0}^{n-1} \cos^{2m}\left(\frac{k\pi}{n}\right) \quad \text{and} \quad S_{2m}(n) = \sum_{k=0}^{n-1} \sin^{2m}\left(\frac{k\pi}{n}\right),$$

and their extensions. In [31] da Fonseca and Kowalenko studied the sums of the form

$$\sum_{k=1}^{n} (-1)^k \cos^{2m}\left(\frac{k\pi}{2n+2}\right),$$

where n and m are arbitrary positive integers.

Recently da Fonseca, Glasser, and Kowalenko [33] have presented an elegant integral approach for computing the so-called Gardner-Fisher trigonometric inverse power sum

$$S_{m,2}(n) = \left(\frac{\pi}{2n}\right)^{2m} \sum_{k=1}^{n-1} \sec^{2m}\left(\frac{k\pi}{2n}\right), \quad n, m \in \mathbb{N}.$$

For example,

$$S_{1,2}(n) = \frac{\pi^2}{6}\left(1 - \frac{1}{n^2}\right) \quad \text{and} \quad S_{2,2}(n) = \frac{\pi^4}{90}\left(1 + \frac{5}{2n^2} - \frac{7}{2n^4}\right).$$

By using contour integrals and residues, similar results for secant and cosecant sums were also obtained by Grabner and Prodinger [41] in terms of Bernoulli numbers and central factorial numbers.

Recently Chu [21] has used the partial fraction decomposition method to get a general reciprocal theorem on trigonometric sums. Several interesting trigonometric reciprocities and summation formulae are derived as consequences.

In this chapter, we mainly give an overview of the Dedekind and Hardy sums and several their generalizations, as well as the trigonometric sums obtained from the quadrature formulas with the highest (algebraic or trigonometric) degree of exactness.

The chapter is organized as follows. In Section 2 we introduce Lambert and Eisenstein series and give some remarks and observations for Eisenstein series. Sections 3 and 4 are dedicated to Dedekind and Hardy sums. In Section 5 we consider the Dedekind type Daehee-Changhee (DC) sums. Their trigonometric representations and connections with some special functions are presented in Sections 6 and 7, respectively. The Section 8 is devoted to the reciprocity law of the previous mentioned sums. Finally, in Sections 9 and 10 we consider trigonometric sums obtained from Gauss-Chebyshev quadrature formulas, as well as ones obtained from the so-called trigonometric quadrature rules.

2 Lambert and Eisenstein Series

Lambert series $G_p(x)$ is defined by

$$G_p(x) = \sum_{m=1}^{\infty} m^{-p} \frac{x^m}{1 - x^m} = \sum_{m,n=1}^{\infty} m^{-p} x^{mn},$$

where $p \geq 1$. These functions are regular for $|x| < 1$. The special case $p = 1$ gives

$$G_1(x) = -\log \prod_{m=1}^{\infty} (1 - x^m).$$

For odd integer values of p, Apostol [2] gave the behavior of these functions in the neighborhood of singularities, using a technique developed by Rademacher [72] in treating the case $p = 1$.

The following series

$$\sum_{(m,n) \in \mathbb{Z}^2} (m + nz)^{-s},$$

for $\operatorname{Im} z > 0$ and $\operatorname{Re} s > 2$, has an analytic continuation to all values of s. In the paper [56] by Lewittes, it is well known this series has transformation formulae for the analytic continuation of very large class of the Eisenstein series. These transformation formulae are related to large class of functions which generalized the case of the Dedekind eta-function, which is given as follows:

Let $z = x + iy$ and $s = \sigma + it$ with x, y, σ, t be real. For any complex number w, branch of $\log w$ with $-\pi \leq \arg w < \pi$. Let

$$V(z) = \frac{az + b}{cz + d}$$

be an arbitrary modular transformation. Let $\mathbb{H}$ denote the upper half-plane,

$$\mathbb{H} = \{z : \operatorname{Im}(z) > 0\}.$$

For $z \in \mathbb{H}$ and $\sigma > 2$, the Eisenstein series, $G(z, s, r_1, r_2)$ is defined by

$$G(z, s, r, h) = \sum_{r \neq (m,n) \in \mathbb{Z}^2} \frac{e^{2\pi i(mh_1 + nh_2)}}{((m + r_1) z + n + r_2)^s}, \tag{1}$$

where $r_1, r_2 \in \mathbb{R}$.

Substituting $r_1 = r_2 = 0$ into Eq. (1), we have

$$G(z, s) = \sum_{r \neq (m,n) \in \mathbb{Z}^2} \frac{1}{(m + nz)^s} \tag{2}$$

(for details see [55, 56]). Let r_1 and r_2 be arbitrary real numbers. For $z \in \mathbb{H}$ and arbitrary s, generalization of Dedekind's eta-function is given by

$$A(z, s, r_1, r_2) = \sum_{m > -r_1} \sum_{k=1}^{\infty} k^{s-1} e^{2\pi i k r_2 + 2\pi i k(m+r_1)z}.$$

For a real and $\sigma > 1$, Lewittes [55, 56] define $\zeta(s, a)$ by

$$\zeta(s, a) = \sum_{n > -a} (n + a)^{-s}.$$

Observe that

$$\zeta(s, a) = \zeta(s, \{a\} + \chi(a)),$$

where $\{a\}$ denotes the fractional part of a, and $\chi(a)$ denotes the characteristic function of integers. Since $0 < \chi(a) + \{a\} \le 1$, $\zeta(s, \{a\} + \chi(a))$ denotes the classical Hurwitz zeta-function. Lewittes ([56, Eq-(18)]) showed a connection between $G(z, s, r_1, r_2)$ and $A(z, s, r_1, r_2)$ as follows

$$G(z, s, r_1, r_2) = \chi(r_1) \left(\zeta(s, r_2) + e^{\pi i s} \zeta(s, -r_2) \right)$$
$$+ \frac{(-2\pi i)^s}{\Gamma(s)} \left(A(z, s, r_1, r_2) + e^{\pi i s} A(z, s, -r_1, -r_2) \right).$$

The above equation was proved by Berndt [10]. He proved transformation formula under modular substitutions which is derived for very large class of generalized Eisenstein series. Berndt's results easily converted into a transformation formula for a large class of functions that includes and generalizes the Dedekind eta-function and the Dedekind sums.

A transformation formula of the function $A(z, s, r_1, r_2)$ is given by Apostol [2] as follows: Let $m (> 0)$ is an even integer. Then

$$(cz + d)^m A(V(z), -m) = A(z, -m) + \frac{1}{2} \zeta(m + 1) \left(1 - (cz + d)^m \right)$$
$$+ \frac{(2\pi i)^{m+1}}{2(m+2)!} \sum_{j=1}^{c} \sum_{k=0}^{m+2} \frac{\binom{m+2}{k} B_k(\frac{j}{c}) \overline{B}_{m+2-k}(\frac{jd}{c})}{(-(cz+d))^{1-k}}. \tag{3}$$

However, due to a miscalculation of residue, the term $\frac{1}{2} \zeta(m + 1)(1 - (cz + d)^m)$ was omitted. The result was also misstated by Carlitz [16]. The proof of this transformation is also given by Lewittes [56] and after that by Berndt [10]. A special value of the function $A(z, s, r_1, r_2)$ is given by

$$\log \eta(z) = \frac{\pi i z}{12} - A(z).$$

Hence, the transformation formula for $A(z)$ is given as follows (cf. [5, 44, 53, 54]):

Theorem 1 *For $z \in \mathbb{H}$ we have*

$$\eta\left(-\frac{1}{z}\right) = \sqrt{(-iz)}\,\eta(z).$$

2.1 Further Remarks and Observations for Eisenstein Series

Now we give some standard results about Eisenstein series.

For $2 \leq k \in \mathbb{N}$ and $z \in \mathbb{H}$,

$$\sum_{m \in \mathbb{Z}} \frac{1}{(z+m)^k} = \frac{(-2\pi i)^k}{(k-1)!} \sum_{n=1}^{\infty} n^{k-1} e^{2\pi i n z}$$

is the known Lipschitz formula.

Apostol-Eisenstein series are given as follows:

If $2 \leq k \in \mathbb{N}$ and $z \in \mathbb{H}$, the Eisenstein series $G(z, 2k)$ is defined by

$$G(z, 2k) = \sum_{0 \neq (m,n) \in \mathbb{Z}^2} \frac{1}{(mz+n)^{2k}}. \tag{4}$$

It converges absolutely and has the Fourier expansion

$$G(z, 2k) = 2\zeta(2k) + \frac{2(2\pi i)^{2k}}{(2k-1)!} \sum_{n=1}^{\infty} \sigma_{2k-1}(n) e^{2\pi i n z}$$

where, as usual, $\sigma_c(n) = \sum_{d|n} d^c$ and $\zeta(z)$ denotes Riemann zeta function.

For $k = 1$, the series in (4); $G(z, 2)$ is no longer absolutely convergent. $G(z, 2)$ is an even function,

$$G(z, 2) = 2\zeta(2) + 2(2\pi i)^2 \sum_{n=1}^{\infty} \sigma(n) e^{2\pi i n z} \tag{5}$$

for $z \in \mathbb{H}$.

For $x = e^{2\pi i z}$ the series in (5) is an absolutely convergent power series for $|x| < 1$ so that $G(z, 2)$ is analytic in $\mathbb{H}$. The behavior of $G(z, 2)$ under the modular group is given by (cf. [5])

$$G\left(-\frac{1}{z}, 2\right) = z^2 G\left(z, 2\right) - 2\pi i z.$$

The well-known Lipschitz formula is given by the following lemma:

Lemma 1 (Lipschitz formula) *Let* $2 \le k \in \mathbb{N}$ *and* $z \in \mathbb{H}$. *Then*

$$\sum_{m \in \mathbb{Z}} \frac{1}{(z+m)^k} = \frac{(-2\pi i)^k}{(k-1)!} \sum_{n=1}^{\infty} n^{k-1} e^{2\pi i n z}.$$

By using this lemma, the Fourier expansion of the Eisenstein series is given by:

Theorem 2 *If* k *is an integer with* $k \ge 2$ *and* $z \in \mathbb{H}$, *then*

$$G\left(z, k\right) = 2\zeta(k) + \frac{2(-2\pi i)^k}{(k-1)!} \sum_{m=1}^{\infty} \sum_{n=1}^{\infty} n^{k-1} e^{2\pi i n m z}.$$

Proof We give only brief sketch of the proof since the method is well-known. Now replacing z by az, where $a > 0$, substituting in Lemma 1 and summing over all $a \ge 1$, we get

$$\sum_{a=1}^{\infty} \sum_{m \in \mathbb{Z}} \frac{1}{(az+m)^k} = \frac{(-2\pi i)^k}{(k-1)!} \sum_{a,n=1}^{\infty} n^{k-1} e^{2\pi i n a z}.$$

We rearrange the terms right member of the above equation, we have

$$\frac{1}{2} \sum_{0 \ne a \in \mathbb{Z}} \sum_{m \in \mathbb{Z}} \frac{1}{(az+m)^k} - 2\zeta(k) = \frac{(-2\pi i)^k}{(k-1)!} \sum_{a,n=1}^{\infty} n^{k-1} e^{2\pi i n a z}.$$

After a further little rearrange and use of (2) we obtain the desired result.

Remark 1 Putting $r_1 = r_2 = 0$, we have

$$G\left(z, s, 0, 0\right) = \chi(0)\left(\zeta(s, 0) + e^{\pi i s} \zeta(s, 0)\right)$$

$$+ \frac{(-2\pi i)^s}{\Gamma(s)}\left(A(z, s, 0, 0) + e^{\pi i s} A(z, s, 0, 0)\right).$$

Replacing s by k (k is an integer with $k > 1$) in the above, we have

$$G\left(z, k\right) = 2\zeta(k) + \frac{2(2\pi i)^k}{(k-1)!} A(z, k).$$

After a number of straightforward calculations we arrive at the desired result.

The Fourier expansion of the function $G(z, k, r, h)$ is given by:

Corollary 1 *Let* $2 \leq k \in \mathbb{N}$, r, h *be rational numbers and* $z \in \mathbb{H}$. *Then*

$$G(z, k, r, h) = 2\zeta(k, h) + \frac{2(-2\pi i)^k}{(k-1)!} \sum_{a,n=1}^{\infty} a^{k-1} e^{2\pi i(n+r)az}.$$

3 Dedekind Sums

The history of the Dedekind sums can be traced back to famous German mathematician Julius Wilhelm Richard Dedekind (1831–1916), who did important work in abstract algebra in particularly including ring theory, algebraic and analytic number theory and the foundations of the real numbers. After Dedekind, Hans Adolph Rademacher (1892–1969), who was one of the most famous German mathematicians, worked the most deeply the Dedekind sums. Rademacher also studied important work in mathematical analysis and its applications and analytic number theory. It is well-known that, the Dedekind sums, named after Dedekind, are certain finite sums of products of a sawtooth function. The Dedekind sums are found in the functional equation that emerges from the action of the Dedekind eta function under modular groups. The Dedekind sums have occurred in analytic number theory, in some problems of topology and also in the other branches of Mathematics. Although two-dimensional Dedekind sums have been around since the nineteenth century and higher-dimensional Dedekind sums have been explored since the 1950s, it is only recently that such sums have figured flashily in so many different areas. The Dedekind sums have also many applications in some areas such as analytic number theory, modular forms, random numbers, the Riemann-Roch theorem, the Atiyah-Singer index theorem, and the family of zeta functions.

In many applications of elliptic modular functions to analytic number theory, and theory of elliptic curves, the Dedekind eta function plays a central role. It was introduced by Dedekind in 1877 by Dedekind. This function is defined on the upper helf-plane as follows:

$$\eta(\tau) = e^{\pi i \tau/12} \prod_{m=1}^{\infty} \left(1 - e^{2\pi i m \tau}\right),$$

The infinite product has the form $\prod_{n=1}^{\infty}(1 - x^n)$, where $x = e^{2\pi i \tau}$. If $\tau \in \mathbb{H}$, then $|x| < 1$ so the product converges absolutely and it is nonzero. Furthermore, since the convergence is uniform on compact subsets of $\mathbb{H}$, $\eta(\tau)$ is analytic on $\mathbb{H}$. The function $\eta(\tau)$ is related to analysis, number theory, combinatorics, q-series, Weierstrass elliptic functions, modular forms, Kronecker limit formula, etc.

The behavior of this function under the modular group $\Gamma(1)$, defined by

$$\Gamma(1) = \left\{ A = \begin{bmatrix} a & b \\ c & d \end{bmatrix} : ad - bc = 1, \ a, b, c, d \in \mathbb{Z} \right\},$$

we note that

$$Az = \frac{az + b}{cz + d}.$$

It is well-known that the Dedekind sums $s(h, k)$ first arose in the transformation formula of the logarithm of the Dedekind-eta function which is given by Apostol,

$$\log \eta(Az) = \log \eta(z) + \frac{\pi i(a + d)}{12c} - \pi i \left(s(d, c) - \frac{1}{4} \right) + \frac{1}{2} \log(cz + d),$$

where $z \in \mathbb{H}$ and $s(d, c)$ denotes the Dedekind sum which defined by

$$s(d, c) = \sum_{\mu \bmod c} \left(\left(\frac{\mu}{c} \right) \right) \left(\left(\frac{d\mu}{c} \right) \right),$$

where $(d, c) = 1, c > 0$, and

$$((x)) = \begin{cases} x - [x] - \frac{1}{2}, & x \notin \mathbb{Z}, \\ 0, & \text{otherwise}, \end{cases}$$

where $[x]$ is the largest integer $\leq x$. The arithmetical function $((x))$ has a period 1 and can thus be expressed by a Fourier series as follows:

$$((x)) = -\frac{1}{\pi} \sum_{n=1}^{\infty} \frac{\sin(2\pi nx)}{n}.$$

For basic properties of the Dedekind sums see monograph of Rademacher and Grosswald [76].

The most important property of Dedekind sums is the reciprocity law. Namely, if $(h, k) = 1$ and h and k are positive, then

$$s(h, k) + s(k, h) = \frac{1}{12} \left(\frac{h}{k} + \frac{k}{h} + \frac{1}{hk} \right) - \frac{1}{4}$$

(cf. [2, 100]). This will be discussed in more detail in a separate section.

Apostol [2] defined the generalized Dedekind sums $s_p(h, k)$ as

$$s_p(h, k) = \sum_{n=0}^{k-1} \frac{n}{k} \overline{B}_p\left(\frac{nh}{k}\right),$$

where $\overline{B}_p(x)$ is the p-th Bernoulli function defined by

$$\overline{B}_n(x) = B_n(x - [x]),$$

where $B_n(x)$ denotes the Bernoulli polynomial. These important polynomials are defined by the following generating function

$$\frac{t}{e^t - 1} e^{xt} = \sum_{n=0}^{\infty} B_n(x) \frac{t^n}{n!}.$$

For $x = 0$ these polynomials reduce to the well-known Bernoulli numbers $B_n = B_n(0)$ (cf. [72]). A few first numbers are $1, -1/2, 1/6, 0, -1/30, 0, 1/42, \ldots$.
The functions $\overline{B}_n(x)$ are 1-periodic, and they satisfy

$$\overline{B}_n(x) = B_n(x)$$

for $0 \le x < 1$, and

$$\overline{B}_n(x + 1) = \overline{B}_n(x)$$

for other real x. The Bernoulli function can be expressed by the following Fourier expansion

$$\overline{B}_n(x) = -\frac{n!}{(2\pi i)^n} \sum_{0 \ne m \in \mathbb{Z}} \frac{1}{m^p} e^{2\pi i m x} \tag{6}$$

Observe that $s_1(h, k) = s(h, k)$. A representation of $s_p(h, k)$ as an infinite series has also given by Apostol [2]. Namely, for odd $p \ge 1$, $(h, k) = 1$ as

$$s_p(h, k) = \frac{p!}{(2\pi i)^p} \sum_{\substack{m \in \mathbb{N} \\ m \not\equiv 0 \,(\mathrm{mod}\ k)}} \frac{1}{m^p} \left(\frac{e^{2\pi i m h/k}}{1 - e^{2\pi i m h/k}} - \frac{e^{2\pi i m h/k}}{1 - e^{2\pi i m h/k}}\right). \tag{7}$$

The relation between Dedekind sums $s(h, k)$ and $\cot \pi x$ are given in the lemma below. This lemma is a special case of (7). The following well-known result is easily given:

$$s(h, k) = \frac{1}{4k} \sum_{m=1}^{k-1} \cot\left(\frac{mh\pi}{k}\right) \cot\left(\frac{m\pi}{k}\right).$$

Recently, many authors proved the above nice formulas by different methods [4, 11, 12, 14, 16, 29, 90, 101].

Using contour integration and Cauchy Residue Theorem, Berndt [11] proved the following result:

Lemma 2 *Let $h, k \in \mathbb{N}$ with $(h, k) = 1$. Then*

$$s(h, k) = \frac{1}{2\pi} \sum_{\substack{m \in \mathbb{N} \\ m \not\equiv 0 \,(\mathrm{mod}\, k)}} \frac{1}{m} \cot\left(\frac{mh\pi}{k}\right).$$

The sums $s_p(h, k)$ are related to the Lambert series $G_p(x)$ in the same way that $s(h, k)$ is related to $\eta(z)$, $\log \eta(z)$ being the same as $(\pi i z/12) - G_1\left(e^{2\pi i z}\right)$, respectively. The sums $s_p(h, k)$ are expressible as infinite series related to certain Lambert series and, for odd $p \geq 1$, $s_p(h, k)$ is also seen to be the Abel sum of a divergent series. This relation is given as follows:

Theorem 3 ([85]) *For $(h, k) = 1$, the Abel sum of the divergent series*

$$\sum_{n=1}^{\infty} \sigma_p(n)\, n^{-p} \sin\left(\frac{2\pi n h}{k}\right)$$

for odd p is given by

$$(-1 \mid p)\, (2\pi)^p\, (2p!)^{-1}\, s_p(h, k),$$

where $\sigma_p(n) = \sum_{d \mid n} d^p$.

Apostol [2] gave a proof of this theorem using a contour integral representation of the Lambert series $G_p(x)$, but his proof is very different from that given below (see [85]).

Brief sketch of the proof of Theorem 3 Starting with (6), replacing x by nx ($n \in \mathbb{N}$) and summing over n we get (cf. [85])

$$\sum_{n=1}^{\infty} \overline{B}_p(nx) = -\frac{p!}{(2\pi i)^p} \sum_{0 \neq a \in \mathbb{Z}} \sum_{n=1}^{\infty} \frac{1}{m^p} e^{2\pi imnx}.$$

If we rearrange the above equation, we get

$$\sum_{n=1}^{\infty} \overline{B}_p(nx) = -\frac{p!}{(2\pi i)^p} \left(\sum_{n=1}^{\infty} \sum_{m=-\infty}^{-1} \frac{1}{m^p} e^{2\pi imnx} + \sum_{m,n=1}^{\infty} \frac{1}{m^p} e^{2\pi imnx} \right).$$

After a little calculation, we easily obtain

$$\sum_{n=1}^{\infty} \overline{B}_p(nx) = -\frac{p!}{(2\pi i)^p} \left(\sum_{m,n=1}^{\infty} \frac{1}{m^p} e^{2\pi i m n x} - \sum_{m,n=1}^{\infty} \frac{1}{m^p} e^{-2\pi i m n x} \right). \tag{8}$$

Because of the identity $2i \sin z = e^{iz} - e^{-iz}$ and putting $x = a/b$ in (8), where $a, b \in \mathbb{Z}$ with $(a, b) = 1$, and writing the Lambert series as a power series $G_p(x) = \sum_{n=1}^{\infty} \sigma_p(n) n^{-p} x^n$, we get (cf. [85])

$$\sum_{n=1}^{\infty} \overline{B}_p\left(\frac{na}{b}\right) = -\frac{p!}{(2\pi i)^p} \sum_{m=1}^{\infty} m^{-p} \sigma_p(m) \sin\left(\frac{2\pi m a}{b}\right). \tag{9}$$

Using a definition of the Lambert series $G_p(x) = \sum_{n=1}^{\infty} n^{-p} x^n / (1 - x^n)$ and replacing x by a/b, with $(a, b) = 1$, in (8), we get

$$\sum_{n=1}^{\infty} \overline{B}_p\left(\frac{na}{b}\right) = -\frac{p!}{(2\pi i)^p} \sum_{\substack{m \in \mathbb{N} \\ m \not\equiv 0 \pmod k}} \frac{1}{m^p} \left(\frac{e^{2\pi i m h/k}}{1 - e^{2\pi i m h/k}} - \frac{e^{2\pi i m h/k}}{1 - e^{2\pi i m h/k}} \right).$$

By substituting (7) into the above, we obtain

$$\sum_{n=1}^{\infty} \overline{B}_p\left(\frac{na}{b}\right) = -s_p(a, b). \tag{10}$$

For odd p we get

$$(i)^{1-p} = (-1)^{(1-p)/2} = (-1 \mid p), \tag{11}$$

which is known as Jacobi (Legendre) symbol. Finally, combining (9)–(11), we find the desired result.

Zagier [101] defined the following multiple Dedekind sums

$$d(p; a_1, a_2, \ldots, a_j) = (-1)^{j/2} \sum_{m=1}^{p-1} \cot\left(\frac{\pi m a_1}{p}\right) \cot\left(\frac{\pi m a_2}{p}\right) \cdots \cot\left(\frac{\pi m a_j}{p}\right).$$

The sum $d(p; a_1, a_2, \ldots, a_j)$ vanishes identically when j is odd. In [90], Simsek, Kim and Koo gave various formulas for the above sums and finite trigonometric sums.

3.1   Some Others Formulas for the Dedekind Sums

Theorem 4 ([85]) *Let $a, b \in \mathbb{Z}$ with $(a, b) = 1$ and let p be odd integers. Then*

$$s_p(a, b) = \frac{2p!}{(2\pi b)^p} (-1 \mid p) \sum_{n=1}^{\infty} \sum_{m=1}^{b-1} \sin\left(\frac{2\pi mna}{b}\right) \zeta\left(p, \frac{m}{b}\right),$$

where $\zeta(p, m/b)$ is the Hurwitz zeta function.

Proof By substituting $x = a/b$ into (8), we easily calculate

$$\sum_{n=1}^{\infty} \overline{B}_p\left(\frac{na}{b}\right) = \frac{2p!}{(2\pi b)^p} \sum_{m,n=1}^{\infty} \frac{1}{m^p} \sin\left(\frac{2\pi nma}{b}\right).$$

Writing $m = ub + c$, with $u = 0, 1, 3, \ldots$ and $c = 1, 2, \ldots, b - 1$ in the above, we obtain

$$\sum_{n=1}^{\infty} \overline{B}_p\left(\frac{na}{b}\right) = -\frac{2p!}{(2\pi b)^p} (-1 \mid p) \sum_{n=1}^{\infty} \sum_{c=1}^{b-1} \zeta\left(p, \frac{c}{b}\right) \sin\left(\frac{2\pi na}{b}\right), \qquad (12)$$

where we assume $p > 1$ in order to insure that the series involved should be absolutely convergent and the rearragements valid. Now, after combining Eqs. (10) and (12), the proof is completed.

Lemma 2, as well as corresponding expression for $s_p(h, k)$ in (7), can be obtained without any knowledge of the function η and the finite sum $\sum_{n=1}^{k-1} nx^n$. By using the well-known equality

$$\cot \pi x = -i\left(\frac{e^{2i\pi x}}{1 - e^{2i\pi x}} - \frac{e^{-2i\pi x}}{1 - e^{-2i\pi x}}\right),$$

Theorems 3 and 4, a relation between $s_p(h, k)$ and $\cot(an\pi/b)$ can be obtained as follows:

Theorem 5 ([85]) *Let $(h, k) = 1$. For odd $p \geq 1$ we have*

$$s_p(h, k) = i\frac{p!}{(2\pi i)^p} \sum_{\substack{n \in \mathbb{N} \\ n \not\equiv 0 \,(\mathrm{mod}\, k)}} \frac{1}{n^p} \cot\left(\frac{\pi nh}{k}\right).$$

Proof By substituting (6) in to definition of $s_p(h, k)$, we have

$$s_p\,(h,k) = \sum_{n=1}^{k-1} \frac{n}{k} \overline{B}_p\left(\frac{nh}{k}\right)$$

$$= -\frac{p!}{k\,(2\pi i)^p} \sum_{n=1}^{k-1} n \sum_{0 \neq m \in \mathbb{Z}} \frac{1}{m^p} e^{\frac{2\pi i m n h}{k}}$$

$$= -\frac{p!}{k\,(2\pi i)^p} \sum_{n=1}^{k-1} n \left(\sum_{m=1}^{\infty} \frac{1}{m^p} e^{\frac{2\pi i m n h}{k}} + \sum_{-\infty}^{-1} \frac{1}{m^p} e^{\frac{2\pi i m n h}{k}}\right)$$

$$= -\frac{p!}{k\,(2\pi i)^p} \sum_{n=1}^{k-1} n \left(\sum_{m=1}^{\infty} \frac{1}{m^p} \left(e^{\frac{2\pi i m n h}{k}} - e^{-\frac{2\pi i m n h}{k}}\right)\right).$$

By applying the well-known identity $2 i \sin x = e^{ix} - e^{-ix}$ in the above, we obtain

$$s_p\,(h,k) = -2i\frac{p!}{k\,(2\pi i)^p} \sum_{n=1}^{k-1} n \sum_{m=1}^{\infty} \frac{1}{m^p} \sin\left(\frac{2\pi m n h}{k}\right).$$

Now, by using the following well-known identity

$$\sum_{a \bmod k} a \sin\left(\frac{2\pi n \phi}{k}\right) = -\frac{k}{2} \cot\left(\frac{\pi \phi}{k}\right),$$

where $k \nmid \phi$, $\phi \in \mathbb{Z}$, then we have the desired result.

By using Theorem 5, we arrive at the following result:

Corollary 2 ([85]) *For odd $p > 1$ we have*

$$s_p\,(a,b) = i\frac{p!}{(2\pi i)^p} \sum_{n=1}^{b-1} \cot\left(\frac{\pi n a}{b}\right) \zeta\left(p, \frac{n}{b}\right). \tag{13}$$

The proof this corollary was proved by Apostol [4].

Using a technique developed by Rademacher (Theorem 2, Eq. (5) in [10]) and Lewittes (Eq. (56) in [4]), we can give the behavior of Lambert series and Dedekind eta-function. Namely, substituting $r_1 = r_2 = s = 0$ into Eq. (3) (and also Eq. (5) in [10]), we have (cf. [85])

$$A\,(V\,(z)) = A\,(z) + \frac{\pi i}{4} - \frac{1}{2} \log\,(cz + d) + \pi i s\,(d,c) - \pi i \left(\frac{a+d}{12c}\right).$$

By the definition of $G_1\left(e^{\pi i z}\right)$ and $A\,(z)$, we get (cf. [85])

$$G_1\left(e^{\pi i V(z)}\right) = G_1\left(e^{\pi i z}\right) + \frac{\pi i}{4} - \frac{1}{2}\log\left(cz+d\right) + \pi i s\,(d,c) - \pi i\left(\frac{a+d}{12c}\right). \tag{14}$$

This relation gives modular transformation of $G_1\left(e^{\pi i z}\right)$. Replacing $V(z) = \frac{az+b}{cz+d}$, by $W(z) = \frac{bz-a}{dz-c}$, where $c, d > 0$ in (14), we have

$$G_1\left(e^{\pi i W(z)}\right) = G_1\left(e^{\pi i z}\right) + \frac{\pi i}{4} - \frac{1}{2}\log\left(dz-c\right) + \pi i s\,(-c,d) - \pi i\left(\frac{b-c}{12d}\right). \tag{15}$$

Comparing (14) with (15) and using reciprocity law of Dedekind sums

$$12s\,(d,c) + 12s\,(c,d) = -3 + \frac{d}{c} + \frac{c}{d} + \frac{1}{dc}, \quad (d,c) = 1,$$

we deduce that

$$G_1\left(e^{\pi i V(z)}\right) - G_1\left(e^{\pi i W(z)}\right) = \frac{1}{2}\log\left(\frac{dz-c}{cz+d}\right) + \frac{\pi i\,(cb - da - 3dc + 1)}{12dc}.$$

Thus, we arrive at the following results (see [85]).

Theorem 6 *Let* $V(z) = \frac{az+b}{cz+d}$ *and* $W(z) = \frac{bz-a}{dz-c}$ *be arbitrary modular transformations, with* $c > 0$, *and let*

$$\mathbb{K} = \left\{z : \mathrm{Re}\,(z) > -\frac{d}{c}, \quad \mathrm{Im}\,(z) > 0\right\}.$$

Then, for $z \in \mathbb{K}$ *we have*

$$\sum_{m,n=1}^{\infty} \frac{1}{m}\left(e^{2\pi i n m V(z)} - e^{2\pi i n m W(z)}\right) = \frac{1}{2}\log\left(\frac{dz-c}{cz+d}\right) + \frac{\pi i\,(cb - da - 3dc + 1)}{12dc}.$$

Theorem 7 *Let* $m > 0$ *be an even integer and* $(a,b) = 1$. *We have*

$$s_{m+1}\,(a,b) = \frac{2\,(m+1)!}{\pi^2\,(4\pi i)^m} \sum_{\substack{n \in \mathbb{N} \\ n \not\equiv 0\,(\mathrm{mod}\,b)}} \frac{\lim_{s \to 1}\left(\zeta\left(s, \frac{na}{b}\right) - \zeta\left(s, 1 - \frac{na}{b}\right)\right)}{n^{1+m}}.$$

4   Hardy Sums

Hardy's sums are derived from theta function. Thus, the well known theta-functions, $\vartheta_n(0,q)(n = 2, 3, 4)$ related to infinite products are given by

$$\vartheta_2(0, q) = 2q^{1/2} \prod_{n=1}^{\infty} (1 - q^{2n})(1 + q^{2n})^2,$$

$$\vartheta_3(0, q) = \prod_{n=1}^{\infty} (1 - q^{2n})(1 + q^{2n-1})^2,$$

$$\vartheta_4(0, q) = \prod_{n=1}^{\infty} (1 - q^{2n})(1 - q^{2n-1})^2.$$

In the sequel we denote $\vartheta_2(0, q)$, $\vartheta_3(0, q)$ and $\vartheta_4(0, q)$ as $\vartheta_2(z)$, $\vartheta_3(z)$ and $\vartheta_4(z)$, respectively, where $q = e^{\pi i z}$. The relations between theta-functions and Dedekind eta-function are defined by

$$\vartheta_3(z) = \frac{\eta^5(z)}{\eta^2(2z)\eta^2(z/2)}.$$

These relations, as well as others, are studied by Rademacher [72] (see also [84], as well as the books [99] and [82]).

Let h and k with $k > 0$ be relatively prime integers (i.e., $(h, k) = 1$). The Hardy sums are defined by (see [48])

$$S(h, k) = \sum_{j=1}^{k-1} (-1)^{j+1+[hj/k]},$$

$$s_1(h, k) = \sum_{j=1}^{k} (-1)^{[hj/k]} \left(\left(\frac{j}{k}\right)\right),$$

$$s_2(h, k) = \sum_{j=1}^{k} (-1)^{j} \left(\left(\frac{j}{k}\right)\right)\left(\left(\frac{hj}{k}\right)\right),$$

$$s_3(h, k) = \sum_{j=1}^{k} (-1)^{j} \left(\left(\frac{hj}{k}\right)\right),$$

$$s_4(h, k) = \sum_{j=1}^{k-1} (-1)^{[hj/k]},$$

$$s_5(h, k) = \sum_{j=1}^{k} (-1)^{j+[hj/k]} \left(\left(\frac{j}{k}\right)\right).$$

By using the following well-known trigonometric formulas, we mention some relations including the Hardy sums (for details see [40]).

If $2m - 1 \not\equiv 0 \,(\mathrm{mod}\ k)$, $(h, k) = 1$, then

$$\sum_{j=1}^{k-1} j \sin\left(\frac{\pi(2m-1)hj}{k}\right) = -\frac{k}{2}\cot\left(\frac{\pi h(2m-1)}{2k}\right).$$

If $m = (2n - 1)h$, $2n - 1 \not\equiv 0 \,(\mathrm{mod}\ k)$, and h and k are of opposite parity, an elementary calculation gives

$$\sum_{j=1}^{k-1}(-1)^j \sin\left(\frac{\pi mj}{k}\right) = -\tan\left(\frac{\pi m}{2k}\right).$$

If $2m \not\equiv 0 \,(\mathrm{mod}\ k)$, and h and k are of opposite parity, an elementary calculation gives

$$\sum_{j=1}^{k-1}(-1)^j j \sin\left(\frac{\pi mj}{k}\right) = \frac{k}{2}\tan\left(\frac{\pi m}{2k}\right).$$

If k is odd, then

$$\sum_{j=1}^{k-1}(-1)^j \sin\left(\frac{2h\pi mj}{k}\right) = -\tan\left(\frac{\pi hm}{k}\right);$$

if k is even, then

$$\sum_{j=1}^{k-1} \sin\left(\frac{2h\pi mj}{k}\right) = 0.$$

Also (cf. [15])

$$\sum_{j=1}^{k-1}\cot^2\left(\frac{\pi j}{k}\right) = \frac{(k-1)(k-3)}{3}.$$

The Fourier series of the function $f(x) = (-1)^{[x]}$ is given by (cf. [40])

$$f(x) = \frac{1}{2\pi}\sum_{n=1}^{\infty}\frac{\sin((2n-1)x\pi)}{2n-1}.$$

Combining the above finite trigonometric sums and Fourier series of the function $f(x) = (-1)^{[x]}$ with definitions of the Hardy sums, some relations between Hardy sums and trigonometric functions can be given. Such results were obtained by Goldberg [40] and Berndt and Goldberg [14]:

Theorem 8 *Let h and k denote relatively prime integers with k > 0.*

1° *If h + k is odd, then*

$$S(h, k) = \frac{4}{\pi} \sum_{n=1}^{\infty} \frac{1}{2n - 1} \tan\left(\frac{\pi h(2n - 1)}{2k}\right); \tag{16}$$

2° *If h is even and k is odd, then*

$$s_1(h, k) = -\frac{2}{\pi} \sum_{\substack{n = 1 \\ 2n - 1 \not\equiv 0 \,(\mathrm{mod}\, k)}}^{\infty} \frac{1}{2n - 1} \cot\left(\frac{\pi h(2n - 1)}{2k}\right); \tag{17}$$

3° *If h is odd and k is even, then*

$$s_2(h, k) = -\frac{1}{2\pi} \sum_{\substack{n = 1 \\ 2n \not\equiv 0 \,(\mathrm{mod}\, k)}}^{\infty} \frac{1}{n} \tan\left(\frac{\pi h n}{k}\right); \tag{18}$$

4° *If k is odd, then*

$$s_3(h, k) = \frac{1}{\pi} \sum_{n=1}^{\infty} \frac{1}{n} \tan\left(\frac{\pi h n}{k}\right); \tag{19}$$

5° *If h is odd, then*

$$s_4(h, k) = \frac{4}{\pi} \sum_{n=1}^{\infty} \frac{1}{2n - 1} \cot\left(\frac{\pi h(2n - 1)}{2k}\right); \tag{20}$$

6° *If h and k are odd, then*

$$s_5(h, k) = \frac{2}{\pi} \sum_{\substack{n = 1 \\ 2n - 1 \not\equiv 0 \,(\mathrm{mod}\, k)}}^{\infty} \frac{1}{2n - 1} \tan\left(\frac{\pi h(2n - 1)}{2k}\right). \tag{21}$$

Using the well-known sum

$$\sum_{n=-\infty}^{\infty} \frac{1}{n + y} = \pi \cot \pi y$$

in (16) through (21), the relations between Hardy sums and finite trigonometric sums can be also obtained [14, 40]:

Theorem 9 *Let h and k be coprime integers with k > 0.*

1° *If h + k is odd, then*

$$S(h, k) = \frac{1}{k} \sum_{m=1}^{k} \tan\left(\frac{\pi h(2m-1)}{2k}\right) \cot\left(\frac{\pi(2m-1)}{2k}\right);$$

2° *If h is even and k is odd, then*

$$s_1(h, k) = -\frac{1}{2k} \sum_{\substack{m=1 \\ m \neq (k+1)/2}}^{k} \cot\left(\frac{\pi h(2m-1)}{2k}\right) \cot\left(\frac{\pi(2m-1)}{2k}\right);$$

3° *If h is odd and k is even, then*

$$s_2(h, k) = -\frac{1}{4k} \sum_{\substack{m=1 \\ m \neq k/2}}^{k-1} \tan\left(\frac{\pi hm}{k}\right) \cot\left(\frac{\pi m}{k}\right);$$

4° *If k is odd, then*

$$s_3(h, k) = \frac{1}{2k} \sum_{m=1}^{k-1} \tan\left(\frac{\pi hm}{k}\right) \cot\left(\frac{\pi m}{k}\right);$$

5° *If h is odd, then*

$$s_4(h, k) = \frac{1}{k} \sum_{m=1}^{k} \cot\left(\frac{\pi h(2m-1)}{2k}\right) \cot\left(\frac{\pi(2m-1)}{2k}\right);$$

6° *If h and k are odd, then*

$$s_5(h, k) = \frac{1}{2k} \sum_{\substack{m=1 \\ m \neq (k+1)/2}}^{k} \tan\left(\frac{\pi h(2m-1)}{2k}\right) \cot\left(\frac{\pi(2m-1)}{2k}\right).$$

Using elementary methods, the previous identities were also obtained by Sitaramachandrarao [91]. Some new higher dimensional generalizations of the Dedekind sums associated with the Bernoulli functions, as well as ones of Hardy sums, have been recently introduced by Rassias and Tóth [77]. They derived the so-called Zagier-type identities for these higher dimensional sums, as well as a sequence of corollaries with interesting particular sums.

5 Dedekind Type Daehee-Changhee (DC) Sums

The first kind n-th Euler function $\overline{E}_m(x)$ is defined by

$$\overline{E}_n(x) = E_n(x)$$

for $0 \leq x < 1$, and by

$$\overline{E}_n(x+1) = -\overline{E}_n(x)$$

for other real x. This function can be expressed by the following Fourier expansion

$$\overline{E}_m(x) = \frac{2m!}{(\pi i)^{m+1}} \sum_{n=-\infty}^{\infty} \frac{e^{(2n-1)\pi i x}}{(2n-1)^{m+1}}, \tag{22}$$

where $m \in \mathbb{N}$ (for details on Euler polynomials and functions and their Fourier series see [1, 16, 49–51, 89, 94, 96]). Hoffman [49] studied the Fourier series of Euler polynomials and expressed the values of Euler polynomials at any rational argument in terms of $\tan x$ and $\sec x$. Suslov [96] considered explicit expansions of some elementary and q-functions in basic Fourier series of the q-extensions of the Bernoulli and Euler polynomials and numbers.

Observe that if $0 \leq x < 1$, then (22) reduces to the first kind n-th Euler polynomials $E_n(x)$ which are defined by means of the following generating function

$$\frac{2e^{tx}}{e^t + 1} = \sum_{n=0}^{\infty} E_n(x)\frac{t^n}{n!}, \quad |t| < \pi. \tag{23}$$

Observe that $E_n(0) = E_n$ denotes the first kind Euler number which is given by the following recurrence formula

$$E_0 = 1 \quad \text{and} \quad E_n = -\sum_{k=0}^{n} \binom{n}{k} E_k. \tag{24}$$

Some of them are given by $1, -1/2, 0, 1/4, \ldots,$ $E_n = 2^n E_n(1/2)$ and $E_{2n} = 0$ ($n \in \mathbb{N}$).

In [50] and [51], by using Fourier transform for the Euler function, Kim derived some formulae related to infinite series and the first kind Euler numbers. From (22) it is easy to see that (cf. [1, 49, 51, 89])

$$\sum_{n=1}^{\infty} \frac{1}{(2n-1)^{2m+2}} = \frac{(-1)^{m+1}\pi^{2m+2}E_{2m+1}}{4(2m+1)!}. \tag{25}$$

By using the first kind n-th Euler function and above infinite series, we can construct infinite series representation of the Dedekind type Daehee-Changhee-sum (DC-sum) and reciprocity law of this sum. We also can give relations between the Dedekind type DC-sum and some special functions.

The second kind Euler numbers, E_m^* are defined by means of the following generating functions (cf. [1, 75, 89])

$$\operatorname{sech} x = \frac{1}{\cosh x} = \frac{2e^x}{e^{2x}+1} = \sum_{n=0}^{\infty} E_n^* \frac{x^n}{n!}, \quad |x| < \frac{\pi}{2}. \tag{26}$$

Kim [51] studied the second kind Euler numbers and polynomials in details. By (23) and (26), it is easy to see that

$$E_m^* = \sum_{n=0}^{m} \binom{m}{n} 2^n E_n \quad \text{and} \quad E_{2m}^* = -\sum_{n=0}^{m-1} \binom{2m}{2n} E_{2n}^*.$$

From the above $E_0^* = 1$, $E_1^* = 0$, $E_2^* = -1$, $E_3^* = 0$, $E_4^* = 5, \ldots$, and $E_{2m+1}^* = 0$ ($m \in \mathbb{N}$).

The first and the second kind Euler numbers are also related to $\tan z$ and $\sec z$,

$$\tan z = -i\frac{e^{iz}-e^{-iz}}{e^{iz}+e^{-iz}} = \frac{e^{2iz}}{2i}\left(\frac{2}{e^{2iz}+1}\right) - \frac{e^{-2iz}}{2i}\left(\frac{2}{e^{-2iz}+1}\right).$$

By using (23) and Cauchy product, we have (cf. [89])

$$\tan z = \frac{1}{2i}\sum_{n=0}^{\infty} E_n \frac{(2iz)^n}{n!} \sum_{n=0}^{\infty} \frac{(2iz)^n}{n!} - \frac{1}{2i}\sum_{n=0}^{\infty} E_n \frac{(-2iz)^n}{n!} \sum_{n=0}^{\infty} \frac{(-2iz)^n}{n!}$$

$$= \frac{1}{2i}\sum_{n=0}^{\infty}\sum_{k=0}^{n} E_k \frac{(2iz)^k}{k!}\frac{(2iz)^{n-k}}{(n-k)!} - \frac{1}{2i}\sum_{n=0}^{\infty}\sum_{k=0}^{n} E_k \frac{(-2iz)^k}{k!}\frac{(-2iz)^{n-k}}{(n-k)!}$$

$$= \frac{1}{2i}\sum_{n=0}^{\infty}\sum_{k=0}^{n} \frac{E_k}{k!(n-k)!}(2i)^n z^n - \frac{1}{2i}\sum_{n=0}^{\infty}\sum_{k=0}^{n} \frac{E_k}{k!(n-k)!}(-2i)^n z^n$$

$$= \sum_{j=0}^{\infty}(-1)^n 2^{2j+1}\left(\sum_{k=0}^{2j+1}\binom{2j+1}{k}E_k\right)\frac{z^{2j+1}}{(2j+1)!}.$$

Finally, using (24) we obtain that (cf. [1, 75, 89])

$$\tan z = \sum_{n=0}^{\infty}(-1)^{n+1}\frac{2^{2n+1}E_{2n+1}}{(2n+1)!}z^{2n+1}, \quad |z| < \frac{\pi}{2}. \tag{27}$$

Remark 2 There are several proofs of (27). For example, Kim [51] used

$$i\tan z = \frac{e^{iz} - e^{-iz}}{e^{iz} + e^{-iz}} = 1 - \frac{2}{e^{2iz} - 1} + \frac{4}{e^{4iz} - 1}$$

to get

$$z\tan z = \sum_{n=1}^{\infty}(-1)^{n}\frac{4^{n}(1 - 4^{n})B_{2n}}{(2n)!}z^{2n},$$

i.e., (27). Similarly, Kim [51] proved the following relation for the secant function (see also [1, 75, 89, 94, 96])

$$\sec z = \sum_{n=0}^{\infty}(-1)^{n}\frac{E_{2n}^{*}}{(2n)!}z^{2n}, \quad |z| < \frac{\pi}{2}.$$

Kim [52] defined *the Dedekind type Daehee-Changhee (DC) sums* as follows:

Definition 1 Let h and k be coprime integers with $k > 0$. Then

$$T_{p}(h, k) = 2\sum_{j=1}^{k-1}(-1)^{j-1}\frac{j}{k}\overline{E}_{p}\left(\frac{hj}{k}\right), \tag{28}$$

where $\overline{E}_{p}(x)$ denotes the p-th Euler function of the first kind.

The behavior of these sum $T_{p}(h, k)$ is similar to that of the Dedekind sums. Several properties and identities of the sum $T_{p}(h, k)$ and Euler polynomials, as well as some other interesting results, were derived in [52]. The most fundamental result in the theory of the Dedekind sums, Hardy-Berndt sums, Dedekind type DC and the other arithmetical sums is the reciprocity law and it can be used as an aid in calculating these sums (see Section 8).

6 Trigonometric Representation of the DC-Sums

In this section we can give relations between trigonometric functions and the sum $T_{p}(h, k)$. We establish analytic properties of the sum $T_{p}(h, k)$ and give their trigonometric representation.

Starting from (22) in the form

$$\frac{(\pi i)^{m+1}}{2\,m!}\overline{E}_m(x) = \sum_{n=-\infty}^{0} \frac{e^{(2n-1)\pi ix}}{(2n-1)^{m+1}} + \sum_{n=1}^{\infty} \frac{e^{(2n-1)\pi ix}}{(2n-1)^{m+1}}, \tag{29}$$

we can get the following auxiliary result (see [89]):

Lemma 3 *Let* $m \in \mathbb{N}$ *and* $0 \le x \le 1$, *except for* $m = 1$ *when* $0 < x < 1$. *Then we have*

$$\overline{E}_{2m-1}(x) = \frac{(-1)^m 4(2m-1)!}{\pi^{2m}} \sum_{n=1}^{\infty} \frac{\cos((2n-1)\pi x)}{(2n-1)^{2m}}, \tag{30}$$

and

$$\overline{E}_{2m}(x) = \frac{(-1)^m 4(2m)!}{\pi^{2m+1}} \sum_{n=1}^{\infty} \frac{\sin((2n-1)\pi x)}{(2n-1)^{2m+1}}. \tag{31}$$

For $0 \le x < 1$, $\overline{E}_{2m-1}(x)$ and $\overline{E}_{2m}(x)$ reduce to the Euler polynomials, which are related to Clausen functions (see Section 7).

We now modify the sums $T_p(h, k)$ for odd and even integer p. Thus, by (28), we define $T_{2m-1}(h, k)$ and $T_{2m}(h, k)$ sums as follows:

Definition 1′ ([89]) Let h and k be coprime integers with $k > 0$. Then

$$T_{2m-1}(h, k) = 2 \sum_{j=1}^{k-1} (-1)^{j-1} \frac{j}{k} \overline{E}_{2m-1}\left(\frac{hj}{k}\right) \tag{32}$$

and

$$T_{2m}(h, k) = 2 \sum_{j=1}^{k-1} (-1)^{j-1} \frac{j}{k} \overline{E}_{2m}\left(\frac{hj}{k}\right), \tag{33}$$

where $\overline{E}_{2m-1}(x)$ and $\overline{E}_{2m}(x)$ denote the Euler functions.

By substituting (30) into (32), we get (cf. [89])

$$T_{2m-1}(h, k) = -\frac{8(-1)^m(2m-1)!}{k\pi^{2m}} \sum_{j=1}^{k-1}(-1)^j j \sum_{n=1}^{\infty} \frac{\cos(\frac{(2n-1)\pi hj}{k})}{(2n-1)^{2m}},$$

i.e.,

$$T_{2m-1}(h,k) = -\frac{8(-1)^m(2m-1)!}{k\pi^{2m}} \sum_{n=1}^{\infty} \frac{1}{(2n-1)^{2m}} \sum_{j=1}^{k-1} (-1)^j j \cos\left(\frac{(2n-1)\pi hj}{k}\right).$$

Since (cf. [14] and [40])

$$\sum_{j=1}^{k-1} j\, e^{(2n-1)\pi ihj/k} = \begin{cases} \dfrac{k}{e^{(2n-1)\pi ih/k} - 1}, & \text{if } 2n-1 \not\equiv 0 \,(\text{mod } k), \\[2mm] \frac{1}{2}k(k-1), & \text{if } 2n-1 \equiv 0 \,(\text{mod } k), \end{cases}$$

we conclude that

$$\sum_{j=1}^{k-1} (-1)^j j\, e^{(2n-1)\pi ihj/k} = \frac{k}{e^{(k+(2n-1)h)\pi i/k} - 1},$$

from which, by some elementary calculations, the following sums follow

$$\sum_{j=1}^{k-1} (-1)^j j \cos\left(\frac{(2n-1)\pi hj}{k}\right) = -\frac{k}{2} \tag{34}$$

and

$$\sum_{j=1}^{k-1} (-1)^j j \sin\left(\frac{(2n-1)\pi hj}{k}\right) = \frac{k}{2} \tan\left(\frac{(2n-1)\pi h}{2k}\right), \tag{35}$$

where $2n-1 \not\equiv 0 \,(\text{mod } k)$.

Using (34) and (25) we obtain the following result:

Theorem 10 *Let h and k be coprime positive integers and $m \in \mathbb{N}$. Then*

$$T_{2m-1}(h,k) = \left(1 - \frac{1}{k^{2m}}\right) E_{2m-1}.$$

Indeed, here we have

$$T_{2m-1}(h,k) = \frac{4(-1)^m(2m-1)!}{\pi^{2m}} \sum_{\substack{n=1 \\ 2n-1 \not\equiv 0 \,(\text{mod } k)}}^{\infty} \frac{1}{(2n-1)^{2m}},$$

i.e.,

$$T_{2m-1}(h, k) = \frac{4(-1)^m (2m-1)!}{\pi^{2m}} \left\{ \sum_{n=1}^{\infty} \frac{1}{(2n-1)^{2m}} - \sum_{\substack{n=1 \\ 2n-1 \equiv 0 \,(\mathrm{mod}\, k)}}^{\infty} \frac{1}{(2n-1)^{2m}} \right\}$$

$$= \frac{4(-1)^m (2m-1)!}{\pi^{2m}} \left\{ \sum_{n=1}^{\infty} \frac{1}{(2n-1)^{2m}} - \sum_{j=1}^{\infty} \frac{1}{k^{2m} (2j-1)^{2m}} \right\},$$

where we put $2n - 1 = (2j - 1)k$ in order to calculate the second sum when $2n - 1 \equiv 0 \,(\mathrm{mod}\, k)$. It proves the statement.

Similarly, by substituting (31) into (33) we get

$$T_{2m}(h, k) = \frac{8(-1)^m (2m)!}{k\pi^{2m+1}} \sum_{j=1}^{k-1} (-1)^j j \sum_{n=1}^{\infty} \frac{\sin\left(\frac{(2n-1)hj\pi}{k}\right)}{(2n-1)^{2m+1}},$$

and then, using (35), we arrive at the following theorem.

Theorem 11 ([89]) *Let h and k be coprime positive integers and $m \in \mathbb{N}$. Then*

$$T_{2m}(h, k) = \frac{4(-1)^m (2m)!}{\pi^{2m+1}} \sum_{\substack{n=1 \\ 2n-1 \not\equiv 0 \,(\mathrm{mod}\, k)}}^{\infty} \frac{\tan\left(\frac{(2n-1)\pi h}{2k}\right)}{(2n-1)^{2m+1}}. \tag{36}$$

7 DC-Sums Related to Special Functions

In this section, we give relations between DC-sums and some special functions.

In [94], Srivastava and Choi gave many applications of the Riemann zeta function, Hurwitz zeta function, Lerch zeta function, Dirichlet series for the polylogarithm function and Dirichlet's eta function. In [45], Guillera and Sandow obtained double integral and infinite product representations of many classical constants, as well as a generalization to Lerch's transcendent of Hadjicostas's double integral formula for the Riemann zeta function, and logarithmic series for the digamma and Euler beta functions. They also gave many applications. The *Lerch transcendent $\Phi(z, s, a)$* (cf. [94, p. 121 et seq.], [45]) is the analytic continuation of the series

$$\Phi(z, s, a) = \frac{1}{a^s} + \frac{z}{(a+1)^s} + \frac{z}{(a+2)^s} + \cdots = \sum_{n=0}^{\infty} \frac{z^n}{(n+a)^s},$$

which converges for $a \in \mathbb{C} \setminus \mathbb{Z}_0^-$, $s \in \mathbb{C}$ when $|z| < 1$, and Re $(s) > 1$ when $|z| = 1$, where as usual, $\mathbb{Z}_0^- = \mathbb{Z}^- \cup \{0\}$ and $\mathbb{Z}^- = \{-1, -2, \ldots\}$. Φ denotes the familiar Hurwitz-Lerch zeta function. Here, we mention some relations between this function Φ and other special functions (cf. [45]).

Special cases include the analytic continuations of the Riemann zeta function

$$\Phi(1, s, 1) = \zeta(s) = \sum_{n=1}^{\infty} \frac{1}{n^s}, \quad \text{Re } (s) > 1,$$

the Hurwitz zeta function

$$\Phi(1, s, a) = \zeta(s, a) = \sum_{n=0}^{\infty} \frac{1}{(n+a)^s}, \quad \text{Re } (s) > 1,$$

the alternating zeta function (also called Dirichlet's eta function $\eta(s)$)

$$\Phi(-1, s, 1) = \zeta^*(s) = \sum_{n=1}^{\infty} \frac{(-1)^{n-1}}{n^s},$$

the Dirichlet beta function

$$2^{-s} \Phi\left(-1, s, \frac{1}{2}\right) = \beta(s) = \sum_{n=0}^{\infty} \frac{(-1)^n}{(2n+1)^s},$$

the Legendre chi function

$$2^{-s} z \, \Phi\left(z^2, s, \frac{1}{2}\right) = \chi_s(z) = \sum_{n=0}^{\infty} \frac{z^{2n+1}}{(2n+1)^s}, \quad |z| \leq 1, \text{ Re } (s) > 1,$$

the polylogarithm

$$z \Phi(z, n, 1) = \mathrm{Li}_m(z) = \sum_{n=0}^{\infty} \frac{z^k}{n^m}$$

and the Lerch zeta function (sometimes called the Hurwitz-Lerch zeta function)

$$L(\lambda, \alpha, s) = \Phi(e^{2\pi i \lambda}, s, \alpha),$$

which is a generalization of the Hurwitz zeta function and polylogarithm (cf. [3, 9, 19, 20, 22, 25, 45, 92–94]).

By using (29), we can give a relation between the Legendre chi function $\chi_s(z)$ and the function $\overline{E}_m(x)$:

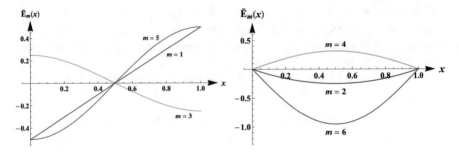

Fig. 1 Graphics of the functions $\overline{E}_m(x)$ for $m = 1, 3, 5$ (left) and $m = 2, 4, 6$ (right)

Corollary 3 ([89]) *Let $m \in \mathbb{N}$. Then we have*

$$\overline{E}_m(x) = \frac{2(m!)}{(\pi i)^{m+1}} \left((-1)^{m+1} \chi_{m+1}(e^{-\pi i x}) + \chi_{m+1}(e^{\pi i x}) \right).$$

The graphics of the functions $x \mapsto \overline{E}_m(x)$ on $(0, 1)$ for $m = 1, 2, \ldots, 6$ are presented in Fig. 1. Most of the aforementioned functions are implemented in Mathematica and Matlab software packages, and the values of these functions can be calculated with arbitrary precision.

Choi et al. gave relations between the Clausen function, multiple gamma function and other functions. The higher-order *Clausen function* $\text{Cl}_n(t)$ is defined for all $n \in \mathbb{N} \setminus \{1\}$ by (see [94])

$$\text{Cl}_n(t) = \begin{cases} \displaystyle\sum_{k=1}^{\infty} \frac{\sin(kt)}{k^n}, & \text{if } n \text{ is even,} \\ \displaystyle\sum_{k=1}^{\infty} \frac{\cos(kt)}{k^n}, & \text{if } n \text{ is odd.} \end{cases}$$

The following functions are related to the higher-order Clausen function (cf. [92], [22, Eq. (5) and Eq. (6)])

$$S(s, x) = \sum_{n=1}^{\infty} \frac{\sin((2n+1)x)}{(2n+1)^s} \quad \text{and} \quad C(s, x) = \sum_{n=1}^{\infty} \frac{\cos((2n+1)x)}{(2n+1)^s}.$$

8 Reciprocity Law

As we mentioned in Section 3, for positive h and k and $(h, k) = 1$ the reciprocity law

$$s(h, k) + s(k, h) = \frac{1}{12} \left(\frac{h}{k} + \frac{k}{h} + \frac{1}{hk} \right) - \frac{1}{4}$$

holds. For the sums $s_p(a, b)$, defined by (13) in Corollary 2, the reciprocity law is given by

$$hk^n s_n(h, k) + kh^n s_n(k, h)$$

$$= \frac{1}{n+1} \sum_{j=0}^{n+1} \binom{n+1}{j} (-1)^j B_j h^j B_{n+1-j} k^{n+1-j} + \frac{n}{n+1} B_{n+1},$$

where $(h, k) = 1$ and B_n is the nth Bernoulli number (cf. [2, 4, 5]).

In the sequel we mention some reciprocity theorems for Hardy sums:

Theorem 12 *Let h and k be coprime positive integers. Then if $h + k$ is odd,*

$$S(h, k) + S(k, h) = 1. \tag{37}$$

Theorem 13 *Let h and k be coprime integers. Then if h and k are odd,*

$$s_5(h, k) + s_5(k, h) = \frac{1}{2} - \frac{1}{2hk}. \tag{38}$$

Theorem 14 *Let h and k be coprime integers. If h is even, then*

$$s_1(h, k) - 2s_2(k, h) = \frac{1}{2} - \frac{1}{2} \left(\frac{1}{hk} + \frac{h}{k} \right). \tag{39}$$

Theorem 15 *Let h and k be coprime integers. If k is odd, then*

$$2s_3(h, k) - s_4(k, h) = 1 - \frac{h}{k}. \tag{40}$$

These reciprocity theorems appear in Hardy's list [48], as Eqs. (viii), (vii), (vi), (vi') and (ix) on pages 122–123. Berndt [13] deduced (37), (39), and (40), and Goldberg [40] deduced (38) from Berndt's transformation formulae. For other proofs which do not depend on transformation theory, we refer to Sitaramachandrarao [91]. Otherwise, all reciprocity theorems can be proved by using contour integration and Cauchy Residue Theorem.

The reciprocity law of the sums $T_p(h, k)$, defined by (28) in Definition 1, is proved in [52]:

Theorem 16 *Let $(h, k) = 1$ and $h, k \in \mathbb{N}$ with $h \equiv 1 \pmod 2$ and $k \equiv 1 \pmod 2$. Then we have*

$$k^p T_p(h, k) + h^p T_p(k, h)$$

$$= 2 \sum_{\substack{j=0 \\ j - \lceil \frac{hj}{k} \rceil \equiv 1 \bmod 2}}^{k-1} \left(kh \left(E + \frac{j}{k} \right) + k \left(E + h - \left[\frac{hj}{k} \right] \right) \right)^p + (hE + kE)^p + (p+2) E_p,$$

where

$$(hE + kE)^{n+1} = \sum_{j=1}^{n+1} \binom{n+1}{j} h^j E_j k^{n+1-j} E_{n+1-j}.$$

The first proof of reciprocity law of the Dedekind sums does not contain the theory of the Dedekind eta function related to Rademacher [73]. The other proofs of the reciprocity law of the Dedekind sums were given by Rademacher and Grosswald [76]. Berndt [11] and Berndt and Goldberg [14] gave various types of Dedekind sums and their reciprocity laws. Berndt's methods are of three types. The first method uses contour integration which was first given by Rademacher [73]. This method has been used by many authors for example Grosswald [43], Hardy [48], his method is a different technique in contour integration. The second method is the Riemann-Stieltjes integral, which was invented by Rademacher [74]. The third method of Berndt is (periodic) Poisson summation formula. For the method and technique see also the references cited in each of these earlier works.

The famous property of the all arithmetic sums is the reciprocity law. In the sequel we prove reciprocity law of (36), by using contour integration.

Theorem 17 *Let h, k, $m \in \mathbb{N}$ with $h \equiv 1$ (mod 2) and $k \equiv 1$ (mod 2) and $(h, k) = 1$. Then we have*

$$hk^{2m+1} T_{2m}(h, k) + kh^{2m+1} T_{2m}(k, h)$$

$$= 2E_{2m+1} - \sum_{j=0}^{m-1} \binom{2m}{2j+1} E_{2j+1} E_{2m-2j-1} h^{2j+2} k^{2m-2j},$$

where E_n are Euler numbers of the first kind.

Proof Following [11, Theorem 4.2], [14, Theorem 3], [43, 44, 76], we give the proof of this theorem. We use the contour integration method with the function

$$F_m(z) = \frac{\tan \pi h z \ \tan \pi k z}{z^{2m+1}}$$

over the contour C_N,

$$I_N = \frac{1}{2\pi i} \int_{C_N} F_m(z) \, dz,$$

where C_N is a positively oriented circle of radius R_N, with $1 \leq N < \infty$, centered at the origin. The sequence of radii R_N is increasing to ∞ and is chosen so that the poles of $F_m(z)$ are at a distance from C_N greater than some fixed positive number for each N.

From the above, we have

$$I_N = \frac{1}{2\pi R_N^{2m}} \int_0^{2\pi} e^{-i2m\theta} \tan\left(\pi h R_N e^{i\theta}\right) \tan\left(\pi k R_N e^{i\theta}\right) d\theta.$$

By the condition on C_N, if $R_N \to \infty$, then $\tan\left(R_N e^{i\theta}\right)$ is bounded, and therefore $\lim_{N\to\infty} I_N = 0$ as $R_N \to \infty$ for each $m \in \mathbb{N}$.

The function $F_m(z)$ has a pole of order $2m-1$ at the origin, whose the residue can be determined from the corresponding Laurent series at $z = 0$. Using the expansion

$$\tan z = \sum_{v=0}^{\infty} \tau_v z^{2v-1}, \quad |z| < \frac{\pi}{2},$$

where

$$\tau_v = (-1)^{v-1}\frac{4^v(4^v - 1)B_{2v}}{(2v)!} = (-1)^v\frac{4^v E_{2v-1}}{2(2v-1)!} \quad \text{(cf. Eq. (27))}$$

we can get the Laurent expansion of $F_m(z)$ at $z = 0$ in the following form

$$F_m(z) = \frac{hk\pi^2}{z^{2m-1}} \sum_{n=0}^{\infty} (-1)^n \pi^{2n} 4^{n+1} f_n z^{2n}, \tag{41}$$

where

$$f_n = \sum_{j=0}^{n} \frac{E_{2j+1}E_{2n-2j+1}h^{2j}k^{2n-2j}}{(2j+1)!(2n-2j+1)!}.$$

Then, from (41) for $n = m - 1$, we get the residue of the function $F_m(z)$ at the pole $z = 0$ as

$$\operatorname*{Res}_{z=0} F_m(z) = hk\pi^2(-1)^{m-1}\pi^{2m-2}4^m f_{m-1}$$

$$= -\frac{(-1)^m(2\pi)^{2m}}{(2m)!\,hk} \sum_{j=0}^{m-1}\binom{2m}{2j+1} E_{2j+1}E_{2m-2j-1}h^{2j+2}k^{2m-2j}. \tag{42}$$

The other singularities of the function $F_m(z)$ in the interior of the contour C_N are points of the sets

$$X_N = \left\{ \xi_j = \frac{2j-1}{2h} : |\xi_j| < R_N, \ j \in \mathbb{Z} \right\}$$

and

$$Y_N = \left\{ \eta_\ell = \frac{2\ell-1}{2k} : |\eta_\ell| < R_N, \ \ell \in \mathbb{Z} \right\}.$$

Since h and k are odd positive integers, then the sets X_N and Y_N has an intersection

$$Z_N = X_N \cap Y_N = \left\{ \zeta_j = \frac{2j-1}{2} : |\zeta_j| < R_N, \ j \in \mathbb{Z} \right\},$$

with double poles of the function $F_m(z)$ in the interior of C_N. Their residues are

$$\operatorname*{Res}_{z=\zeta_j} F_m(z) = \lim_{z\to\zeta_j} \frac{d}{dz}\left[(z-\zeta_j)^2 F_m(z) \right] = -\frac{(2m+1)4^{m+1}}{(2j-1)^{2m+2}hk\pi^2}. \tag{43}$$

Te residues of the simple poles $z = \xi_j \in X_N \setminus Z_N$ and $\eta_\ell \in Y_N \setminus Z_N$ are easily found to be

$$\operatorname*{Res}_{z=\xi_j} F_m(z) = \lim_{z\to\xi_j} (z-\xi_j) F_m(z) = -\frac{2^{2m+1}h^{2m}}{\pi(2j-1)^{2m+1}} \tan\left(\frac{(2j-1)\pi k}{2h} \right) \tag{44}$$

and

$$\operatorname*{Res}_{z=\eta_\ell} F_m(z) = \lim_{z\to\eta_\ell} (z-\eta_\ell) F_m(z) = -\frac{2^{2m+1}k^{2m}}{\pi(2\ell-1)^{2m+1}} \tan\left(\frac{(2\ell-1)\pi h}{2k} \right), \tag{45}$$

respectively. Now, by the Cauchy residue theorem, we have

$$I_N = \sum_{\xi_j \in X_N\setminus Z_N} \operatorname*{Res}_{z=\xi_j} F_m(z) + \sum_{\eta_\ell \in Y_N\setminus Z_N} \operatorname*{Res}_{z=\eta_\ell} F_m(z) + \operatorname*{Res}_{z=0} F_m(z) + \sum_{\zeta_j \in Z_N} \operatorname*{Res}_{z=\zeta_j} F_m(z). \tag{46}$$

Using the residues (45), (44), (42), (43), and letting $N \to \infty$ in (46), we have that $I_N \to 0$ and

$$\frac{4^{m+1}k^{2m}}{\pi} \sum_{\substack{\ell=1 \\ 2\ell-1 \not\equiv 0 \,(\mathrm{mod}\, k)}}^{\infty} \frac{\tan\left(\frac{(2\ell-1)\pi h}{2k} \right)}{(2\ell-1)^{2m+1}} + \frac{4^{m+1}h^{2m}}{\pi} \sum_{\substack{j=1 \\ 2j-1 \not\equiv 0 \,(\mathrm{mod}\, h)}}^{\infty} \frac{\tan\left(\frac{(2j-1)\pi k}{2h} \right)}{(2j-1)^{2m+1}}$$

$$= -\frac{(-1)^m (2\pi)^{2m}}{(2m)! \, hk} \sum_{j=0}^{m-1} \binom{2m}{2j+1} E_{2j+1} E_{2m-2j-1} h^{2j+2} k^{2m-2j}$$

$$- \frac{(2m+1)4^{m+2}}{2hk\pi^2} \sum_{j=1}^{\infty} \frac{1}{(2j-1)^{2m+2}},$$

where the last sum is given by (25).

Multiplying this equality by the factor $hk(-1)^m (2m)!/(2\pi)^{2m}$, we arrive at the desired result.

Corollary 4 *Let For each $m \in \mathbb{N}$ and each positive odd integer $k \geq 3$, we have*

$$k^{2m+1} T_{2m}(1, k) = 2E_{2m+1} - \sum_{j=0}^{m-1} \binom{2m}{2j+1} E_{2j+1} E_{2m-2j-1} h^{2j+2} k^{2m-2j}.$$

This result can be proved in a similar way as Theorem 17. For the sets of singularities here $Z_N = X_N$, so that the first term on the right hand side in (46) vanishes.

We now give a relation between Hurwitz zeta function, $\tan z$ and the sum $T_{2m}(h, k)$.

Hence, substituting $n = rk + j, \, 0 \leq r \leq \infty, \, 1 \leq j \leq k \, (j \neq (k+1)/2)$ into (36), and recalling that $\tan(\pi + \alpha) = \tan \alpha$, then we have

$$T_{2m}(h, k) = \frac{4(-1)^m (2m)!}{\pi^{2m+1}} \sum_{\substack{j=1 \\ j \neq (k+1)/2}}^{k} \sum_{r=0}^{\infty} \frac{\tan\left(\frac{\pi h 2(rk+j)+1}{2k}\right)}{(2(rk+j)+1)^{2m+1}}$$

$$= \frac{4(-1)^m (2m)!}{\pi^{2m+1} (2k)^{2m+1}} \sum_{\substack{j=1 \\ j \neq (k+1)/2}}^{k} \tan\left(\frac{(2j-1)\pi h}{2k}\right) \sum_{r=0}^{\infty} \frac{1}{\left(r + \frac{2j-1}{2k}\right)^{2m+1}},$$

where the last sum can be identified as the Hurwitz zeta function $\zeta(s, x)$ at $s = 2m + 1$ and $x = (2j - 1)/2$. Note that $j \neq (k + 1)/2$ provides the condition $2n - 1 \not\equiv 0 \pmod{k}$ in the summation process in (36).

In this way, we now arrive at the following result:

Theorem 18 *Let h and k be coprime positive integers and $m \in \mathbb{N}$. Then*

$$T_{2m}(h, k) = \frac{4(-1)^m (2m)!}{(2k\pi)^{2m+1}} \sum_{\substack{j=1 \\ j \neq (k+1)/2}}^{k} \tan\left(\frac{(2j-1)\pi h}{2k}\right) \zeta\left(2m+1, \frac{2j-1}{2k}\right).$$

Remark 3 Finally, we mention the sums $Y(h, k)$ defined by Simsek [88] (see also [86] and [65, p. 211]) as $Y(h, k) = 4ks_5(h, k)$, where h and k are odd with $(h, k) = 1$. If we integrate the function $F(z) = \cot(\pi z) \tan(\pi h z) \tan(\pi k z)$ over a contour, obtained from the rectangle with vertices at $\pm iB$, $\frac{1}{2} \pm iB$ $(B > 0)$, we see that $F(z)$ has the poles $z = 0$ and $z = 1/2$ on this contour; therefore, reciprocity of the sums $Y(h, k)$ is given by

$$hY(h, k) + kY(k, h) = 2hk - 2.$$

Remark 4 Using the definitions of the q-analogues of some classical arithmetic functions (Riemann zeta functions, Dirichlet L-functions, Hurwitz zeta functions, Dedekind sums), Simsek [87] defined q-analogues of these functions and gave some relations among them.

Remark 5 The higher multiple elliptic Dedekind sums and the reciprocity law have been introduced and considered by Bayad and Simsek [7] (see also [6, 8, 29, 46, 83]).

9 Sums Obtained from Gauss-Chebyshev Quadratures

It is well-known that Gaussian quadrature formulas with respect to the Chebyshev weight functions of the first, second, third, and fourth kind,

$$w_1(t) = \frac{1}{\sqrt{1 - t^2}}, \quad w_2(t) = \sqrt{1 - t^2}, \quad w_3(t) = \sqrt{\frac{1 + t}{1 - t}}, \quad w_4(t) = \sqrt{\frac{1 - t}{1 + t}},$$

respectively (cf. [18, 35], [58, p. 122]) have nodes expressible by trigonometric functions. In a short note in 1884 Stieltjes [95] gave the explicit expressions for these quadrature formulas for the weights w_1, w_2, and w_4,

$$\int_{-1}^{1} w_1(t) f(t)\, dt = \frac{\pi}{n} \sum_{k=1}^{n} f\left(\cos \frac{(2k-1)\pi}{2n}\right) + R_{n,1}[f], \tag{47}$$

$$\int_{-1}^{1} w_2(t) f(t)\, dt = \frac{\pi}{n+1} \sum_{k=1}^{n} \sin^2 \frac{k\pi}{n+1} f\left(\cos \frac{k\pi}{n+1}\right) + R_{n,2}[f], \tag{48}$$

$$\int_{-1}^{1} w_4(t) f(t)\, dt = \frac{4\pi}{2n+1} \sum_{k=1}^{n} \sin^2 \frac{k\pi}{2n+1} f\left(\cos \frac{2k\pi}{2n+1}\right) + R_{n,4}[f], \tag{49}$$

where $R_{n,\nu}(f) = 0$, $\nu = 1, 2, 4$, for all algebraic polynomials of degree at most $2n - 1$. In the class of functions $C^{2n}[-1, 1]$, the remainder term of these Gaussian formulas can be done in the form (cf. [58, p. 333])

$$R_{n,v}[f] = \frac{\|\pi_{n,v}\|^2}{(2n)!} f^{(2n)}(\xi), \quad -1 < \xi < 1,$$

where the norms of the corresponding orthogonal polynomials $\pi_{n,v}$ can be expressed by the coefficients $\beta_{k,v}$ in their three-term recurrence relations

$$\pi_{k+1,v}(t) = (t - \alpha_{k,v})\pi_{k,v}(t) - \beta_{k,v}\pi_{k-1,v}(t), \quad k = 1, 2, \dots,$$

$$\pi_{0,v}(t) = 1, \quad \pi_{-1,v}(t) = 0,$$

as $\|\pi_{n,v}\|^2 = \beta_{0,v}\beta_{1,v}\cdots\beta_{n,v}$, with

$$\beta_{0,v} = \mu_{0,v} = \int_{-1}^{1} w_v(t)\,dt.$$

The recurrence coefficients for these Chebyshev weights, as well as the corresponding values of $\|\pi_{n,v}\|^2$ are presented in Table 1 (cf. [35, p. 29]). For completeness, we also give parameters for the Chebyshev weight of the third kind w_3. The corresponding quadrature formula for w_3 (cf. [61]),

$$\int_{-1}^{1} w_3(t) f(t)\,dt = \frac{4\pi}{2n+1} \sum_{k=1}^{n} \cos^2 \frac{(2k-1)\pi}{2(2n+1)} f\left(\cos \frac{(2k-1)\pi}{2n+1}\right) + R_{n,3}[f],$$

(50)

Table 1 Recurrence coefficients for different kind of Chebyshev polynomials

Weight function	Recurrence coefficients	$\|\pi_{n,v}\|^2$
Chebyshev I	$\alpha_{k,1} = 0 \ (k \geq 0)$	$\dfrac{2\pi}{4^n}$
$(v = 1)$	$\beta_{0,1} = \pi, \ \beta_{1,1} = \dfrac{1}{2}, \ \beta_{k,1} = \dfrac{1}{4} \ (k \geq 1)$	
Chebyshev II	$\alpha_{k,2} = 0 \ (k \geq 0)$	$\dfrac{\pi}{2^{2n+1}}$
$(v = 2)$	$\beta_{0,2} = \dfrac{\pi}{2}, \ \beta_{k,2} = \dfrac{1}{4} \ (k \geq 1)$	
Chebyshev III	$\alpha_{0,3} = \dfrac{1}{2}, \ \alpha_{k,3} = 0 \ (k \geq 1)$	$\dfrac{\pi}{4^n}$
$(v = 3)$	$\beta_{0,3} = \pi, \ \beta_{k,3} = \dfrac{1}{4} \ (k \geq 1)$	
Chebyshev IV	$\alpha_{0,4} = -\dfrac{1}{2}, \ \alpha_{k,4} = 0 \ (k \geq 1)$	$\dfrac{\pi}{4^n}$
$(v = 4)$	$\beta_{0,4} = \pi, \ \beta_{k,4} = \dfrac{1}{4} \ (k \geq 1)$	

can be obtained by changing $t := -t$ and using (49), as well as

$$\int_{-1}^{1} w_3(t) f(t)\, dt = \int_{-1}^{1} w_4(t) f(-t)\, dt = Q_{n,4}(f(-\cdot)) + R_{n,3}[f].$$

Taking certain algebraic polynomials of degree at most $2n$ and integrating them by using some of quadrature formulas (47)–(49) and (50), we can obtain some trigonometric sums. For example, by a polynomial $p_m(t)$ of degree $m < 2n$, using the quadrature formula (47) we get

$$\sum_{k=1}^{n} p_m\left(\cos\frac{(2k-1)\pi}{2n}\right) = \frac{n}{\pi} \int_{-1}^{1} \frac{p_m(t)}{\sqrt{1-t^2}}\, dt,$$

but if $p_{2n}(t) = At^{2n} +$ terms of lower degree, then

$$\sum_{k=1}^{n} p_{2n}\left(\cos\frac{(2k-1)\pi}{2n}\right) = \frac{n}{\pi} \int_{-1}^{1} \frac{p_{2n}(t)}{\sqrt{1-t^2}}\, dt - \frac{2nA}{4^n}.$$

Thus, we need to compute only integrals of the forms $\int_{-1}^{1} p_m(t) w_\nu(t)\, dt$ for $\nu = 1, 2, 3, 4$. As usual $(s)_n$ is the well known Pochhammer symbol defined by

$$(s)_n = s(s+1)\cdots(s+n-1) = \frac{\Gamma(s+n)}{\Gamma(s)} \qquad (\Gamma \text{ is the gamma function}).$$

If we take

$$p_m(t) = U_m(c\,t) = 2^m c^m t^m + \text{ terms of lower degree}, \quad c \in \mathbb{R},$$

where U_m is the Chebyshev polynomial of the second kind and degree m and use the integral (cf. [71, p. 456])

$$\int_{-1}^{1} \frac{U_{2m}(cx)}{\sqrt{1-x^2}}\, dx = \pi P_m(2c^2 - 1),$$

where P_m is the Legendre polynomial of degree m, we get the following statement:

Theorem 19 *Let U_m be Chebyshev polynomial of the second kind and degree m and $c \in \mathbb{R}$. We have*

$$\sum_{\nu=1}^{n} U_m\left(c\cos\frac{(2\nu-1)\pi}{2n}\right) = \begin{cases} 0, & m \text{ is odd } (m \geq 1), \\ n\, P_{m/2}(2c^2-1), & m \text{ is even } (0 \leq m < 2n), \\ n\left(P_n(2c^2-1) - 2c^{2n}\right), & m = 2n, \end{cases}$$

where P_n is the Legendre polynomial of degree n.

Similarly, using the Legendre polynomials

$$P_m(t) = \frac{(n+1)_n}{2^n n!} t^n + \text{ terms of lower degree,}$$

defined by the generating function (cf. [58, p. 129])

$$\frac{1}{\sqrt{1 - 2xt + x^2}} = \sum_{m=0}^{\infty} P_m(t) x^m,$$

for which the following integrals are true (see [71, p. 423])

$$\int_{-1}^{1} \frac{P_{2m}(x)}{\sqrt{1 - x^2}} \, dx = \pi \left(\frac{(1/2)_m}{m!} \right)^2$$

and

$$\int_{-1}^{1} \frac{x \, P_{2m+1}(x)}{\sqrt{1 - x^2}} \, dx = \pi \left(\frac{(1/2)_m}{m!} \right)^2 \frac{2m+1}{2m+2},$$

we can prove the following statement:

Theorem 20 *Let P_m be the Legendre polynomial of degree n. Then*

$$\sum_{v=1}^{n} P_m \left(\cos \frac{(2v-1)\pi}{2n} \right) = \begin{cases} 0, & m \text{ is odd } (m \geq 1), \\ \frac{n}{4^m} \binom{m}{m/2}^2, & m \text{ is even } (0 \leq m < 2n), \\ \frac{n}{4^{2n}} \left(\binom{2n}{n}^2 - 2\binom{4n}{2n} \right), & m = 2n, \end{cases}$$

and

$$\sum_{v=1}^{n} \cos \frac{(2v-1)\pi}{2n} P_{m-1} \left(\cos \frac{(2v-1)\pi}{2n} \right)$$

$$= \begin{cases} 0, & m \text{ is odd } (m \geq 1), \\ \frac{n}{4^{m-2}} \frac{m-1}{m} \binom{m-2}{(m-1)/2}^2, & m \text{ is even } (0 < m < 2n), \\ \frac{1}{4^{2n-1}} \left(2(2n-1) \binom{2n-2}{n-1}^2 - n\binom{4n-2}{2n-1} \right), & m = 2n. \end{cases}$$

Now, we use the following weighted integral for Legendre polynomials, expressed in terms of the hypergeometric function (see [71, p. 422]),

$$I_\beta(m) = \int_{-1}^{1} (1-x^2)^{\beta-1} P_m(x)\, dx = (-1)^m \sqrt{\pi}\, \frac{\Gamma(\beta)}{\Gamma(\beta+\frac{1}{2})}\, {}_3F_2\left[\begin{matrix} -m,\, m+1,\, \beta \\ 2\beta,\, 1 \end{matrix}\,;\, 1\right].$$

For odd m this integral vanishes. We need now its value for $\beta = 3/2$ and even m, i.e.,

$$I_{3/2}(m) = \frac{\pi}{2}\, {}_3F_2\left[\begin{matrix} -m,\, m+1,\, 3/2 \\ 3,\, 1 \end{matrix}\,;\, 1\right]$$

$$= \frac{\pi}{2} \cdot \frac{\pi}{2\Gamma\left(\frac{1}{2}-\frac{m}{2}\right)\Gamma\left(\frac{3}{2}-\frac{m}{2}\right)\Gamma\left(\frac{m}{2}+1\right)\Gamma\left(\frac{m}{2}+2\right)}.$$

Since $\Gamma(z) = \Gamma(z+k)/(z)_k$ $(k \in \mathbb{N})$, we have

$$\Gamma\left(\frac{1}{2}-\frac{m}{2}\right) = \frac{\Gamma\left(\frac{1}{2}\right)}{\left(\frac{1}{2}-\frac{m}{2}\right)_{m/2}} = (-1)^{m/2}\frac{2^m\sqrt{\pi}}{m!}\left(\frac{m}{2}\right)!,$$

so that the previous integral becomes

$$I_{3/2}(m) = -\frac{\pi}{4^m(m-1)(m+2)}\binom{m}{m/2}^2.$$

Setting $p_m(t) = P_m(t)$ in (48) we get the following result:

Theorem 21 *Let P_m be the Legendre polynomial of degree n. Then*

$$\sum_{\nu=1}^{n} \sin^2\frac{k\pi}{n+1} P_m\left(\cos\frac{k\pi}{n+1}\right)$$

$$= \begin{cases} 0, & m \text{ is odd } (m \geq 1), \\[2mm] -\dfrac{n+1}{4^m(m+2)(m-1)}\dbinom{m}{m/2}^2, & m \text{ is even } (0 \leq m < 2n), \\[3mm] -\dfrac{2}{4^{2n+1}}\left(\dfrac{1}{2n-1}\dbinom{2n}{n}^2 + (n+1)\dbinom{4n}{2n}\right), & m = 2n. \end{cases}$$

In the sequel we give trigonometric sums obtained by monomials $p_m(t) = t^m$ in the quadrature formulas (47)–(49). Because of that, we need the moments

$$\mu_{m,\nu} = \int_{-1}^{1} w_\nu(t) t^m\, dt, \quad \nu = 1, 2, 3, 4.$$

These moments are

$$\mu_{m,1} = \frac{\sqrt{\pi}\,((-1)^m + 1)\,\Gamma\left(\frac{m+1}{2}\right)}{m\Gamma\left(\frac{m}{2}\right)} = \begin{cases} 0, & m \text{ is odd,} \\[2mm] \dfrac{\pi}{2^m}\dbinom{m}{m/2}, & m \text{ is even;} \end{cases}$$

$$\mu_{m,2} = \frac{\sqrt{\pi}\,((-1)^m + 1)\,\Gamma\left(\frac{m+1}{2}\right)}{4\Gamma\left(\frac{m}{2} + 2\right)} = \begin{cases} 0, & m \text{ is odd,} \\[2mm] \dfrac{\pi}{2^m(m+2)}\dbinom{m}{m/2}, & m \text{ is even;} \end{cases}$$

$$\mu_{m,4} = \frac{2^m\left(((-1)^m - 1)\,\Gamma\left(\frac{m}{2}+1\right)^2 + ((-1)^m + 1)\,\Gamma\left(\frac{m+1}{2}\right)\Gamma\left(\frac{m+3}{2}\right)\right)}{\Gamma(m+2)}$$

$$= \begin{cases} -\dfrac{\pi}{2^{m+1}}\dbinom{m+1}{(m+1)/2}, & m \text{ is odd,} \\[3mm] \dfrac{\pi}{2^m}\dbinom{m}{m/2}, & m \text{ is even.} \end{cases}$$

Note that $\mu_{m,3} = (-1)^m \mu_{m,4}$.

According to (47)–(49) we conclude that the following results hold.

Theorem 22 *We have*

$$\sum_{v=1}^{n} \cos^m \frac{(2v-1)\pi}{2n} = \begin{cases} 0, & m \text{ is odd } (m \geq 1), \\[2mm] \dfrac{n}{2^m}\dbinom{m}{m/2}, & m \text{ is even } (0 \leq m < 2n), \\[2mm] \dfrac{n}{4^n}\left(\dbinom{2n}{n} - 2\right), & m = 2n; \end{cases}$$

$$\sum_{v=1}^{n} \sin^2 \frac{v\pi}{n+1}\cos^m \frac{v\pi}{n+1} = \begin{cases} 0, & m \text{ is odd } (m \geq 1), \\[2mm] \dfrac{n+1}{(m+2)2^m}\dbinom{m}{m/2}, & m \text{ is even } (0 \leq m < 2n), \\[2mm] \dfrac{1}{2^{2n+1}}\left(\dbinom{2n}{n} - n - 1\right), & m = 2n; \end{cases}$$

$$\sum_{v=1}^{n} \sin^2 \frac{v\pi}{2n+1}\cos^m \frac{2v\pi}{2n+1} = \begin{cases} \dfrac{2n+1}{2^{m+2}}\dbinom{m}{m/2}, & m \text{ is even } (0 \leq m < 2n), \\[2mm] -\dfrac{2n+1}{2^{m+3}}\dbinom{m+1}{(m+1)/2}, & m \text{ is odd } (0 \leq m < 2n), \\[2mm] \dfrac{2n+1}{4^{n+1}}\left(\dbinom{2n}{n} - 1\right), & m = 2n. \end{cases}$$

Trigonometric sums can be also obtained in a similar way using quadrature formulas of Radau and Lobatto type with respect to Chebyshev weights. We mention that shortly after Stieltjes' results [95], Markov [57] (see also [61]) obtained the explicit expressions for Gauss-Radau and Gauss-Lobatto formulas, with respect to the Chebyshev weight of the first kind $w_1(t)$ (for both endpoints),

$$\int_{-1}^{1} w_1(t) f(t)\, dt = \frac{2\pi}{2n+1} \left[\frac{1}{2} f(-1) + \sum_{k=1}^{n} f\left(\cos \frac{(2k-1)\pi}{2n+1} \right) \right] + R_{n+1,1}^{(-1)}[f],$$

$$\int_{-1}^{1} w_1(t) f(t)\, dt = \frac{2\pi}{2n+1} \left[\frac{1}{2} f(1) + \sum_{k=1}^{n} f\left(\cos \frac{2k\pi}{2n+1} \right) \right] + R_{n+1,1}^{(+1)}[f],$$

and

$$\int_{-1}^{1} w_1(t) f(t)\, dt = \frac{\pi}{n+1} \left[\frac{1}{2} f(-1) + \sum_{k=1}^{n} f\left(\cos \frac{k\pi}{n+1} \right) + \frac{1}{2} f(1) \right] + R_{n+2,1}^{L}[f],$$

respectively, where

$$R_{n+1,1}^{(\mp 1)}[f] = \pm \frac{\pi f^{(2n+1)}(\xi)}{(2n+1)! 2^{2n}} \qquad (f \in C^{2n+1}[-1,1])$$

and

$$R_{n+2,1}^{L}[f] = -\frac{\pi f^{(2n+2)}(\xi)}{(2n+2)! 2^{2n+1}}, \qquad (f \in C^{2n+2}[-1,1])$$

and $\xi \in (-1,1)$. These Gauss-Radau formulas are exact for all algebraic polynomials of degree at most $2n$, while the Gauss-Lobatto is exact for polynomials up to degree $2n+1$, so that we can obtain trigonometric sums taking monomials x^m in the previous formulas for all $m \le 2n+1$ and $m \le 2n+2$, respectively.

There are also similar quadrature formulas for other Chebyshev weights $w_\nu(t)$, $\nu = 2, 3, 4$. For details see [34, 36, 37, 67, 69, 78]. A general approach in construction of Gauss-Radau and Gauss-Lobatto formulas can be found in [58, pp. 328–332]. In some of these cases the nodes of quadratures can be expressed in terms of trigonometric functions, and such quadratures can be used for getting trigonometric sums.

Some more complex trigonometric sums can be obtained using quadrature formulas of Turán type (quadrature with multiple nodes) with respect to the Chebyshev weight functions. For some details on such quadrature formulas see [38, 39, 59, 60, 64, 79–81, 97].

10 Sums Obtained from Trigonometric Quadrature Rules

Another way for getting trigonometric sums is based on quadrature rules with maximal trigonometric degree of exactness. With $\mathscr{T}_d$ we denote the linear space of all trigonometric polynomials of degree less than or equal to d,

$$
t_d(x) = \frac{a_0}{2} + \sum_{k=1}^{d} (a_k \cos kx + \sin kx) \qquad (a_k, b_k \in \mathbb{R}).
$$

If $|a_d| + |b_d| > 0$ the degree of t_d is strictly d.

We say that a quadrature formula of the form

$$
\int_0^{2\pi} f(x)w(x)\,dx = \sum_{v=0}^{n} w_v f(x_v) + R_n(f), \quad 0 \le x_0 < x_1 < \cdots < x_n < 2\pi,
$$

has trigonometric degree of exactness equal to d if the remainder term $R_n(f) = 0$ for all $f \in \mathscr{T}_d$ and there exists some g in $\mathscr{T}_{d+1}$ such that $R_n(g) \ne 0$.

Quadrature rules with a maximal trigonometric degree of exactness are known in the literature as *trigonometric quadrature rules of Gaussian type*. Maximal trigonometric degree of exactness for quadrature rule with $n + 1$ nodes is n.

A brief historical survey of available approaches for the construction of quadrature rules with maximal trigonometric degree of exactness has been given in [61] (see also [62, 68]). Here, we consider only a simple case of the $(2n + 1)$-point trigonometric quadrature formula

$$
\int_0^{2\pi} f(x)\,dx = \frac{2\pi}{2n+1} \sum_{k=0}^{2n} f(x_k) + R_{2n+1}[f], \tag{51}
$$

with the nodes

$$
x_k = \theta + \frac{2k\pi}{2n+1}, \qquad k = 0, 1, \ldots, 2n,
$$

where $0 \le \theta < 2\pi/(2n+1)$. Formula (51) is exact for every trigonometric polynomial of degree at most $2n$ (cf. [98]). Such kind of quadratures have applications in numerical integration of 2π-periodic functions. Two special cases of the quadrature formula (51) for which $\theta = 0$ and $\theta = \pi/(2n+1)$ are very interesting in applications. Their quadrature sums are

$$
Q_{2n+1}^T(f) = \frac{2\pi}{2n+1} \sum_{k=0}^{2n} f\left(\frac{2k\pi}{2n+1}\right) \tag{52}
$$

and

$$Q_{2n+1}^M(f) = \frac{2\pi}{2n+1} \sum_{k=0}^{2n} f\left(\frac{(2k+1)\pi}{2n+1}\right),$$ (53)

respectively. Some details on $Q_{2n+1}^T(f)$ and its applications in the trigonometric approximation can be found in [58, Chap. 3]. The second formula $Q_{2n+1}^M(f)$ has been analyzed in [62].

Remark 6 Putting $h = 2\pi/(2n+1)$ and $f_\alpha \equiv f(\alpha h)$, we can write the formulas (52) and (53) in the forms

$$Q_{2n+1}^T(f) = h\left\{\frac{1}{2}f_0 + f_1 + \cdots + f_{2n} + \frac{1}{2}f_{2n+1}\right\}$$

and

$$Q_{2n+1}^M(f) = h\left\{f_{1/2} + f_{3/2} + \cdots + f_{2n} + f_{2n+1/2}\right\},$$

where, because of periodicity, we introduced $f_{2n+1} = f(2\pi) = f(0) = f_0$. These quadratures (52) and (53) are symmetric with respect to the point $x = \pi$, and they are, in fact, the composite *trapezoidal* and *midpoint* rules, respectively. Also, they are equivalent to the trigonometric version of the Gauss-Radau formulas with respect to the Chebyshev weight of the first kind on $(-1, 1)$ (cf. [61]).

Taking $f(x) = \cos^{2m-2\nu} x \sin^{2\nu} x$ in (51), where $0 \le m \le n$ and $0 \le \nu \le m$, we get the following result:

Theorem 23 *If $n, m, \nu \in \mathbb{N}$ and $0 \le \nu \le m \le n$, we have*

$$\sum_{k=0}^{2n} \cos^{2m-2\nu}\left(\theta + \frac{2k\pi}{2n+1}\right) \sin^{2\nu}\left(\theta + \frac{2k\pi}{2n+1}\right) = \frac{2n+1}{4^m} \frac{\binom{2m}{m}\binom{m}{\nu}}{\binom{2m}{2\nu}}$$

for each real θ.

Here, we see that $f \in \mathcal{T}_{2m}$, as well as that

$$I_{m,\nu} = \int_0^{2\pi} \cos^{2m-2\nu} x \sin^{2\nu} x \, dx = 4\int_0^{\pi/2} \cos^{2m-2\nu} x \sin^{2\nu} x \, dx$$

$$= 2\int_0^1 t^{\nu-1/2}(1-t)^{m-\nu-1/2} \, dt$$

$$= \frac{2}{m!}\Gamma\left(\nu+\frac{1}{2}\right)\Gamma\left(m-\nu+\frac{1}{2}\right),$$

after the change of variables $t = \sin^2 x$. Since $\Gamma(\nu + 1/2) = (2\nu - 1)!!\sqrt{\pi}/2^\nu$, by using (51), we obtain the desired result.

Selecting other functions, such that $f \in \mathscr{T}_{2n}$, we can get similar results as in Section 9.

Acknowledgements The authors have been supported by the Serbian Academy of Sciences and Arts, Φ-96 (G. V. Milovanović) and by the Scientific Research Project Administration of Akdeniz University (Y. Simsek).

References

1. M. Abramowitz, I.A. Stegun, *Handbook of Mathematical Functions with Formulas, Graphs, and Mathematical Tables*. National Bureau of Standards Applied Mathematics Series, vol. 55 (Dover, New York, 1965)
2. T.M. Apostol, Generalized dedekind sums and transformation formulae of certain Lambert series. Duke Math. J. **17**, 147–157 (1950)
3. T.M. Apostol, On the Lerch zeta function. Pac. J. Math. **1**, 161–167 (1951)
4. T.M. Apostol, Theorems on generalized Dedekind sums. Pac. J. Math. **2**, 1–9 (1952)
5. T.M. Apostol, *Modular Functions and Dirichlet Series in Number Theory* (Springer, New York, 1976)
6. A. Bayad, Sommes de Dedekind elliptiques et formes de Jacobi. Ann. Instit. Fourier **51**(1), 29–42 (2001)
7. A. Bayad, Y. Simsek, Dedekind sums involving Jacobi modular forms and special values of Barnes zeta functions. Ann. Inst. Fourier, Grenoble **61**(5), 1977–1993 (2011)
8. M. Beck, Dedekind cotangent sums. Acta Arithmetica **109**(2), 109–130 (2003)
9. B.C. Berndt, On the Hurwitz zeta-function. Rocky Mt. J. Math. **2**(1), 151–157 (1972)
10. B.C. Berndt, Generalized Dedekind eta-Functions and generalized Dedekind sums. Trans. Am. Math. Soc. **178**, 495–508 (1973)
11. B.C. Berndt, Dedekind sums and a paper of G. H. Hardy. J. Lond. Math. Soc. (2). **13**(1), 129–137 (1976)
12. B.C. Berndt, Reciprocity theorems for Dedekind sums and generalizations. Adv. Math. **23**(3), 285–316 (1977)
13. B.C. Berndt, Analytic Eisenstein series, theta-functions and series relations in the spirit of Ramanujan. J. Reine Angew. Math. **303/304**, 332–365 (1978)
14. B.C. Berndt, L.A. Goldberg, Analytic properties of arithmetic sums arising in the theory of the classical theta-functions. SIAM J. Math. Anal. **15**, 143–150 (1984)
15. B.C. Berndt, B.P. Yeap, Explicit evaluations and reciprocity theorems for finite trigonometric sums. Adv. Appl. Math. **29**(3), 358–385 (2002)
16. L. Carlitz, Some theorems on generalized Dedekind sums. Pac. J. Math. **3**, 513–522 (1953)
17. H. Chen, On some trigonometric power sums. Int. J. Math. Math. Sci. **30**(3), 185–191 (2002)
18. T.S. Chihara, *An Introduction to Orthogonal Polynomials* (Gordon and Breach, New York, 1978)
19. J. Choi, Some Identities involving the Legendre's chi-function. Commun. Korean Math. Soc. **22**(2), 219–225 (2007)
20. J. Choi, D.S. Jang, H.M. Srivastava, A generalization of the Hurwitz-Lerch zeta function. Integral Transform. Spec. Funct. **19**(1–2), 65–79 (2008)
21. W. Chu, Reciprocal relations for trigonometric sums. Rocky Mt. J. Math. **48**(1), 121–140 (2018)

22. D. Cvijović, Integral representations of the Legendre chi function. J. Math. Anal. Appl. **332**, 1056–1062 (2007)
23. D. Cvijović, Summation formulae for finite cotangent sums. Appl. Math. Comput. **215**, 1135–1140 (2009)
24. D. Cvijović, Summation formulae for finite tangent and secant sums. Appl. Math. Comput. **218**, 741–745 (2011)
25. D. Cvijović, J. Klinkovski, Values of the Legendre chi and Hurwitz zeta functions at rational arguments. Math. Comp. **68**, 1623–1630 (1999)
26. D. Cvijović, J. Klinkovski, Finite cotangent sums and the Riemann zeta function. Math. Slovaca **50**(2), 149–157 (2000)
27. D. Cvijović, H.S. Srivastava, Summation of a family of finite secant sums. Appl. Math. Copmput. **190**, 590–598 (2007)
28. D. Cvijović, H.S. Srivastava, Closed-form summations of Dowker's and related trigonometric sums. J. Phys. A **45**(37), 374015 (2012)
29. U. Dieter, Cotangent sums, a further generalization of Dedekind sums. J. Number Theory **18**, 289–305 (1984)
30. J.S. Dowker, On Verlinde's formula for the dimensions of vector bundles on moduli spaces. J. Phys. A **25**(9), 2641–2648 (1992)
31. C M. da Fonseca, V. Kowalenko, On a finite sum with powers of cosines. Appl. Anal. Discrete Math. **7**, 354–377 (2013)
32. C.M. da Fonseca, M.L. Glasser, V. Kowalenko, Basic trigonometric power sums with applications. Ramanujan J. **42**(2), 401–428 (2017)
33. C.M. da Fonseca, M.L. Glasser, V. Kowalenko, Generalized cosecant numbers and trigonometric inverse power sums. Appl. Anal. Discrete Math. **42**(2), 70–109 (2018)
34. W. Gautschi, On the remainder term for analytic functions of Gauss-Lobatto and Gauss-Radau quadratures. Rocky Mt. J. Math. **21**, 209–226 (1991)
35. W. Gautschi, *Orthogonal Polynomials: Computation and Approximation* (Clarendon Press, Oxford, 2004)
36. W. Gautschi, S. Li, The remainder term for analytic functions of Gauss-Radau and Gauss-Lobatto quadrature rules with multiple end points. J. Comput. Appl. Math. **33**, 315–329 (1990)
37. W. Gautschi, S. Li, Gauss-Radau and Gauss-Lobatto quadratures with double end points. J. Comput. Appl. Math. **34**, 343–360 (1991)
38. W. Gautschi, G.V. Milovanović, s-orthogonality and construction of Gauss-Turán-type quadrature formulae. J. Comput. Appl. Math. **86**, 205–218 (1997)
39. A. Ghizzetti, A. Ossicini, *Quadrature Formulae* (Akademie Verlag, Berlin, 1970)
40. L.A. Goldberg, Transformation of Theta-Functions and Analogues of Dedekind Sums, Thesis, University of Illinois Urbana (1981)
41. P.J. Grabner, H. Prodinger, Secant and cosecant sums and Bernoulli-Nörlund polynomials. Quaest. Math. **30**, 159–165 (2007)
42. I.S. Gradshteyn, I.M. Ryzhik, *Table of Integrals, Series, and Products*, 7th edn. (Elsevier/Academic Press, Amsterdam, 2007)
43. E. Grosswald, Dedekind-Rademacher sums. Am. Math. Mon. **78**, 639–644 (1971)
44. E. Grosswald, Dedekind-Rademacher sums and their reciprocity formula. J. Reine. Angew. Math. **25**(1), 161–173 (1971)
45. J. Guillera, J. Sondow, Double integrals and infinite products for some classical constants via analytic continuations of Lerch's transcendent. Ramanujan J. **16**(3), 247–270 (2008)
46. R.R. Hall, J.C. Wilson, D. Zagier, Reciprocity formulae for general Dedekind-Rademacher sums. Acta Arith. **73**(4), 389–396 (1995)
47. E.R. Hansen, *A Table of Series and Products* (Prentice-Hall, Englewood Cliffs, 1975)
48. G.H. Hardy, On certain series of discontinous functions connected with the modular functions. Quart. J. Math. **36**, 93–123 (1905); [Collected Papers, vol. IV (Clarendon Press, Oxford, 1969), pp. 362–392]
49. M.E. Hoffman, Derivative polynomials and associated integer sequences. Electron. J. Combin. **6**, 13 (1999), Research Paper 21

50. T. Kim, Note on the Euler numbers and polynomials. Adv. Stud. Contemp. Math. **17**, 131–136 (2008)
51. T. Kim, Euler numbers and polynomials associated with zeta functions. Abstr. Appl. Anal. **2008**, 11 (2008), Art. ID 581582
52. T. Kim, Note on Dedekind type DC sums. Adv. Stud. Contemp. Math. (Kyungshang) **18**(2), 249–260 (2009)
53. M.I. Knoop, *Modular Functions in Analytic Number Theory* (Markham Publishing Company, Chicago, 1970)
54. N. Koblitz, *Introduction to Elliptic Curves and Modular Forms* (Springer, New York, 1993)
55. J. Lewittes, Analytic continuation of the series $\sum (m + nz)^{-s}$. Trans. Am. Math. Soc. **159**, 505–509 (1971)
56. J. Lewittes, Analytic continuation of Eisenstein series. Trans. Am. Math. Soc. **177**, 469–490 (1972)
57. A. Markov, Sur la méthode de Gauss pour le calcul approché des intégrales. Math. Ann. **25**, 427–432 (1885)
58. G. Mastroianni, G.V. Milovanović, *Interpolation Processes – Basic Theory and Applications*. Springer Monographs in Mathematics (Springer, Berlin/Heidelberg/New York, 2008)
59. G.V. Milovanović, Construction of s-orthogonal polynomials and Turán quadrature formulae, in *Numerical Methods and Approximation Theory III*, ed. by G.V. Milovanović (Niš, 1987) (Univ. Niš, Niš, 1988), pp. 311–328
60. G.V. Milovanović, Quadrature with multiple nodes, power orthogonality, and moment-preserving spline approximation. J. Comput. Appl. Math. **127**, 267–286 (2001)
61. G.V. Milovanović, D. Joksimović, On a connection between some trigonometric quadrature rules and Gauss-Radau formulas with respect to the Chebyshev weight. Bull. Cl. Sci. Math. Nat. Sci. Math. **39**, 79–88 (2014)
62. G.V. Milovanović, A.S. Cvetković, M.P. Stanić, Trigonometric orthogonal systems and quadrature formulae. Comput. Math. Appl. **56**, 2915–2931 (2008)
63. G.V. Milovanović, D.S. Mitrinović, T.M. Rassias, *Topics in Polynomials: Extremal Problems, Inequalities, Zeros* (World Scientific Publishing Co., Inc., River Edge, 1994)
64. G.V. Milovanović, M.S. Pranić, M.M. Spalević, Quadrature with multiple nodes, power orthogonality, and moment-preserving spline approximation, Part II. Appl. Anal. Discrete Math. **13**, 1–27 (2019)
65. G.V. Milovanović, T.M. Rassias (eds.), *Analytic Number Theory, Approximation Theory, and Special Functions*. In Honor of Hari M. Srivastava (Springer, New York, 2014)
66. G.V. Milovanović, T.M. Rassias, Inequalities connected with trigonometric sums, in *Constantin Carathéodory: an International Tribute*, vol. II (World Science Publications, Teaneck, 1991), pp. 875–941
67. G.V. Milovanović, M.M. Spalević, M.S. Pranić, On the remainder term of Gauss-Radau quadratures for analytic functions. J. Comput. Appl. Math. **218**, 281–289 (2008)
68. G.V. Milovanović, M.P. Stanić, Quadrature rules with multiple nodes, in *Mathematical Analysis, Approximation Theory and Their Applications*, ed. by T.M. Rassias, V. Gupta (Springer, Cham, 2016), pp. 435–462
69. S.E. Notaris, The error norm of Gauss-Radau quadrature formulae for Chebyshev weight functions. BIT Numer. Math. **50**, 123–147 (2010)
70. A.P. Prudnikov, Y.A. Brychkov, O.I. Marichev, *Integrals and Series*, vol. 1. Elementary functions (Gordon and Breach Science Publishers, New York, 1986)
71. A.P. Prudnikov, Y.A. Brychkov, O.I. Marichev, *Integrals and Series*, vol. 2 (Gordon and Breach, New York, 1986)
72. H. Rademacher, Zur Theorie der Modulfunktionen. J. Reine Angew. Math. **167**, 312–336 (1932)
73. H. Rademacher, Über eine Reziprozitätsformel aus der Theorie der Modulfunktionen. Mat. Fiz. Lapok **40**, 24–34 (1933) (Hungarian)
74. H. Rademacher, Die reziprozitatsformel für Dedekindsche Summen. Acta Sci. Math. (Szeged) **12**(B), 57–60 (1950)

75. H. Rademacher, Topics in Analytic Number Theory, Die Grundlehren der Math. Wissenschaften, Band 169 (Springer, Berlin, 1973)
76. H. Rademacher, E. Grosswald, *Dedekind Sums, The Carus Mathematical Monographs*, vol. 16 (The Mathematical Association of America, Washington, DC, 1972)
77. M.T. Rassias, L. Toth, Trigonometric representations of generalized Dedekind and Hardy sums via the discrete Fourier transform, in *Analytic Number Theory: In Honor of Helmut Maier's 60th Birthday*, ed. by C. Pomerance, M.T. Rassias (Springer, Cham, 2015), pp. 329–343
78. T. Schira, The remainder term for analytic functions of Gauss-Lobatto quadratures. J. Comput. Appl. Math. **76**, 171–193 (1996)
79. Y.G. Shi, On Turán quadrature formulas for the Chebyshev weight. J. Approx. Theory **96**, 101–110 (1999)
80. Y.G. Shi, On Gaussian quadrature formulas for the Chebyshev weight. J. Approx. Theory **98**, 183–195 (1999)
81. Y.G. Shi, Generalized Gaussian quadrature formulas with Chebyshev nodes. J. Comput. Math. **17**(2), 171–178 (1999)
82. B. Schoeneberg, *Elliptic Modular Functions: An Introduction*, Die Grundlehren der mathematischen Wissenschaften, Band 203 (Springer, New York/Heidelberg, 1974)
83. R. Sczech, Dedekind summen mit elliptischen Funktionen. Invent. Math. **76**, 523–551 (1984)
84. Y. Simsek, Relations between theta-functions Hardy sums Eisenstein and Lambert series in the transformation formula of $\log \eta_{g,h}(z)$. J. Number Theory **99**, 338–360 (2003)
85. Y. Simsek, Generalized Dedekind sums associated with the Abel sum and the Eisenstein and Lambert series. Adv. Stud. Contemp. Math. (Kyungshang) **9**(2), 125–137 (2004)
86. Y. Simsek, On generalized Hardy sums $S_5(h, k)$. Ukrainian Math. J. **56**(10), 1434–1440 (2004)
87. Y. Simsek, q-Dedekind type sums related to q-zeta function and basic L-series. J. Math. Anal. Appl. **318**(1), 333–351 (2006)
88. Y. Simsek, On analytic properties and character analogs of Hardy sums. Taiwanese J. Math. **13**(1), 253–268 (2009)
89. Y. Simsek, Special functions related to Dedekind-type DC-sums and their applications. Russ. J. Math. Phys. **17**(4), 495–508 (2010)
90. Y. Simsek, D. Kim, J.K. Koo, Oon elliptic analogue to the Hardy sums. Bull. Korean Math. Soc. **46**(1), 1–10 (2009)
91. R. Sitaramachandrarao, Dedekind and Hardy sums. Acta Arith. **48**, 325–340 (1987)
92. H.M. Srivastava, A note on the closed-form summation of some trigonometric series. Kobe J. Math. **16**(2), 177–182 (1999)
93. H.M. Srivastava, A. Pinter, Remarks on some relationships between the Bernoulli and Euler polynomials. Appl. Math. Lett. **17**(4), 375–380 (2004)
94. H.M. Srivastava, J. Choi, *Series Associated with the Zeta and Related Functions* (Kluwer Academic Publishers, Dordrecht/Boston/London, 2001)
95. T.J. Stijeltes, Note sur quelques formules pour l'évaluation de certaines intégrales, Bul. Astr. Paris 1, 568–569 (1884) [Oeuvres I, 426–427]
96. S.K. Suslov, Some Expansions in basic Fourier series and related topics. J. Approx. Theory **115**(2), 289–353 (2002)
97. P. Turán, On the theory of the mechanical quadrature. Acta Sci. Math. Szeged **12**, 30–37 (1950)
98. A.H. Turetzkii, On quadrature formulae that are exact for trigonometric polynomials. East J. Approx. 11(3), 337–335 (2005). [Translation in English from Uchenye Zapiski, Vypusk 1 (149). Seria Math. Theory of Functions, Collection of papers, Izdatel'stvo Belgosuniversiteta imeni V.I. Lenina, Minsk (1959), pp. 31–54]
99. E.T. Wittaker, G.N. Watson, A Course of Modern Analysis, 4th edn. (Cambridge University Press, Cambridge, 1962)
100. M. Waldschmidt, P. Moussa, J.M. Luck, C. Itzykson, *From Number Theory to Physics* (Springer, New York, 1995)
101. D. Zagier, Higher dimensional Dedekind sums. Math. Ann. **202**, 149–172 (1973)

On a Half-Discrete Hilbert-Type Inequality in the Whole Plane with the Kernel of Hyperbolic Secant Function Related to the Hurwitz Zeta Function

Michael Th. Rassias, Bicheng Yang, and Andrei Raigorodskii

Abstract Using weight functions, we obtain a half-discrete Hilbert-type inequality in the whole plane with the kernel of hyperbolic secant function and multi-parameters. The constant factor related to the Hurwitz zeta function is proved to be the best possible. We also consider equivalent forms, two kinds of particular inequalities, the operator expressions and some reverses.

Keywords Half-discrete Hilbert-type inequality · Weight function · Equivalent form · Operator expression · Hurwitz zeta function

2000 Mathematics Subject Classification 26D15, 30A10, 47A05

M. T. Rassias (✉)
Institute of Mathematics, University of Zurich, Zurich, Switzerland

Moscow Institute of Physics and Technology, Dolgoprudny, Russia

Institute for Advanced Study, Program in Interdisciplinary Studies, Princeton, NJ, USA
e-mail: michail.rassias@math.uzh.ch

B. Yang
Department of Mathematics, Guangdong University of Education,
Guangzhou, Guangdong, P. R. China
e-mail: bcyang@gdei.edu.cn; bcyang818@163.com

A. Raigorodskii
Moscow Institute of Physics and Technology, Dolgoprudny, Russia

Moscow State University, Moscow, Russia

Buryat State University, Ulan-Ude, Russia

Caucasus Mathematical Center, Adyghe State University, Maykop, Russia
e-mail: raigorodsky@yandex-team.ru

1 Introduction

Assuming that $p > 1$, $\frac{1}{p} + \frac{1}{q} = 1$, $a_m, b_n > 0$,

$$0 < \sum_{m=1}^{\infty} a_m^p < \infty, \ \ 0 < \sum_{n=1}^{\infty} b_n^q < \infty,$$

we have the following discrete Hardy-Hilbert inequality (cf. [1]):

$$\sum_{n=1}^{\infty} \sum_{m=1}^{\infty} \frac{a_m b_n}{m+n} < \frac{\pi}{\sin(\pi/p)} \left(\sum_{m=1}^{\infty} a_m^p \right)^{\frac{1}{p}} \left(\sum_{n=1}^{\infty} b_n^q \right)^{\frac{1}{q}}, \tag{1}$$

where the constant factor

$$\frac{\pi}{\sin(\pi/p)}$$

is the best possible.

If $f(x), g(y) \geq 0$,

$$0 < \int_0^{\infty} f^p(x)dx < \infty \ \text{ and } \ 0 < \int_0^{\infty} g^q(y)dy < \infty,$$

then we have the following Hardy-Hilbert integral inequality with the same best possible constant factor $\frac{\pi}{\sin(\pi/p)}$ (cf. [2]):

$$\int_0^{\infty} \int_0^{\infty} \frac{f(x)g(y)}{x+y}dxdy < \frac{\pi}{\sin(\pi/p)} \left(\int_0^{\infty} f^p(x)dx \right)^{\frac{1}{p}} \left(\int_0^{\infty} g^q(y)dy \right)^{\frac{1}{q}}. \tag{2}$$

Recently, the following half-discrete Hardy-Hilbert inequality with the same best possible constant factor was established (cf. [3]):

$$\sum_{n=1}^{\infty} \int_0^{\infty} \frac{b_n f(x)}{x+n}dx < \frac{\pi}{\sin(\pi/p)} \left(\int_0^{\infty} f^p(x)dx \right)^{\frac{1}{p}} \left(\sum_{n=1}^{\infty} b_n^q \right)^{\frac{1}{q}}. \tag{3}$$

Inequalities (1), (2) and (3) are fairly applicable in various domains of mathematical analysis (cf. [2, 4–6]).

We notice that the inequalities (1)–(3) involve a homogenous kernel of degree -1. In 2009, a survey of the study of Hilbert-type inequalities with homogeneous kernels having negative numbers as a degree was presented in [7]. A few inequalities with homogenous kernels of degree 0 and non-homogeneous kernels have been studied in [8–10]. Some other kinds of Hilbert-type inequalities are provided in [11–21, 24]. It is worth mentioning that all of the above inequalities are built in the quarter plane of the first quadrant.

In 2007, a Hilbert-type integral inequality in the whole plane was proved by Yang [22]. Additionally, the following Hilbert-type integral inequality in the whole plane was established in [23]:

$$\int_{-\infty}^{\infty} \int_{-\infty}^{\infty} \frac{1}{|1+xy|^{\lambda}} f(x)g(y)dxdy$$

$$< k_{\lambda} \left[\int_{-\infty}^{\infty} |x|^{p(1-\frac{\lambda}{2})-1} f^{p}(x)dx \right]^{\frac{1}{p}} \left[\int_{-\infty}^{\infty} |y|^{q(1-\frac{\lambda}{2})-1} g^{q}(y)dy \right]^{\frac{1}{q}}, \quad (4)$$

where the constant factor

$$k_{\lambda} = B\left(\frac{\lambda}{2}, \frac{\lambda}{2}\right) + 2B\left(1-\lambda, \frac{\lambda}{2}\right) \quad (0 < \lambda < 1)$$

is the best possible.

Furthermore He et al. [25–29] also proved some integral and discrete Hilbert-type inequalities in the whole plane.

In this chapter, using weight functions, we prove the following new half-discrete Hilbert-type inequality in the whole plane with the kernel of hyperbolic secant function

$$\sec h(u) := \frac{2}{e^{u} + e^{-u}} \quad (u \ge 0)$$

and a best possible constant factor:

$$\sum_{|n|=1}^{\infty} \int_{-\infty}^{\infty} sech\left(\rho\left(\frac{|n|}{|x|}\right)^{\gamma}\right) f(x)b_{n}dx$$

$$< \frac{4}{\gamma(2\rho)^{\sigma/\gamma}} \Gamma(\frac{\sigma}{\gamma}) \left(\zeta(\frac{\sigma}{\gamma}, \frac{1}{2}) - \frac{2}{2^{\sigma/\gamma}} \zeta(\frac{\sigma}{\gamma}, \frac{3}{4}) \right)$$

$$\times \left[\int_{-\infty}^{\infty} |x|^{p(1+\sigma)-1} f^{p}(x)dx \right]^{\frac{1}{p}} \left[\sum_{|n|=1}^{\infty} |n|^{q(1-\sigma)-1} b_{n}^{q} \right]^{\frac{1}{q}}, \quad (5)$$

where, $\rho > 0, 0 < \gamma < \sigma \le 1$, and

$$\zeta(s, a) := \sum_{k=0}^{\infty} \frac{1}{(k+a)^{s}} \quad (\text{Re}s > 1; 0 < a \le 1)$$

is the Hurwitz zeta function. Note that $\zeta(s) := \zeta(s, 1)$ is the Reimann zeta function (cf. [30]).

Moreover, an extension of (5) with multi-parameters is given in Theorem 1. The equivalent forms, two kinds of particular inequalities, the operator expressions and some reverses are also considered.

2 Weight Functions and Some Lemmas

In what follows, we assume that $p \in \mathbf{R}\backslash\{0, 1\}$, $\frac{1}{p} + \frac{1}{q} = 1$, $\delta \in \{-1, 1\}$, $\alpha, \beta \in (0, \pi)$, $\rho > 0$, $0 < \gamma < \sigma \leq 1$.

We set

$$h(x, y) := sech\left(\rho\left[\frac{|y| + y\cos\beta}{(|x| + x\cos\alpha)^\delta}\right]^\gamma\right)$$

$$= \frac{2}{e^{\rho\left[\frac{|y|+y\cos\beta}{(|x|+x\cos\alpha)^\delta}\right]^\gamma} + e^{-\rho\left[\frac{|y|+y\cos\beta}{(|x|+x\cos\alpha)^\delta}\right]^\gamma}} \quad (x \neq 0, y \neq 0), \qquad (6)$$

wherefrom,

$$h(x, y) = sech\left(\rho\left[\frac{y(1 + \cos\beta)}{(|x| + x\cos\alpha)^\delta}\right]^\gamma\right) \quad (y > 0),$$

$$h(x, y) = \sec h\left(\rho\left\{\frac{|y| + y\cos\beta}{[x(1 + \cos\alpha)]^\delta}\right\}^\gamma\right) \quad (x > 0),$$

$$h(-x, y) = sech\left(\rho\left\{\frac{|y| + y\cos\beta}{[x(1 - \cos\alpha)]^\delta}\right\}^\gamma\right) \quad (x > 0),$$

$$h(x, -y) = sech\left(\rho\left[\frac{y(1 - \cos\beta)}{(|x| + x\cos\alpha)^\delta}\right]^\gamma\right) \quad (y > 0).$$

Lemma 1 *We define two weight functions $\omega(\sigma, n)$ and $\varpi(\sigma, x)$ as follows:*

$$\omega(\sigma, n) := \int_{-\infty}^{\infty} h(x, n)\frac{(|n| + n\cos\beta)^\sigma}{(|x| + x\cos\alpha)^{1+\delta\sigma}}dx \quad (|n| \in \mathbf{N}), \qquad (7)$$

$$\varpi(\sigma, x) := \sum_{|n|=1}^{\infty} h(x, n)\frac{(|x| + x\cos\alpha)^{-\delta\sigma}}{(|n| + n\cos\beta)^{1-\sigma}} \quad (x \in \mathbf{R}\backslash\{0\}). \qquad (8)$$

Then we have

(i)

$$\omega(\sigma, n) = k_\alpha(\sigma) := \frac{4\csc^2\alpha}{\gamma(2\rho)^{\sigma/\gamma}}\Gamma(\frac{\sigma}{\gamma})$$

$$\times\left((\zeta(\frac{\sigma}{\gamma},\frac{1}{2}) - \frac{2}{2^{\sigma/\gamma}}\zeta(\frac{\sigma}{\gamma},\frac{3}{4}))\right) \in \mathbf{R}_+ \ (|n| \in \mathbf{N}); \tag{9}$$

(ii)

$$k_\beta(\sigma)(1 - \theta(\sigma, x)) < \varpi(\sigma, x) < k_\beta(\sigma) \ (x \in \mathbf{R}\backslash\{0\}), \tag{10}$$

where

$$\theta(\sigma, x) := \frac{2^{\sigma/\gamma}}{2\Gamma(\frac{\sigma}{\gamma})(\zeta(\frac{\sigma}{\gamma},\frac{1}{2}) - \frac{2}{2^{\sigma/\gamma}}\zeta(\frac{\sigma}{\gamma},\frac{3}{4}))}$$

$$\times \int_0^{\rho\left[\frac{1+\cos\beta}{(|x|+x\cos\alpha)^\delta}\right]^\gamma} u^{\frac{\sigma}{\gamma}-1}sech(u)du$$

$$= O\left(\frac{1}{(|x| + x\cos\alpha)^{\delta\sigma}}\right) \in (0, 1). \tag{11}$$

Proof We obtain

$$\omega(\sigma, n) = \int_{-\infty}^0 h(x, n)\frac{(|n| + n\cos\beta)^\sigma}{[x(\cos\alpha - 1)]^{1+\delta\sigma}}dx$$

$$+ \int_0^\infty h(x, n)\frac{(|n| + n\cos\beta)^\sigma}{[x(\cos\alpha + 1)]^{1+\delta\sigma}}dx$$

$$= \int_0^\infty h(-x, n)\frac{(|n| + n\cos\beta)^\sigma}{[x(1 - \cos\alpha)]^{1+\delta\sigma}}dx$$

$$+ \int_0^\infty h(x, n)\frac{(|n| + n\cos\beta)^\sigma}{[x(1 + \cos\alpha)]^{1+\delta\sigma}}dx.$$

Setting $u = \rho\left\{\frac{|n|+n\cos\beta}{[x(1-\cos\alpha)]^\delta}\right\}^\gamma$ $\left(u = \rho\left\{\frac{|n|+n\cos\beta}{[x(1+\cos\alpha)]^\delta}\right\}^\gamma\right)$ in the above first (respectively second) integral, by Lebesgue term by term integrations (cf. [31]), we derive that

$$\omega(\sigma, n) = \left(\frac{1}{1 - \cos\alpha} + \frac{1}{1 + \cos\alpha}\right)\frac{1}{\gamma\rho^{\sigma/\gamma}}\int_0^\infty u^{\frac{\sigma}{\gamma}-1}sech(u)du$$

$$= \frac{4\csc^2\alpha}{\gamma\rho^{\sigma/\gamma}}\int_0^\infty \frac{e^{-u}u^{\frac{\sigma}{\gamma}-1}du}{1+e^{-2u}} = \frac{4\csc^2\alpha}{\gamma\rho^{\sigma/\gamma}}\int_0^\infty \sum_{k=0}^\infty \frac{(-1)^k u^{\frac{\sigma}{\gamma}-1}}{e^{(2k+1)u}}du$$

$$= \frac{4 \csc^2 \alpha}{\gamma \rho^{\sigma/\gamma}} \int_0^\infty \sum_{k=0}^\infty [e^{-(4k+1)u} - e^{-(4k+3)u}] u^{\frac{\sigma}{\gamma}-1} du$$

$$= \frac{4 \csc^2 \alpha}{\gamma \rho^{\sigma/\gamma}} \sum_{k=0}^\infty \int_0^\infty [e^{-(4k+1)u} - e^{-(4k+3)u}] u^{\frac{\sigma}{\gamma}-1} du$$

$$= \frac{4 \csc^2 \alpha}{\gamma \rho^{\sigma/\gamma}} \sum_{k=0}^\infty \int_0^\infty (-1)^k e^{-(2k+1)u} u^{\frac{\sigma}{\gamma}-1} du \ (v = (2k+1)u)$$

$$= \frac{4 \csc^2 \alpha}{\gamma \rho^{\sigma/\gamma}} \int_0^\infty e^{-v} v^{\frac{\sigma}{\gamma}-1} dv \sum_{k=0}^\infty \frac{(-1)^k}{(2k+1)^{\sigma/\gamma}}$$

$$= \frac{4 \csc^2 \alpha}{\gamma \rho^{\sigma/\gamma}} \Gamma(\frac{\sigma}{\gamma}) \sum_{k=0}^\infty \frac{(-1)^k}{(2k+1)^{\sigma/\gamma}},$$

where

$$\int_0^\infty e^{-v} v^{a-1} dv = \Gamma(a) \ (a > 0)$$

is the gamma function (cf. [30]).

Since $\frac{\sigma}{\gamma} > 1$, we obtain that

$$\sum_{k=0}^\infty \frac{(-1)^k}{(2k+1)^{\sigma/\gamma}} = \sum_{k=0}^\infty \frac{1}{(2k+1)^{\sigma/\gamma}} - 2 \sum_{k=0}^\infty \frac{1}{(4k+3)^{\sigma/\gamma}}$$

$$= \frac{1}{2^{\sigma/\gamma}} \left(\zeta(\frac{\sigma}{\gamma}, \frac{1}{2}) - \frac{2}{2^{\sigma/\gamma}} \zeta(\frac{\sigma}{\gamma}, \frac{3}{4}) \right),$$

from which we deduce (9).

We obtain

$$\varpi(\sigma, x) = \sum_{n=-1}^{-\infty} h(x, n) \frac{(|x| + x \cos \alpha)^{-\delta\sigma}}{(|n| + n \cos \beta)^{1-\sigma}}$$

$$+ \sum_{n=1}^\infty h(x, n) \frac{(|x| + x \cos \alpha)^{-\delta\sigma}}{(|n| + n \cos \beta)^{1-\sigma}}$$

$$= \frac{(|x| + x \cos \alpha)^{-\delta\sigma}}{(1 - \cos \beta)^{1-\sigma}} \sum_{n=1}^\infty \frac{h(x, -n)}{n^{1-\sigma}}$$

$$+ \frac{(|x| + x \cos \alpha)^{-\delta\sigma}}{(1 + \cos \beta)^{1-\sigma}} \sum_{n=1}^\infty \frac{h(x, n)}{n^{1-\sigma}}. \qquad (12)$$

Since

$$\frac{d}{du}sech(u^\gamma) = \frac{-2\gamma u^{\gamma-1}(e^{u^\gamma} - e^{-u^\gamma})}{(e^{u^\gamma} + e^{-u^\gamma})^2} < 0 \ (\gamma > 0),$$

we observe that for $0 < \sigma \le 1$, both

$$\frac{h(x, -y)}{y^{1-\sigma}} \quad \text{and} \quad \frac{h(x, y)}{y^{1-\sigma}}$$

are strictly decreasing in $y \in (0, \infty)$. By (12) and this decreasing property, we have

$$\varpi(\sigma, x) < \frac{(|x| + x\cos\alpha)^{-\delta\sigma}}{(1 - \cos\beta)^{1-\sigma}} \int_0^\infty \frac{h(x, -y)}{y^{1-\sigma}}dy$$

$$+ \frac{(|x| + x\cos\alpha)^{-\delta\sigma}}{(1 + \cos\beta)^{1-\sigma}} \int_0^\infty \frac{h(x, y)}{y^{1-\sigma}}dy.$$

Setting $u = \rho\left[\frac{y(1-\cos\beta)}{(|x|+x\cos\alpha)^\delta}\right]^\gamma$ $\left(u = \rho\left[\frac{y(1+\cos\beta)}{(|x|+x\cos\alpha)^\delta}\right]^\gamma\right)$ in the above first (respectively second) integral, and by carrying out the corresponding simplifications, we obtain

$$\varpi(\sigma, x) < \frac{4\csc^2\beta}{\gamma(2\rho)^{\sigma/\gamma}}\Gamma(\frac{\sigma}{\gamma})\left(\zeta(\frac{\sigma}{\gamma}, \frac{1}{2}) - \frac{2}{2^{\sigma/\gamma}}\zeta(\frac{\sigma}{\gamma}, \frac{3}{4})\right)$$

$$= k_\beta(\sigma).$$

By (12) and the decreasing property, we obtain that

$$\varpi(\sigma, x) > \frac{(|x| + x\cos\alpha)^{-\delta\sigma}}{(1 - \cos\beta)^{1-\sigma}} \int_1^\infty \frac{h(x, -y)dy}{y^{1-\sigma}}$$

$$+ \frac{(|x| + x\cos\alpha)^{-\delta\sigma}}{(1 + \cos\beta)^{1-\sigma}} \int_1^\infty \frac{h(x, y)dy}{y^{1-\sigma}}.$$

Still setting $u = \rho\left[\frac{y(1-\cos\beta)}{(|x|+x\cos\alpha)^\delta}\right]^\gamma$ $\left(u = \rho\left[\frac{y(1+\cos\beta)}{(|x|+x\cos\alpha)^\delta}\right]^\gamma\right)$ in the above first (respectively second) integral, and by carrying out the corresponding simplifications, we have

$$\varpi(\sigma, x) > \frac{1}{\gamma\rho^{\sigma/\gamma}(1 - \cos\beta)} \int_{\rho\left[\frac{1-\cos\beta}{(|x|+x\cos\alpha)^\delta}\right]^\gamma}^\infty u^{\frac{\sigma}{\gamma}-1}sech(u)du$$

$$+ \frac{1}{\gamma\rho^{\sigma/\gamma}(1 + \cos\beta)} \int_{\rho\left[\frac{1+\cos\beta}{(|x|+x\cos\alpha)^\delta}\right]^\gamma}^\infty u^{\frac{\sigma}{\gamma}-1}sech(u)du$$

$$\geq \frac{2\csc^2\beta}{\gamma\rho^{\sigma/\gamma}}\int_\rho^\infty \left[\frac{1+\cos\beta}{(|x|+x\cos\alpha)^\delta}\right]^\gamma u^{\frac{\sigma}{\gamma}-1}sech(u)du$$

$$= k_\beta(\sigma)(1-\theta(\sigma,x)) > 0.$$

We deduce that

$$\lim_{u\to 0} sech(u) = 1, \ \lim_{u\to\infty} sech(u) = 0$$

and then $0 < sech(u) \leq 1$ ($u \in (0,\infty)$). Hence, we have

$$0 < \theta(\sigma,x) = \frac{2^{\sigma/\gamma}}{2\Gamma(\frac{\sigma}{\gamma})(\zeta(\frac{\sigma}{\gamma},\frac{1}{2}) - \frac{2}{2^{\sigma/\gamma}}\zeta(\frac{\sigma}{\gamma},\frac{3}{4}))}$$

$$\times \int_0^\rho \left[\frac{1+\cos\beta}{(|x|+x\cos\alpha)^\delta}\right]^\gamma u^{\frac{\sigma}{\gamma}-1}sech(u)du$$

$$\leq \frac{2^{\sigma/\gamma}}{2\Gamma(\frac{\sigma}{\gamma})(\zeta(\frac{\sigma}{\gamma},\frac{1}{2}) - \frac{2}{2^{\sigma/\gamma}}\zeta(\frac{\sigma}{\gamma},\frac{3}{4}))}\int_0^\rho \left[\frac{1+\cos\beta}{(|x|+x\cos\alpha)^\delta}\right]^\gamma u^{\frac{\sigma}{\gamma}-1}du$$

$$= \frac{\gamma(2\rho)^{\sigma/\gamma}}{2\sigma\Gamma(\frac{\sigma}{\gamma})(\zeta(\frac{\sigma}{\gamma},\frac{1}{2}) - \frac{2}{2^{\sigma/\gamma}}\zeta(\frac{\sigma}{\gamma},\frac{3}{4}))}\left[\frac{1+\cos\beta}{(|x|+x\cos\alpha)^\delta}\right]^\sigma,$$

namely, (10) and (11) follow.

Lemma 2 *If $\varepsilon > 0$,*

$$H_\varepsilon(\beta) := \sum_{|n|=1}^\infty \frac{1}{(|n|+n\cos\beta)^{1+\varepsilon}},$$

then it holds

$$H_\varepsilon(\beta) = \frac{1}{\varepsilon}(2\csc^2\beta + o_1(1))(1+o_2(1)) \ (\varepsilon \to 0^+). \tag{13}$$

Proof We have

$$H_\varepsilon(\beta) = \sum_{n=-1}^{-\infty}\frac{1}{[n(\cos\beta-1)]^{1+\varepsilon}} + \sum_{n=1}^\infty \frac{1}{[n(\cos\beta+1)]^{1+\varepsilon}}$$

$$= \left[\frac{1}{(1-\cos\beta)^{1+\varepsilon}} + \frac{1}{(1+\cos\beta)^{1+\varepsilon}}\right]\sum_{n=1}^\infty \frac{1}{n^{1+\varepsilon}}. \tag{14}$$

By (14) and the decreasing property, we find

$$H_\varepsilon(\beta) = \left[\frac{1}{(1+\cos\beta)^{1+\varepsilon}} + \frac{1}{(1-\cos\beta)^{1+\varepsilon}}\right]\left(1 + \sum_{n=2}^{\infty}\frac{1}{n^{1+\varepsilon}}\right)$$

$$< \left[\frac{1}{(1+\cos\beta)^{1+\varepsilon}} + \frac{1}{(1-\cos\beta)^{1+\varepsilon}}\right]\left(1 + \int_{1}^{\infty}\frac{dy}{y^{1+\varepsilon}}\right)$$

$$= \frac{1}{\varepsilon}(2\csc^2\beta + o_1(1))(1+\varepsilon),$$

$$H_\varepsilon(\beta) > \left[\frac{1}{(1+\cos\beta)^{1+\varepsilon}} + \frac{1}{(1-\cos\beta)^{1+\varepsilon}}\right]\int_{1}^{\infty}\frac{dy}{y^{1+\varepsilon}}$$

$$= \frac{1}{\varepsilon}(2\csc^2\beta + o_1(1)).$$

Hence, we obtain (13).

Lemma 3 *For* $\varepsilon > 0$, *setting*

$$E_\delta := \{x \in \mathbf{R}\backslash\{0\}; \frac{1}{(|x| + x\cos\alpha)^\delta} \geq 1\},$$

we have

$$H_\delta := \int_{E_\delta}\frac{1}{(|x| + x\cos\alpha)^{1+\delta\varepsilon}}dx = \frac{2}{\varepsilon}\csc^2\alpha. \qquad (15)$$

Proof Setting

$$E_\delta^+ := \{x > 0; \frac{1}{[x(1+\cos\alpha)]^\delta} \geq 1\},$$

$$E_\delta^- := \{x < 0; \frac{1}{[(-x)(1-\cos\alpha)]^\delta} \geq 1\},$$

it follows that $E_\delta = E_\delta^+ \cup E_\delta^-$ and

$$H_\delta = \frac{1}{(1+\cos\alpha)^{1+\delta\varepsilon}}\int_{E_\delta^+}\frac{1}{x^{1+\delta\varepsilon}}dx$$

$$+ \frac{1}{(1-\cos\alpha)^{1+\delta\varepsilon}}\int_{E_\delta^-}\frac{1}{(-x)^{1+\delta\varepsilon}}dx.$$

Setting $u = [x(1+\cos\alpha)]^\delta$ $(u = [(-x)(1-\cos\alpha)]^\delta)$ in the above first (respectively second) integral, we obtain

$$H_\delta = \left(\frac{1}{1+\cos\alpha} + \frac{1}{1-\cos\alpha}\right) \int_1^\infty \frac{du}{u^{1+\varepsilon}} = \frac{2}{\varepsilon}\csc^2\alpha,$$

namely, (15) follows.

3   Main Results

Theorem 1 *Suppose that $p > 1$,*

$$K_{\alpha,\beta}(\sigma) := k_\alpha^{\frac{1}{q}}(\sigma)k_\beta^{\frac{1}{p}}(\sigma) = \frac{4}{\gamma(2\rho)^{\sigma/\gamma}}\Gamma\left(\frac{\sigma}{\gamma}\right)$$

$$\times\left(\zeta\left(\frac{\sigma}{\gamma},\frac{1}{2}\right) - \frac{2}{2^{\sigma/\gamma}}\zeta\left(\frac{\sigma}{\gamma},\frac{3}{4}\right)\right)\csc^{\frac{2}{q}}\alpha\csc^{\frac{2}{p}}\beta. \qquad (16)$$

If $f(x), b_n \geq 0$, satisfying

$$0 < \int_{-\infty}^\infty (|x| + x\cos\alpha)^{p(1+\delta\sigma)-1}f^p(x)dx < \infty,$$

$$0 < \sum_{|n|=1}^\infty (|n| + n\cos\beta)^{q(1-\sigma)-1}b_n^q < \infty,$$

then we have the following equivalent inequalities:

$$I := \sum_{|n|=1}^\infty \int_{-\infty}^\infty \text{sech}\left(\rho\left[\frac{|n| + n\cos\beta}{(|x| + x\cos\alpha)^\delta}\right]^\gamma\right) f(x)b_n dx$$

$$< K_{\alpha,\beta}(\sigma)\left[\int_{-\infty}^\infty (|x| + x\cos\alpha)^{p(1+\delta\sigma)-1}f^p(x)dx\right]^{\frac{1}{p}}$$

$$\times\left[\sum_{|n|=1}^\infty (|n| + n\cos\beta)^{q(1-\sigma)-1}b_n^q\right]^{\frac{1}{q}}, \qquad (17)$$

$$J_1 \; := \; \left\{ \sum_{|n|=1}^{\infty} (|n| + n \cos \beta)^{p\sigma - 1} \right.$$

$$\left. \times \left[\int_{-\infty}^{\infty} sech \left(\rho \left[\frac{|n| + n \cos \beta}{(|x| + x \cos \alpha)^{\delta}} \right]^{\gamma} \right) f(x) dx \right]^p \right\}^{\frac{1}{p}}$$

$$< K_{\alpha,\beta}(\sigma) \left[\int_{-\infty}^{\infty} (|x| + x \cos \alpha)^{p(1+\delta\sigma)-1} f^p(x) dx \right]^{\frac{1}{p}}, \tag{18}$$

$$J_2 \; := \; \left\{ \int_{-\infty}^{\infty} (|x| + x \cos \alpha)^{-q\delta\sigma - 1} \right.$$

$$\left. \times \left[\sum_{|n|=1}^{\infty} sech \left(\rho \left[\frac{|n| + n \cos \beta}{(|x| + x \cos \alpha)^{\delta}} \right]^{\gamma} \right) b_n \right]^q dx \right\}^{\frac{1}{q}}$$

$$< K_{\alpha,\beta}(\sigma) \left[\sum_{|n|=1}^{\infty} (|n| + n \cos \beta)^{q(1-\sigma)-1} b_n^q \right]^{\frac{1}{q}}. \tag{19}$$

In particular, for $\alpha = \beta = \frac{\pi}{2}$, we have the following equivalent inequalities:

$$\sum_{|n|=1}^{\infty} \int_{-\infty}^{\infty} sech \left(\rho \left(\frac{|n|}{|x|^{\delta}} \right)^{\gamma} \right) f(x) b_n dx$$

$$< \frac{4}{\gamma (2\rho)^{\sigma/\gamma}} \Gamma(\frac{\sigma}{\gamma}) \left(\zeta(\frac{\sigma}{\gamma}, \frac{1}{2}) - \frac{2}{2^{\sigma/\gamma}} \zeta(\frac{\sigma}{\gamma}, \frac{3}{4}) \right)$$

$$\times \left(\int_{-\infty}^{\infty} |x|^{p(1+\delta\sigma)-1} f^p(x) dx \right)^{\frac{1}{p}} \left(\sum_{|n|=1}^{\infty} |n|^{q(1-\sigma)-1} b_n^q \right)^{\frac{1}{q}}, \tag{20}$$

$$\left\{ \sum_{|n|=1}^{\infty} |n|^{p\sigma - 1} \left[\int_{-\infty}^{\infty} sech \left(\rho \left(\frac{|n|}{|x|^{\delta}} \right)^{\gamma} \right) f(x) dx \right]^p \right\}^{\frac{1}{p}}$$

$$< \frac{4}{\gamma (2\rho)^{\sigma/\gamma}} \Gamma(\frac{\sigma}{\gamma}) \left(\zeta(\frac{\sigma}{\gamma}, \frac{1}{2}) - \frac{2}{2^{\sigma/\gamma}} \zeta(\frac{\sigma}{\gamma}, \frac{3}{4}) \right)$$

$$\times \left[\int_{-\infty}^{\infty} |x|^{p(1+\delta\sigma)-1} f^p(x) dx \right]^{\frac{1}{p}}, \tag{21}$$

$$\left\{\int_{-\infty}^{\infty} |x|^{-q\delta\sigma-1} \left[\sum_{|n|=1}^{\infty} sech\left(\rho\left(\frac{|n|}{|x|^{\delta}}\right)^{\gamma}\right) b_n\right]^q dx\right\}^{\frac{1}{q}} < \frac{4}{\gamma(2\rho)^{\sigma/\gamma}} \Gamma(\frac{\sigma}{\gamma})$$

$$\times \left(\zeta(\frac{\sigma}{\gamma},\frac{1}{2}) - \frac{2}{2^{\sigma/\gamma}}\zeta(\frac{\sigma}{\gamma},\frac{3}{4})\right) \left[\sum_{|n|=1}^{\infty} |n|^{q(1-\sigma)-1} b_n^q\right]^{\frac{1}{q}}. \tag{22}$$

Proof By Hölder's inequality with weight (cf. [32]) and (7), we obtain

$$\left[\int_{-\infty}^{\infty} h(x,n) f(x) dx\right]^p$$

$$= \left[\int_{-\infty}^{\infty} h(x,n) \frac{(|x|+x\cos\alpha)^{(1+\delta\sigma)/q} f(x)}{(|n|+n\cos\beta)^{(1-\sigma)/p}} \frac{(|n|+n\cos\beta)^{(1-\sigma)/p}}{(|x|+x\cos\alpha)^{(1+\delta\sigma)/q}} dx\right]^p$$

$$\leq \int_{-\infty}^{\infty} h(x,n) \frac{(|x|+x\cos\alpha)^{(1+\delta\sigma)(p-1)}}{(|n|+n\cos\beta)^{1-\sigma}} f^p(x) dx$$

$$\times \left[\int_{-\infty}^{\infty} h(x,n) \frac{(|n|+n\cos\beta)^{(1-\sigma)(q-1)}}{(|x|+x\cos\alpha)^{1+\delta\sigma}} dx\right]^{p-1}$$

$$= \frac{\omega^{p-1}(\sigma,n)}{(|n|+n\cos\beta)^{p\sigma-1}} \int_{-\infty}^{\infty} h(x,n) \frac{(|x|+x\cos\alpha)^{(1+\delta\sigma)(p-1)}}{(|n|+n\cos\beta)^{1-\sigma}} f^p(x) dx.$$

Then by (9) and the Lebesgue term by term integration theorem (cf. [31]), in view of (8), we deduce that

$$J_1 \leq k_\alpha^{\frac{1}{q}}(\sigma) \left[\sum_{|n|=1}^{\infty} \int_{-\infty}^{\infty} h(x,n) \frac{(|x|+x\cos\alpha)^{(1+\delta\sigma)(p-1)}}{(|n|+n\cos\beta)^{1-\sigma}} f^p(x) dx\right]^{\frac{1}{p}}$$

$$= k_\alpha^{\frac{1}{q}}(\sigma) \left[\int_{-\infty}^{\infty} \sum_{|n|=1}^{\infty} h(x,n) \frac{(|x|+x\cos\alpha)^{(1+\delta\sigma)(p-1)}}{(|n|+n\cos\beta)^{1-\sigma}} f^p(x) dx\right]^{\frac{1}{p}}$$

$$= k_\alpha^{\frac{1}{q}}(\sigma) \left[\int_{-\infty}^{\infty} \varpi(\sigma,x)(|x|+x\cos\alpha)^{p(1+\delta\sigma)-1} f^p(x) dx\right]^{\frac{1}{p}}. \tag{23}$$

Hence, by (10), since

$$K_{\alpha,\beta}(\sigma) = k_\alpha^{\frac{1}{q}}(\sigma) k_\beta^{\frac{1}{p}}(\sigma),$$

we derive (18).

By Hölder's inequality (cf. [32]), we have

$$I = \sum_{|n|=1}^{\infty} \left[(|n| + n\cos\beta)^{\frac{-1}{p}+\sigma} \int_{-\infty}^{\infty} h(x,n) f(x) dx \right] \left[(|n| + n\cos\beta)^{\frac{1}{p}-\sigma} b_n \right]$$

$$\leq J_1 \left[\sum_{|n|=1}^{\infty} (|n| + n\cos\beta)^{q(1-\sigma)-1} b_n^q \right]^{\frac{1}{q}}. \tag{24}$$

Then by (18), we derive (17). On the other hand, assuming that (17) is valid, we set

$$b_n := (|n| + n\cos\beta)^{p\sigma-1} \left[\int_{-\infty}^{\infty} h(x,n) f(x) dx \right]^{p-1} \quad (|n| \in \mathbf{N}).$$

Then we obtain

$$J_1 = \left[\sum_{|n|=1}^{\infty} (|n| + n\cos\beta)^{q(1-\sigma)-1} b_n^q \right]^{\frac{1}{p}}.$$

In view of (23), it follows that $J_1 < \infty$. If $J_1 = 0$, then (18) is trivially valid; if $J_1 > 0$, then by (17), we have

$$\sum_{|n|=1}^{\infty} (|n| + n\cos\beta)^{q(1-\sigma)-1} b_n^q$$

$$= J_1^p = I < K_{\alpha,\beta}(\sigma) \left[\int_{-\infty}^{\infty} (|x| + x\cos\alpha)^{p(1+\delta\sigma)-1} f^p(x) dx \right]^{\frac{1}{p}}$$

$$\times \left[\sum_{|n|=1}^{\infty} (|n| + n\cos\beta)^{q(1-\sigma)-1} b_n^q \right]^{\frac{1}{q}},$$

$$J_1 = \left[\sum_{|n|=1}^{\infty} (|n| + n\cos\beta)^{q(1-\sigma)-1} b_n^q \right]^{1-\frac{1}{q}}$$

$$< K_{\alpha,\beta}(\sigma) \left[\int_{-\infty}^{\infty} (|x| + x\cos\alpha)^{p(1+\delta\sigma)-1} f^p(x) dx \right]^{\frac{1}{p}},$$

namely, (18) follows, which is equivalent to (17).

By Hölder's inequality with weight, we still obtain

$$
\left[\sum_{|n|=1}^{\infty} h(x,n) b_n \right]^q
$$

$$
= \left[\sum_{|n|=1}^{\infty} h(x,n) \frac{(|x| + x\cos\alpha)^{(1+\delta\sigma)/q}}{(|n| + n\cos\beta)^{(1-\sigma)/p}} \frac{(|n| + n\cos\beta)^{(1-\sigma)/p}}{(|x| + x\cos\alpha)^{(1+\delta\sigma)/q}} b_n \right]^q
$$

$$
\leq \left[\sum_{|n|=1}^{\infty} h(x,n) \frac{(|x| + x\cos\alpha)^{(1+\delta\sigma)(p-1)}}{(|n| + n\cos\beta)^{1-\sigma}} \right]^{q-1}
$$

$$
\times \sum_{|n|=1}^{\infty} h(x,n) \frac{(|n| + n\cos\beta)^{(1-\sigma)(q-1)}}{(|x| + x\cos\alpha)^{1+\delta\sigma}} b_n^q
$$

$$
= \frac{(\varpi(\sigma,x))^{q-1}}{(|x| + x\cos\alpha)^{-q\delta\sigma-1}} \sum_{|n|=1}^{\infty} h(x,n) \frac{(|n| + n\cos\beta)^{(1-\sigma)(q-1)}}{(|x| + x\cos\alpha)^{1+\delta\sigma}} b_n^q.
$$

By (10) and the Lebesgue term by term theorem, we have

$$
J_2 < k_\alpha^{\frac{1}{p}}(\sigma) \left[\int_{-\infty}^{\infty} \sum_{|n|=1}^{\infty} h(x,n) \frac{(|n| + n\cos\beta)^{(1-\sigma)(q-1)}}{(|x| + x\cos\alpha)^{1+\delta\sigma}} b_n^q dx \right]^{\frac{1}{q}}
$$

$$
= k_\alpha^{\frac{1}{p}}(\sigma) \left[\sum_{|n|=1}^{\infty} \omega(\sigma,n)(|n| + n\cos\beta)^{q(1-\sigma)-1} b_n^q \right]^{\frac{1}{q}}. \tag{25}
$$

Hence, by (9), we deduce (19).

We have proved that (17) is valid. Setting

$$
f(x) := (|x| + x\cos\alpha)^{-q\delta\sigma-1} \left[\sum_{|n|=1}^{\infty} h(x,n) b_n \right]^{q-1} \quad (x \in \mathbf{R}\backslash\{0\}),
$$

it follows that

$$
J_2 = \left[\int_{-\infty}^{\infty} (|x| + x\cos\alpha)^{p(1+\delta\sigma)-1} f^p(x) dx \right]^{\frac{1}{q}},
$$

and in view of (25), we find $J_2 < \infty$. If $J_2 = 0$, then (19) is trivially valid; if $J_2 > 0$, then by (17), we have

$$\int_{-\infty}^{\infty} (|x| + x \cos \alpha)^{p(1+\delta\sigma)-1} f^p(x) dx$$

$$= J_2^q = I < K_{\alpha,\beta}(\sigma) \left[\int_{-\infty}^{\infty} (|x| + x \cos \alpha)^{p(1+\delta\sigma)-1} f^p(x) dx \right]^{\frac{1}{p}}$$

$$\times \left[\sum_{|n|=1}^{\infty} (|n| + n \cos \beta)^{q(1-\sigma)-1} b_n^q \right]^{\frac{1}{q}},$$

$$J_2 = \left[\int_{-\infty}^{\infty} (|x| + x \cos \alpha)^{p(1+\delta\sigma)-1} f^p(x) dx \right]^{1-\frac{1}{p}}$$

$$< K_{\alpha,\beta}(\sigma) \left[\sum_{|n|=1}^{\infty} (|n| + n \cos \beta)^{q(1-\sigma)-1} b_n^q \right]^{\frac{1}{q}},$$

namely, (19) follows.

On the other hand, assuming that (19) is valid, by Hölder's inequality and similarly to as we derived (24), we have

$$I \leq \left[\int_{-\infty}^{\infty} (|x| + x \cos \alpha)^{p(1+\delta\sigma)-1} f^p(x) dx \right]^{\frac{1}{p}} J_2. \tag{26}$$

Then by (19), we obtain (17), which is equivalent to (19).

Therefore, inequalities (17), (18) and (19) are equivalent.

Theorem 2 *With respect to the assumptions of Theorem 1, the constant factor $K_{\alpha,\beta}(\sigma)$ in (17), (18) and (19) is the best possible.*

Proof For $0 < \varepsilon < q\sigma$, we set $\tilde{\sigma} = \sigma - \frac{\varepsilon}{q}$ $(\in (0, 1))$,

$$\tilde{f}(x) := \begin{cases} \dfrac{1}{(|x|+x \cos \alpha)^{\delta(\sigma+\frac{\varepsilon}{p})+1}}, & x \in E_\delta, \\ 0, & x \in \mathbf{R} \backslash E_\delta, \end{cases}$$

and

$$\tilde{b}_n := (|n| + n \cos \beta)^{(\sigma-\frac{\varepsilon}{q})-1} \quad (|n| \in \mathbf{N}).$$

By (13) and (15), we obtain that

$$\widetilde{I}_1 := \left[\int_{-\infty}^{\infty} (|x| + x \cos \alpha)^{p(1+\delta\sigma)-1} \widetilde{f}^p(x) dx \right]^{\frac{1}{p}}$$

$$\times \left[\sum_{|n|=1}^{\infty} (|n| + n \cos \beta)^{q(1-\sigma)-1} \widetilde{b}_n^q \right]^{\frac{1}{q}}$$

$$= \left[\int_{-\infty}^{\infty} \frac{dx}{(|x| + x \cos \alpha)^{\delta\varepsilon+1}} \right]^{\frac{1}{p}} \left[\sum_{|n|=1}^{\infty} \frac{1}{(|n| + n \cos \beta)^{\varepsilon+1}} \right]^{\frac{1}{q}}$$

$$\le \frac{1}{\varepsilon} \left(2 \csc^2 \alpha \right)^{\frac{1}{p}} \left[(2 \csc^2 \beta + o_1(1))(1 + o_2(1)) \right]^{\frac{1}{q}}.$$

By (10), we have

$$\widetilde{I} : = \sum_{|n|=1}^{\infty} \int_{-\infty}^{\infty} h(x, n) \widetilde{f}(x) \widetilde{b}_n dx$$

$$= \int_{E_\delta} \sum_{|n|=1}^{\infty} h(x, n) \frac{(|x| + x \cos \alpha)^{-\delta(\widetilde{\sigma}+\varepsilon)-1}}{(|n| + n \cos \beta)^{1-\widetilde{\sigma}}} dx$$

$$= \int_{E_\delta} \frac{\varpi(\widetilde{\sigma}, x)}{(|x| + x \cos \alpha)^{\delta\varepsilon+1}} dx \ge k_\beta(\widetilde{\sigma}) \int_{E_\delta} \frac{1 - \theta(\widetilde{\sigma}, x)}{(|x| + x \cos \alpha)^{\delta\varepsilon+1}} dx$$

$$= k_\beta(\widetilde{\sigma}) \left[\int_{E_\delta} \frac{dx}{(|x| + x \cos \alpha)^{\delta\varepsilon+1}} - \int_{E_\delta} \frac{dx}{O((|x| + x \cos \alpha)^{\delta(\sigma+\frac{\varepsilon}{p})+1})} \right]$$

$$= \frac{1}{\varepsilon} k_\beta(\sigma - \frac{\varepsilon}{q})(2 \csc^2 \alpha - \varepsilon O(1)).$$

If the constant factor $K_{\alpha,\beta}(\sigma)$ in (17) is not the best possible, then, there exists a positive number k, with $K_{\alpha,\beta}(\sigma) > k$, such that (17) is valid when replacing $K_{\alpha,\beta}(\sigma)$ by k. Thus, in particular, we have $\varepsilon \widetilde{I} < \varepsilon k \widetilde{I}_1$, namely,

$$k_\beta(\sigma - \frac{\varepsilon}{q})(2 \csc^2 \alpha - \varepsilon O(1))$$

$$< k \cdot \left(2 \csc^2 \alpha \right)^{\frac{1}{p}} \left[(2 \csc^2 \beta + o_1(1))(1 + o_2(1)) \right]^{\frac{1}{q}}.$$

It follows that

$$2 k_\beta(\sigma) \csc^2 \alpha \le 2k \csc^{\frac{2}{p}} \alpha \csc^{\frac{2}{q}} \beta \ (\varepsilon \to 0^+),$$

namely,

$$K_{\alpha,\beta}(\sigma) = \frac{4}{\gamma(2\rho)^{\sigma/\gamma}} \Gamma(\frac{\sigma}{\gamma}) \left(\zeta(\frac{\sigma}{\gamma},\frac{1}{2}) - \frac{2}{2^{\sigma/\gamma}} \zeta(\frac{\sigma}{\gamma},\frac{3}{4}) \right) \csc^{\frac{2}{q}} \alpha \csc^{\frac{2}{p}} \beta$$

$$\leq k.$$

This is a contradiction. Hence, the constant factor $K_{\alpha,\beta}(\sigma)$ in (17) is the best possible.

The constant factor $K_{\alpha,\beta}(\sigma)$ in (18) (respectively (19)) is still the best possible. Otherwise, we would reach a contradiction by (24) (or (26)) that the constant factor $K_{\alpha,\beta}(\sigma)$ in (17) is not the best possible.

4 Operator Expressions

Suppose that $p > 1$. We set the following functions:

$$\Phi(x) := (|x| + x \cos\alpha)^{p(1+\delta\sigma)-1} \quad \text{and} \quad \Psi(n) := (|n| + n\cos\beta)^{q(1-\sigma)-1},$$

wherefrom

$$\Phi^{1-q}(x) = (|x| + x\cos\alpha)^{-q\delta\sigma-1},$$

$$\Psi^{1-p}(n) = (|n| + n\cos\beta)^{p\sigma-1} \quad (x \in \mathbf{R}\backslash\{0\}, |n| \in \mathbf{N}).$$

Define the following real weight normed linear spaces:

$$L_{p,\Phi}(\mathbf{R}) := \left\{ f; \|f\|_{p,\Phi} := \left(\int_{-\infty}^{\infty} \Phi(x)|f(x)|^p dx \right)^{\frac{1}{p}} < \infty \right\},$$

$$L_{q,\Phi^{1-q}}(\mathbf{R}) := \left\{ h; \|h\|_{q,\Phi^{1-q}} := \left(\int_{-\infty}^{\infty} \Phi^{1-q}(x)|h(x)|^q dx \right)^{\frac{1}{q}} < \infty \right\},$$

$$l_{q,\Psi} := \left\{ b = \{b_n\}_{|n|=1}^{\infty}; \|b\|_{q,\Psi} := \left(\sum_{|n|=1}^{\infty} \Psi(n)|b_n|^q \right)^{\frac{1}{q}} < \infty \right\}$$

$$l_{p,\Psi^{1-p}} := \left\{ c = \{c_n\}_{|n|=1}^{\infty}; \|c\|_{p,\Psi^{1-p}} := \left(\sum_{|n|=1}^{\infty} \Psi^{1-p}(n)|c_n|^p \right)^{\frac{1}{p}} < \infty \right\}.$$

(a) In view of Theorem 1, for $f \in L_{p,\Phi}(\mathbf{R})$, setting

$$H^{(1)}(n) := \int_{-\infty}^{\infty} h(x,n)|f(x)|dx \ (|n| \in \mathbf{N}),$$

by (18), we have

$$||H^{(1)}||_{p,\psi^{1-p}} = \left[\sum_{|n|=1}^{\infty} \psi^{1-p}(n)(H^{(1)}(n))^p \right]^{\frac{1}{p}} < K_{\alpha,\beta}(\sigma)||f||_{p,\Phi} < \infty, \tag{27}$$

namely, $H^{(1)} \in l_{p,\psi^{1-p}}$.

Definition 1 Define a Hilbert-type operator in the whole plane

$$T^{(1)} : L_{p,\Phi}(\mathbf{R}) \to l_{p,\psi^{1-p}}$$

as follows: For any $f \in L_{p,\Phi}(\mathbf{R})$, there exists a unique representation

$$T^{(1)}f = H^{(1)} \in l_{p,\psi^{1-p}},$$

satisfying

$$(T^{(1)}f)(n) = H^{(1)}(n),$$

for any $|n| \in \mathbf{N}$.

In view of (27), it follows that

$$||T^{(1)}f||_{p,\psi^{1-p}} = ||H^{(1)}||_{p,\psi^{1-p}} \leq K_{\alpha,\beta}||f||_{p,\Phi},$$

and then the operator $T^{(1)}$ is bounded, satisfying

$$||T^{(1)}|| = \sup_{f(\neq\theta)\in L_{p,\Phi}(\mathbf{R})} \frac{||T^{(1)}f||_{p,\psi^{1-p}}}{||f||_{p,\Phi}} \leq K_{\alpha,\beta}(\sigma).$$

In virtue of the fact that the constant factor $K_{\alpha,\beta}(\sigma)$ in (27) is the best possible, we have

$$||T^{(1)}|| = K_{\alpha,\beta}(\sigma) = \frac{4}{\gamma(2\rho)^{\sigma/\gamma}} \Gamma(\frac{\sigma}{\gamma})$$

$$\times \left(\zeta(\frac{\sigma}{\gamma}, \frac{1}{2}) - \frac{2}{2^{\sigma/\gamma}} \zeta(\frac{\sigma}{\gamma}, \frac{3}{4}) \right) \csc^{\frac{2}{q}} \alpha \csc^{\frac{2}{p}} \beta. \tag{28}$$

If we define the formal inner product of $T^{(1)}f$ and b $(\in l_{q,\psi})$ as follows:

$$(T^{(1)}f, b) := \sum_{|n|=1}^{\infty} (\int_{-\infty}^{\infty} h(x, n)f(x)dx)b_n$$

then we can rewrite the equivalent forms (17) and (18) in the following manner:

$$(T^{(1)}f, b) < ||T^{(1)}||\cdot||f||_{p,\psi}||b||_{q,\Phi}, \; ||T^{(1)}f||_{p,\psi^{1-p}} < ||T^{(1)}||\cdot||f||_{p,\Phi}. \quad (29)$$

(b) In view of Theorem 1, for $b \in l_{q,\psi}$, setting

$$H^{(2)}(x) := \sum_{|n|=1}^{\infty} h(x, n)b_n \; (x \in \mathbf{R}\backslash\{0\}),$$

by (19) we derive that

$$||H^{(2)}||_{q,\Phi^{1-q}} = \left[\int_{-\infty}^{\infty} \Phi^{1-q}(x)(H^{(2)}(x))^q dx\right]^{\frac{1}{q}} < K_{\alpha,\beta}(\sigma)||b||_{q,\psi} < \infty, \quad (30)$$

namely, $H^{(2)} \in L_{q,\psi^{1-q}}(\mathbf{R})$.

Definition 2 Define a Hilbert-type operator in the whole plane

$$T^{(2)} : l_{q,\psi} \to L_{q,\psi^{1-q}}(\mathbf{R})$$

as follows: For any $b \in l_{q,\psi}$, there exists a unique representation

$$T^{(2)}b = H^{(2)} \in L_{q,\psi^{1-q}}(\mathbf{R}),$$

satisfying

$$(T^{(2)}b)(x) = H^{(2)}(x),$$

for any $x \in \mathbf{R}$.

In view of (30), we have

$$||T^{(2)}b||_{q,\Phi^{1-q}} = ||H^{(2)}||_{q,\Phi^{1-q}} \leq K_{\alpha,\beta}(\sigma)||b||_{q,\psi},$$

and then the operator $T^{(2)}$ is bounded, satisfying

$$||T^{(2)}|| = \sup_{b(\neq\theta)\in l_{q,\psi}} \frac{||T^{(2)}b||_{q,\Phi^{1-q}}}{||b||_{q,\psi}} \leq K_{\alpha,\beta}(\sigma).$$

By the fact that the constant factor $K_{\alpha,\beta}(\sigma)$ in (30) is the best possible, we have

$$||T^{(2)}|| = K_{\alpha,\beta}(\sigma) = ||T^{(1)}||. \tag{31}$$

If we define the formal inner product of $T^{(2)}b$ and f $(\in L_{p,\Phi}(\mathbf{R}))$ as follows:

$$(T^{(2)}b, f) := \int_{-\infty}^{\infty} \sum_{|n|=1}^{\infty} h(x,n)b_n f(x)dx,$$

we can then rewrite the equivalent forms (17) and (19) as follows:

$$(T^{(2)}b, f) < ||T^{(2)}|| \cdot ||f||_{p,\Psi}||b||_{q,\Phi}, ||T^{(2)}b||_{q,\Phi^{1-q}} < ||T^{(2)}|| \cdot ||b||_{q,\Psi}. \tag{32}$$

Remark 1 (i) For $\delta = 1$, (20) reduces to (5). If $f(-x) = f(x)$ $(x > 0)$, $b_{-n} = b_n$ $(n \in \mathbf{N})$, then (5) reduces to the following half-discrete Hilbert-type inequality (cf. [6]):

$$\sum_{n=1}^{\infty} \int_0^{\infty} \operatorname{sech}\left(\rho\left(\frac{n}{x}\right)^\gamma\right) f(x)b_n dx$$

$$< \frac{2}{\gamma(2\rho)^{\sigma/\gamma}}\Gamma(\frac{\sigma}{\gamma})\left(\zeta(\frac{\sigma}{\gamma},\frac{1}{2}) - \frac{2}{2^{\sigma/\gamma}}\zeta(\frac{\sigma}{\gamma},\frac{3}{4})\right)$$

$$\times \left[\int_0^{\infty} x^{p(1+\sigma)-1}f^p(x)dx\right]^{\frac{1}{p}}\left[\sum_{n=1}^{\infty} n^{q(1-\sigma)-1}b_n^q\right]^{\frac{1}{q}}. \tag{33}$$

(ii) For $\delta = 1$, (17) reduces to the following particular inequality with homogeneous kernel of degree 0:

$$\sum_{|n|=1}^{\infty} \int_{-\infty}^{\infty} \operatorname{sech}\left(\rho\left(\frac{|n| + n\cos\beta}{|x| + x\cos\alpha}\right)^\gamma\right) f(x)b_n dx$$

$$< K_{\alpha,\beta}(\sigma)\left[\int_{-\infty}^{\infty}(|x| + x\cos\alpha)^{p(1+\sigma)-1}f^p(x)dx\right]^{\frac{1}{p}}$$

$$\times \left[\sum_{|n|=1}^{\infty}(|n| + n\cos\beta)^{q(1-\sigma)-1}b_n^q\right]^{\frac{1}{q}}. \tag{34}$$

(iii) For $\delta = -1$, (17) reduces to the following particular inequality with non-homogeneous kernel:

$$\sum_{|n|=1}^{\infty} \int_{-\infty}^{\infty} sech\left(\rho[(|x|+x\cos\alpha)(|n|+n\cos\beta)]^{\gamma}\right) f(x)b_n dx$$

$$< K_{\alpha,\beta}(\sigma) \left[\int_{-\infty}^{\infty} (|x|+x\cos\alpha)^{p(1-\sigma)-1} f^p(x)dx\right]^{\frac{1}{p}}$$

$$\times \left[\sum_{|n|=1}^{\infty} (|n|+n\cos\beta)^{q(1-\sigma)-1} b_n^q\right]^{\frac{1}{q}}. \tag{35}$$

The constant factors in the above inequalities are the best possible.

5 Two Kinds of Equivalent Reverse Inequalities

In the following, for the cases $0 < p < 1$ and $p < 0$, we still use the symbols $\|b\|_{q,\Phi}$ and $\|f\|_{p,\Psi}$.

Theorem 3 *Suppose that* $0 < p < 1$,

$$K_{\alpha,\beta}(\sigma) = \frac{4}{\gamma(2\rho)^{\sigma/\gamma}} \Gamma(\frac{\sigma}{\gamma})$$

$$\times \left(\zeta(\frac{\sigma}{\gamma}, \frac{1}{2}) - \frac{2}{2^{\sigma/\gamma}} \zeta(\frac{\sigma}{\gamma}, \frac{3}{4})\right) \csc^{\frac{2}{q}} \alpha \csc^{\frac{2}{p}} \beta.$$

Let $f(x), b_n \geq 0$, *satisfy* $0 < \|f\|_{p,\Psi} < \infty, 0 < \|b\|_{q,\Phi} < \infty$, *then we have the following equivalent inequalities:*

$$I = \sum_{|n|=1}^{\infty} \int_{-\infty}^{\infty} sech\left(\rho\left[\frac{|n|+n\cos\beta}{(|x|+x\cos\alpha)^{\delta}}\right]^{\gamma}\right) f(x)b_n dx$$

$$> K_{\alpha,\beta}(\sigma) \left[\int_{-\infty}^{\infty} (1-\theta(\sigma,x))(|x|+x\cos\alpha)^{p(1+\delta\sigma)-1} f^p(x)dx\right]^{\frac{1}{p}} \|b\|_{q,\Phi}, \tag{36}$$

$$J_1 = \left\{\sum_{|n|=1}^{\infty} (|n|+n\cos\beta)^{p\sigma-1}\right.$$

$$\times \left.\left[\int_{-\infty}^{\infty} sech\left(\rho\left[\frac{|n|+n\cos\beta}{(|x|+x\cos\alpha)^{\delta}}\right]^{\gamma}\right) f(x)dx\right]^p\right\}^{\frac{1}{p}}$$

$$> K_{\alpha,\beta}(\sigma) \left[\int_{-\infty}^{\infty} (1 - \theta(\sigma, x))(|x| + x \cos \alpha)^{p(1+\delta\sigma)-1} f^p(x) dx \right]^{\frac{1}{p}}, \quad (37)$$

$$\tilde{J}_2 := \left\{ \int_{-\infty}^{\infty} \frac{(1 - \theta(\sigma, x))^{1-q}}{(|x| + x \cos \alpha)^{q\delta\sigma+1}} \right.$$

$$\left. \times \left[\sum_{|n|=1}^{\infty} sech \left(\rho \left[\frac{|n| + n \cos \beta}{(|x| + x \cos \alpha)^\delta} \right]^\gamma \right) b_n \right]^q dx \right\}^{\frac{1}{q}}$$

$$> K_{\alpha,\beta}(\sigma) \|b\|_{q,\Phi}, \quad (38)$$

where the constant factor $K_{\alpha,\beta}(\sigma)$ *is the best possible.*

Proof Similarly, by the reverse Hölder inequality (cf. [32]) and (7), we obtain that

$$\left[\int_{-\infty}^{\infty} h(x, n) f(x) dx \right]^p$$

$$\geq \frac{\omega^{p-1}(\sigma, n)}{(|n| + n \cos \beta)^{p\sigma-1}} \int_{-\infty}^{\infty} h(x, n) \frac{(|x| + x \cos \alpha)^{(1+\delta\sigma)(p-1)}}{(|n| + n \cos \beta)^{1-\sigma}} f^p(x) dx.$$

In view of (9), by the Lebesgue term by term integration theorem (cf. [31]) and (8), we deduce that

$$J_1 \geq k_\alpha^{\frac{1}{q}}(\sigma) \left[\int_{-\infty}^{\infty} \varpi(\sigma, x)(|x| + x \cos \alpha)^{p(1+\delta\sigma)-1} f^p(x) dx \right]^{\frac{1}{p}}. \quad (39)$$

Hence, by (10), we have (37).

By the reverse Hölder inequality (cf. [32]), we still have

$$I \geq J_1 \left[\sum_{|n|=1}^{\infty} (|n| + n \cos \beta)^{q(1-\sigma)-1} b_n^q \right]^{\frac{1}{q}}. \quad (40)$$

In view of (37), we obtain (36).

On the other hand, assuming that (36) is valid, we set

$$b_n := (|n| + n \cos \beta)^{p\sigma-1} \left[\int_{-\infty}^{\infty} h(x, n) f(x) dx \right]^{p-1} \quad (|n| \in \mathbf{N}).$$

Then we get

$$J_1 = \left[\sum_{|n|=1}^{\infty} (|n| + n \cos \beta)^{q(1-\sigma)-1} b_n^q \right]^{\frac{1}{p}}.$$

In view of (39), it follows that $J_1 > 0$. If $J_1 = \infty$, then (37) is trivially valid; if $J_1 < \infty$, then by (36), we have

$$||b||_{q,\Phi}^q = J_1^p = I$$

$$> K_{\alpha,\beta}(\sigma) \left[\int_{-\infty}^{\infty} (1 - \theta(\sigma, x))(|x| + x \cos \alpha)^{p(1+\delta\sigma)-1} f^p(x)dx \right]^{\frac{1}{p}} ||b||_{q,\Phi},$$

$$||b||_{q,\Phi}^{q-1} = J_1 > K_{\alpha,\beta}(\sigma) \left[\int_{-\infty}^{\infty} (1 - \theta(\sigma, x))(|x| + x \cos \alpha)^{p(1+\delta\sigma)-1} f^p(x)dx \right]^{\frac{1}{p}},$$

namely, (37) holds, which is equivalent to (36).

Similarly to as we obtained (39), we deduce that

$$\tilde{J}_2 > k_\alpha^{\frac{1}{p}}(\sigma) \left[\sum_{|n|=1}^{\infty} \omega(\sigma, n)(|n| + n \cos \beta)^{q(1-\sigma)-1} b_n^q \right]^{\frac{1}{q}}. \tag{41}$$

Hence, by (9), we have (38). We have proved that (36) is valid. Setting

$$f(x) := \frac{(1 - \theta(\sigma, x))^{1-q}}{(|x| + x \cos \alpha)^{q\delta\sigma+1}} \left[\sum_{|n|=1}^{\infty} h(x, n)b_n \right]^{q-1} \quad (x \in \mathbf{R} \backslash \{0\}),$$

it then follows that

$$\tilde{J}_2 = \left[\int_{-\infty}^{\infty} (1 - \theta(\sigma, x))(|x| + x \cos \alpha)^{p(1+\delta\sigma)-1} f^p(x)dx \right]^{\frac{1}{q}},$$

and in view of (41), we get that $\tilde{J}_2 > 0$. If $\tilde{J}_2 = \infty$, then (38) is trivially valid; if $\tilde{J}_2 < \infty$, then by (36), we have

$$\int_{-\infty}^{\infty} (1 - \theta(\sigma, x))(|x| + x \cos \alpha)^{p(1+\delta\sigma)-1} f^p(x)dx = \tilde{J}_2^q = I$$

$$> K_{\alpha,\beta}(\sigma) \left[\int_{-\infty}^{\infty} (1 - \theta(\sigma, x))(|x| + x \cos \alpha)^{p(1+\delta\sigma)-1} f^p(x)dx \right]^{\frac{1}{p}} ||b||_{q,\Phi},$$

$$\tilde{J}_2 = \left[\int_{-\infty}^{\infty}(1-\theta(\sigma,x))(|x|+x\cos\alpha)^{p(1+\delta\sigma)-1}f^p(x)dx\right]^{1-\frac{1}{p}} > K_{\alpha,\beta}(\sigma)\|b\|_{q,\Phi},$$

namely (38) follows. On the other hand, assuming that (38) is valid, by the reverse Hölder inequality (cf. [32]), we obtain

$$I \geq \left[\int_{-\infty}^{\infty}(1-\theta(\sigma,x))(|x|+x\cos\alpha)^{p(1+\delta\sigma)-1}f^p(x)dx\right]^{\frac{1}{p}}\tilde{J}_2. \qquad (42)$$

Then by (38), we have (16), which is equivalent to (38).

Therefore, inequalities (36), (37) and (38) are equivalent.

For $\varepsilon > 0$, we set $\tilde{\sigma} = \sigma + \frac{\varepsilon}{p}$ ($> \gamma$),

$$\tilde{f}(x) := \begin{cases} \dfrac{1}{(|x|+x\cos\alpha)^{\delta(\sigma+\frac{\varepsilon}{p})+1}}, x \in E_\delta, \\ 0, x \in \mathbf{R}\backslash E_\delta, \end{cases}$$

and

$$\tilde{b}_n := (|n|+n\cos\beta)^{(\sigma-\frac{\varepsilon}{q})-1} \quad (|n| \in \mathbf{N}).$$

Then by (13) and (15), we find

$$\tilde{I}_1 := \left[\int_{-\infty}^{\infty}(1-\theta(\sigma,x))(|x|+x\cos\alpha)^{p(1+\delta\sigma)-1}\tilde{f}^p(x)dx\right]^{\frac{1}{p}}$$

$$\times \left[\sum_{|n|=1}^{\infty}(|n|+n\cos\beta)^{q(1-\sigma)-1}\tilde{b}_n^q\right]^{\frac{1}{q}}$$

$$= \left[\int_{-\infty}^{\infty}\frac{(1-\theta(\sigma,x))dx}{(|x|+x\cos\alpha)^{\delta\varepsilon+1}}\right]^{\frac{1}{p}}\left[\sum_{|n|=1}^{\infty}\frac{1}{(|n|+n\cos\beta)^{\varepsilon+1}}\right]^{\frac{1}{q}}$$

$$= \left[\frac{2}{\varepsilon}\csc^2\alpha - \int_{-\infty}^{\infty}\frac{dx}{O((|x|+x\cos\alpha)^{\delta(\sigma+\varepsilon)+1})}\right]^{\frac{1}{p}}$$

$$\times \left[\sum_{|n|=1}^{\infty}\frac{1}{(|n|+n\cos\beta)^{\varepsilon+1}}\right]^{\frac{1}{q}}$$

$$= \frac{1}{\varepsilon}\left(2\csc^2\alpha - \varepsilon O(1)\right)^{\frac{1}{p}}\left[(2\csc^2\beta + o_1(1))(1+o_2(1))\right]^{\frac{1}{q}}.$$

By (10), we still have

$$
\begin{aligned}
\widetilde{I} : &= \sum_{|n|=1}^{\infty} \int_{-\infty}^{\infty} h(x,n)\widetilde{f}(x)\widetilde{b}_n dx \\
&= \sum_{|n|=1}^{\infty} \int_{E_\delta} h(x,n)\frac{(|n|+n\cos\beta)^{(\sigma-\frac{\varepsilon}{q})-1}}{(|x|+x\cos\alpha)^{\delta(\sigma+\frac{\varepsilon}{p})+1}}dx \\
&\leq \sum_{|n|=1}^{\infty} \int_{-\infty}^{\infty} h(x,n)\frac{(|n|+n\cos\beta)^{(\widetilde{\sigma}-\varepsilon)-1}}{(|x|+x\cos\alpha)^{\delta\widetilde{\sigma}+1}}dx \\
&= \sum_{|n|=1}^{\infty} \frac{\omega(\widetilde{\sigma},n)}{(|n|+n\cos\beta)^{\varepsilon+1}} = k_\alpha(\widetilde{\sigma})\sum_{|n|=1}^{\infty}\frac{1}{(|n|+n\cos\beta)^{\varepsilon+1}} \\
&= \frac{1}{\varepsilon}k_\alpha(\sigma+\frac{\varepsilon}{p})(2\csc^2\beta+o_1(1))(1+o_2(1)).
\end{aligned}
$$

If the constant factor $K_{\alpha,\beta}(\sigma)$ in (37) is not the best possible, then there exists a positive number k, with $K_{\alpha,\beta}(\sigma) < k$, such that (37) is valid when replacing $K_{\alpha,\beta}(\sigma)$ by k. Thus in particular, we have $\varepsilon\widetilde{I} > \varepsilon k\widetilde{I}_1$, namely,

$$
k_\alpha(\sigma + \frac{\varepsilon}{p})(2\csc^2\beta+o_1(1))(1+o_2(1))
$$

$$
> k \cdot \left(2\csc^2\alpha - \varepsilon O(1)\right)^{\frac{1}{p}}\left[(2\csc^2\beta+o_1(1))(1+o_2(1))\right]^{\frac{1}{q}}.
$$

It follows that

$$
2k_\alpha(\sigma)\csc^2\beta \geq 2k\csc^{\frac{2}{p}}\alpha\csc^{\frac{2}{q}}\beta \ (\varepsilon\to 0^+),
$$

namely

$$
K_{\alpha,\beta}(\sigma) = \frac{4}{\gamma(2\rho)^{\sigma/\gamma}}\Gamma(\frac{\sigma}{\gamma})
$$

$$
\times\left(\zeta(\frac{\sigma}{\gamma},\frac{1}{2}) - \frac{2}{2^{\sigma/\gamma}}\zeta(\frac{\sigma}{\gamma},\frac{3}{4})\right)\csc^{\frac{2}{q}}\alpha\csc^{\frac{2}{p}}\beta \geq k.
$$

This is a contradiction. Hence, the constant factor $K_{\alpha,\beta}(\sigma)$ in (36) is the best possible.

The constant factor $K_{\alpha,\beta}(\sigma)$ in (37) (respectively (38)) is still the best possible. Otherwise, we would reach a contradiction by (40) (or (42)) that the constant factor $K_{\alpha,\beta}(\sigma)$ in (36) is not the best possible.

Theorem 4 *Suppose that $p < 0$,*

$$K_{\alpha,\beta}(\sigma) = \frac{4}{\gamma(2\rho)^{\sigma/\gamma}} \Gamma(\frac{\sigma}{\gamma})$$

$$\times \left(\zeta(\frac{\sigma}{\gamma}, \frac{1}{2}) - \frac{2}{2^{\sigma/\gamma}} \zeta(\frac{\sigma}{\gamma}, \frac{3}{4}) \right) \csc^{\frac{2}{q}} \alpha \, \csc^{\frac{2}{p}} \beta.$$

Let $f(x), b_n \geq 0$, satisfy $0 < \|f\|_{p,\psi}, \|b\|_{q,\Phi} < \infty$, then we have the following equivalent inequalities:

$$I = \sum_{|n|=1}^{\infty} \int_{-\infty}^{\infty} sech\left(\rho \left[\frac{|n| + n\cos\beta}{(|x| + x\cos\alpha)^{\delta}} \right]^{\gamma} \right) f(x)b_n dx > K_{\alpha,\beta}(\sigma)\|f\|_{p,\psi}\|b\|_{q,\Phi},$$

$$\tag{43}$$

$$J_1 = \left\{ \sum_{|n|=1}^{\infty} (|n| + n\cos\beta)^{p\sigma - 1} \right.$$

$$\times \left[\int_{-\infty}^{\infty} sech\left(\rho \left[\frac{|n| + n\cos\beta}{(|x| + x\cos\alpha)^{\delta}} \right]^{\gamma} \right) f(x)dx \right]^{p} \right\}^{\frac{1}{p}} > K_{\alpha,\beta}(\sigma)\|f\|_{p,\psi}, \quad (44)$$

$$J_2 = \left\{ \int_{-\infty}^{\infty} \frac{1}{(|x| + x\cos\alpha)^{q\delta\sigma + 1}} \right.$$

$$\times \left[\sum_{|n|=1}^{\infty} sech\left(\rho \left[\frac{|n| + n\cos\beta}{(|x| + x\cos\alpha)^{\delta}} \right]^{\gamma} \right) b_n \right]^{q} dx \right\}^{\frac{1}{q}} > K_{\alpha,\beta}(\sigma)\|b\|_{q,\Phi}, \quad (45)$$

where the constant factor $K_{\alpha,\beta}(\sigma)$ is the best possible.

Proof For $p < 0$, by the reverse Hölder inequality (cf. [32]) and (7), we obtain that

$$\left[\int_{-\infty}^{\infty} h(x, n) f(x)dx \right]^{p}$$

$$\leq \frac{\omega^{p-1}(\sigma, n)}{(|n| + n\cos\beta)^{p\sigma - 1}} \int_{-\infty}^{\infty} h(x, n)\frac{(|x| + x\cos\alpha)^{(1+\delta\sigma)(p-1)}}{(|n| + n\cos\beta)^{1-\sigma}} f^{p}(x)dx.$$

Then by (9), the Lebesgue term by term integration theorem (cf. [31]) and (8), we deduce that

$$J_1 \geq k_{\alpha}^{\frac{1}{q}}(\sigma) \left[\int_{-\infty}^{\infty} \varpi(\sigma, x)(|x| + x\cos\alpha)^{p(1+\delta\sigma)-1} f^{p}(x)dx \right]^{\frac{1}{p}}. \tag{46}$$

Hence, by (10), we have (44).

By the reverse Hölder inequality (cf. [32]), we obtain that

$$I \geq J_1 \left[\sum_{|n|=1}^{\infty} (|n| + n \cos \beta)^{q(1-\sigma)-1} b_n^q \right]^{\frac{1}{q}}. \qquad (47)$$

Then by (44), we have (43).

On the other hand, assuming that (43) is valid, we set

$$b_n := (|n| + n \cos \beta)^{p\sigma-1} \left(\int_{-\infty}^{\infty} h(x, n) f(x) dx \right)^{p-1} \quad (|n| \in \mathbf{N}),$$

and find $J_1 = ||b||_{q,\Phi}^{\frac{q}{p}}$. In view of (46), it follows that $J_1 > 0$. If $J_1 = \infty$, then (44) is trivially valid; if $J_1 < \infty$, then by (43), we have

$$||b||_{q,\Phi}^q = J_1^p = I > K_{\alpha,\beta}(\sigma)||f||_{p,\psi}||b||_{q,\Phi},$$

$$J_1 = ||b||_{q,\Phi}^{q-1} > K_{\alpha,\beta}(\sigma)||f||_{p,\psi},$$

namely, (44) holds, which is equivalent to (43).

We similarly get that

$$J_2 > k_\beta^{\frac{1}{p}}(\sigma) \left[\sum_{|n|=1}^{\infty} \omega(\sigma, n)(|n| + n \cos \beta)^{q(1-\sigma)-1} b_n^q \right]^{\frac{1}{q}}. \qquad (48)$$

Hence, by (9), we deduce (38). We have proved that (43) is valid. Setting

$$f(x) := \frac{1}{(|x| + x \cos \alpha)^{q\delta\sigma+1}} \left(\sum_{|n|=1}^{\infty} h(x, n) b_n \right)^{q-1} \quad (x \in \mathbf{R}\backslash\{0\}),$$

it follows that $J_2 = ||f||_{p,\psi}^{\frac{p}{q}}$ and in view of (48), we find $J_2 > 0$. If $J_2 = \infty$, then (45) is trivially valid; if $J_2 < \infty$, then by (43), we have

$$||f||_{p,\psi}^p = J_2^q = I > K_{\alpha,\beta}(\sigma)||f||_{p,\psi}||b||_{q,\Phi},$$

$$J_2 = ||f||_{p,\psi}^{p-1} > K_{\alpha,\beta}(\sigma)||b||_{q,\Phi},$$

namely (45) follows.

On the other hand, assuming that (45) is valid, by the reverse Hölder's inequality (cf. [32]), we obtain

$$I \geq \left[\int_{-\infty}^{\infty} (|x| + x \cos \alpha)^{p(1+\delta\sigma)-1} f^p(x) dx \right]^{\frac{1}{p}} J_2. \tag{49}$$

Then by (45), we have (23), which is equivalent to (45).

Therefore, inequalities (43), (44) and (45) are equivalent.

For $0 < \varepsilon < |p|(\sigma - \gamma)$, we set $\tilde{\sigma} = \sigma + \frac{\varepsilon}{p} \ (> \gamma)$,

$$\tilde{f}(x) := \begin{cases} \dfrac{1}{(|x|+x\cos\alpha)^{\delta(\sigma+\frac{\varepsilon}{p})+1}}, & x \in E_\delta, \\ 0, & x \in \mathbf{R}\backslash E_\delta, \end{cases}$$

and

$$\tilde{b}_n := (|n| + n \cos \beta)^{(\sigma - \frac{\varepsilon}{q})-1} (|n| \in \mathbf{N}).$$

Then by (13) and (15), we obtain that

$$\begin{aligned} \tilde{I}_1 :&= \left[\int_{-\infty}^{\infty} (|x| + x \cos \alpha)^{p(1+\delta\sigma)-1} \tilde{f}^p(x) dx \right]^{\frac{1}{p}} \\ &\quad \times \left[\sum_{|n|=1}^{\infty} (|n| + n \cos \beta)^{q(1-\sigma)-1} \tilde{b}_n^q \right]^{\frac{1}{q}} \\ &= \left[\int_{-\infty}^{\infty} \frac{dx}{(|x| + x \cos \alpha)^{\delta\varepsilon+1}} \right]^{\frac{1}{p}} \left[\sum_{|n|=1}^{\infty} \frac{1}{(|n| + n \cos \beta)^{\varepsilon+1}} \right]^{\frac{1}{q}} \\ &= \frac{1}{\varepsilon} \left(2 \csc^2 \alpha \right)^{\frac{1}{p}} \left[(2 \csc^2 \beta + o_1(1))(1 + o_2(1)) \right]^{\frac{1}{q}}. \end{aligned}$$

By (10), we still have

$$\begin{aligned} \tilde{I} :&= \sum_{|n|=1}^{\infty} \int_{-\infty}^{\infty} h(x, n) \tilde{f}(x) \tilde{b}_n dx \\ &= \sum_{|n|=1}^{\infty} \int_{E_\delta} h(x, n) \frac{(|n| + n \cos \beta)^{(\sigma-\frac{\varepsilon}{q})-1}}{(|x| + x \cos \alpha)^{\delta(\sigma+\frac{\varepsilon}{p})+1}} dx \\ &\leq \sum_{|n|=1}^{\infty} \int_{-\infty}^{\infty} h(x, n) \frac{(|n| + n \cos \beta)^{(\tilde{\sigma}-\varepsilon)-1}}{(|x| + x \cos \alpha)^{\delta\tilde{\sigma}+1}} dx \end{aligned}$$

$$= \sum_{|n|=1}^{\infty} \frac{\omega(\tilde{\sigma}, n)}{(|n| + n \cos \beta)^{\varepsilon+1}} = k_\alpha(\tilde{\sigma}) \sum_{|n|=1}^{\infty} \frac{1}{(|n| + n \cos \beta)^{\varepsilon+1}}$$

$$= \frac{1}{\varepsilon} k_\alpha(\sigma + \frac{\varepsilon}{p})(2 \csc^2 \beta + o_1(1))(1 + o_2(1))$$

If the constant factor $K_{\alpha,\beta}(\sigma)$ in (43) is not the best possible, then there exists a positive number k, with $K_{\alpha,\beta}(\sigma) < k$, such that (43) is valid when replacing $K_{\alpha,\beta}(\sigma)$ by k. Hence in particular, we have $\varepsilon \tilde{I} > \varepsilon k \tilde{I}_1$, namely

$$k_\alpha(\sigma + \frac{\varepsilon}{p})(2 \csc^2 \beta + o_1(1))(1 + o_2(1))$$

$$> k \cdot \left(2 \csc^2 \alpha\right)^{\frac{1}{p}} \left[(2 \csc^2 \beta + o_1(1))(1 + o_2(1))\right]^{\frac{1}{q}}.$$

It follows that

$$2 k_\alpha(\sigma) \csc^2 \beta \geq 2k \csc^{\frac{2}{p}} \alpha \csc^{\frac{2}{q}} \beta \ (\varepsilon \to 0^+),$$

namely

$$K_{\alpha,\beta}(\sigma) = \frac{4}{\gamma (2\rho)^{\sigma/\gamma}} \Gamma(\frac{\sigma}{\gamma}) \left(\zeta(\frac{\sigma}{\gamma}, \frac{1}{2}) - \frac{2}{2^{\sigma/\gamma}} \zeta(\frac{\sigma}{\gamma}, \frac{3}{4})\right) \csc^{\frac{2}{q}} \alpha \csc^{\frac{2}{p}} \beta$$

$$\geq k.$$

This is a contradiction. Hence, the constant factor $K_{\alpha,\beta}(\sigma)$ in (43) is the best possible.

The constant factor $K_{\alpha,\beta}(\sigma)$ in (44) (respectively (45)) is still the best possible. Otherwise, we would reach a contradiction by (47) (respectively (49)) that the constant factor $K_{\alpha,\beta}(\sigma)$ in (43) is not the best possible.

6 Conclusions

Using weight functions, a half-discrete Hilbert-type inequality in the whole plane with the kernel of hyperbolic secant function and multi-parameters is established in Theorem 1. The constant factor related to the Hurwitz zeta function is proved to be the best possible in Theorem 2. Equivalent forms, two kinds of particular inequalities, the operator expressions and some reverses are also considered in Theorem 1, Remark 1, and Theorems 3–4. The lemmas and theorems provide and extensive account of this type of inequalities.

Acknowledgements B. Yang: This work is supported by the National Natural Science Foundation (No. 61772140), and Science and Technology Planning Project Item of Guangzhou City (No. 201707010229). I would like to express my gratitude for this support.

References

1. G.H. Hardy, Note on a theorem of Hilbert concerning series of positive terms. Proc. Lond. Math. Soc. **23**(2), xlv–xlvi (1925). Records of Proc
2. G.H. Hardy, J.E. Littlewood, G. Pólya, *Inequalities* (Cambridge University Press, Cambridge, 1934)
3. B.C. Yang, A half-discrete Hilbert's inequality. J. Guangdong Univ. Edu. **31**(3), 1–7 (2011)
4. D.S. Mitrinović, J.E. Peččarić, A.M. Fink, *Inequalities Involving Functions and their Integrals and Deivatives* (Kluwer Academic, Boston, 1991)
5. B.C. Yang, *The Norm of Operator and Hilbert-Type Inequalities* (Science Press, Beijing, 2009)
6. B.C. Yang, L. Debnath, *Half-Discrete Hilbert-Type Inequalitiea* (World Scientific Publishing, Singapore, 2014)
7. B.C. Yang, A survey of the study of Hilbert-type inequalities with parameters. Adv. Math. **38**(3), 257–268 (2009)
8. B.C. Yang, On the norm of an integral operator and applications. J. Math. Anal. Appl. **321**, 182–192 (2006)
9. J.S. Xu, Hardy-Hilbert's inequalities with two parameters. Adv. Math. **36**(2), 63–76 (2007)
10. D.M. Xin, A Hilbert-type integral inequality with the homogeneous Kernel of zero degree. Math. Theory Appl. **30**(2), 70–74 (2010)
11. G.V. Milovanovic, M.T. Rassias, Some properties of a hypergeometric function which appear in an approximation problem. J. Glob. Optim. **57**, 1173–1192 (2013)
12. M. Krnić, J. Pečarić, General Hilbert's and Hardy's inequalities. Math. Inequal. Appl. **8**(1), 29–51 (2005)
13. I. Perić, P. Vuković, Multiple Hilbert's type inequalities with a homogeneous kernel. Banach J. Math. Anal. **5**(2), 33–43 (2011)
14. Q. Huang, A new extension of Hardy-Hilbert-type inequality. J. Inequal. Appl. **2015**, 397 (2015)
15. B. He, A multiple Hilbert-type discrete inequality with a new kernel and best possible constant factor. J. Math. Anal. Appl. **431**, 990–902 (2015)
16. V. Adiyasuren, T. Batbold, M. Krnić, Multiple Hilbert-type inequalities involving some differential operators. Banach J. Math. Anal. **10**(2), 320–337 (2016)
17. M.T. Rassias, B.C. Yang, On half-discrete Hilbert's inequality. Appl. Math. Comput. **220**, 75–93 (2013)
18. M.T. Rassias, B.C. Yang, A multidimensional half – discrete Hilbert – type inequality and the Riemann zeta function. Appl. Math. Comput. **225**, 263–277 (2013)
19. M.T. Rassias, B.C. Yang, A multidimensional Hilbert-type integral inequality related to the Riemann zeta function, in *Applications of Mathematics and Informatics in Science and Engineering* (Springer, New York, 2014), pp. 417–433
20. M.T. Rassias, B.C. Yang, On a multidimensional half – discrete Hilbert – type inequality related to the hyperbolic cotangent function. Appl. Math. Comput. **242**, 800–813 (2014)
21. M.T. Rassias, B.C. Yang, A. Raigorodskii, Two kinds of the reverse Hardy-type integral inequalities with the equivalent forms related to the extended Riemann zeta function. Appl. Anal. Discret. Math. **12**, 273–296 (2018)
22. B.C. Yang, A new Hilbert-type integral inequality. Soochow J. Math. **33**(4), 849–859 (2007)
23. B.C. Yang, A new Hilbert-type integral inequality with some parameters. J. Jilin Univ. (Science Edition) **46**(6), 1085–1090 (2008)

24. B.C. Yang, A Hilbert-type integral inequality with a non-homogeneous kernel. J. Xiamen Univ. (Natural Science) **48**(2), 165–169 (2008)
25. Z. Zeng, Z.T. Xie, On a new Hilbert-type integral inequality with the homogeneous kernel of degree 0 and the integral in whole plane. J. Inequal. Appl. **2010**, 9. Article ID 256796 (2010)
26. Q.L. Huang, S.H. Wu, B.C. Yang, Parameterized Hilbert-type integral inequalities in the whole plane. Sci. World J. **2014**, 8. Article ID 169061 (2014)
27. Z. Zeng, K.R.R. Gandhi, Z.T. Xie, A new Hilbert-type inequality with the homogeneous kernel of degree −2 and with the integral. Bull. Math. Sci. Appl. **3**(1), 11–20 (2014)
28. Z.H. Gu, B.C. Yang, A Hilbert-type integral inequality in the whole plane with a non-homogeneous kernel and a few parameters. J. Inequal. Appl. **2015**, 314 (2015)
29. D.M. Xin, B.C. Yang, Q. Chen, A discrete Hilbert-type inequality in the whole plane. J. Inequal. Appl. **2016**, 133 (2016)
30. Z.Q. Wang, D.R. Guo, *Inteoduction to Special Functions* (Science Press, Beijing, 1979)
31. J.C. Kuang, *Real and Functional Analysis*(Continuation), vol. 2 (Higher Education Press, Beijing, 2015)
32. J.C. Kuang, *Applied Inequalities* (Shangdong Science and Technology Press, Jinan, 2004)

A Remark on Sets with Small Wiener Norm

I. D. Shkredov

Abstract We show that any set with small Wiener norm has small multiplicative energy. It gives some new bounds for Wiener norm for sets with small product set. Also, we prove that any symmetric subset S of an abelian group has a nonzero Fourier coefficient of size $\Omega(|S|^{1/3})$.

Keywords Exponential sums · Wiener norm · Sum-product · Multiplicative subgroups

1 Introduction

We consider the abelian group $\mathbb{F}_p = \mathbb{Z}/p\mathbb{Z}$, where p is a prime number. Denote the Fourier transform of a complex function on $\mathbb{F}_p$ to be a new function

$$\hat{f}(\gamma) = \sum_{x \in \mathbb{F}_p} f(x)e_p(-x\gamma),$$

where $e_p(u) = \exp(2\pi i u/p)$ (we note that e_p is correctly defined for $u \in \mathbb{F}_p$). It is known that the function f can be reconstructed from $\hat{f}$ by the inverse Fourier transform

This work is supported by the Russian Science Foundation under grant 19–11–00001.

I. D. Shkredov (✉)
Steklov Mathematical Institute, Moscow, Russia

IITP RAS, Moscow, Russia

MIPT, Dolgoprudnii, Russia

© Springer Nature Switzerland AG 2020
A. Raigorodskii, M. T. Rassias (eds.), *Trigonometric Sums and Their Applications*,
https://doi.org/10.1007/978-3-030-37904-9_12

$$f(x) = \frac{1}{p} \sum_{\gamma \in \mathbb{F}_p} \hat{f}(\gamma) e_p(x\gamma). \tag{1}$$

We define the Wiener norm of a function f as

$$\|f\|_{A(\mathbb{F}_p)} = p^{-1}\|\hat{f}\|_1 = p^{-1} \sum_{\gamma \in \mathbb{F}_p} |\hat{f}(\gamma)|.$$

By 1_S, $S \subset \mathbb{F}_p$ denote the characteristic function of a certain set S. The problem of finding lower bounds for $\|f\|_{A(\mathbb{F}_p)}$ for f equals 1_S for an arbitrary set $S \subseteq \mathbb{F}_p$, $|S| \leqslant p/2$ is called the *modular Littlewood problem* and many results in this direction were obtained by various authors [1–7, 18] (for the original continues Littlewood problem see [8, 9]). In this paper we find a new property of sets with small Wiener norm. For any set $S \subseteq \mathbb{F}_p$ consider *the multiplicative energy* $\mathsf{E}^{\times}(S)$ of S, namely, $\mathsf{E}^{\times}(S) := |\{ab = cd : a, b, c, d \in S\}|$. We show that if S has small Wiener norm, then it has small multiplicative energy as well.

Theorem 1 *Let $S \subseteq \mathbb{F}_p$ be a set, and $K := \|1_S\|_{A(\mathbb{F}_p)}$. Suppose that $|S|^5 \leqslant K^2 p^3$. Then*

$$\mathsf{E}^{\times}(S) \lesssim K^{4/3}|S|^{8/3}, \tag{2}$$

and in any case

$$\mathsf{E}^{\times}(S) \leqslant \frac{|S|^4}{p-1} + \|S\|_{A(\mathbb{F}_p)}^2 |S|p.$$

All logarithms in this paper are to base 2. The signs $\ll$ and $\gg$ are the usual Vinogradov symbols. If we have a set A, then we will write $a \lesssim b$ or $b \gtrsim a$ if $a = O(b \cdot \log^c |A|)$, $c > 0$.

The proof of Theorem 1 uses some results from the sum–product phenomenon, see [10] and, e.g., [11, 12] and especially the Balog–Wooley decomposition [13]. Theorem 1 implies an interesting statement which separates the prime field from general finite fields (also, see comments after Corollary 10 below).

Corollary 2 *Let $S \subseteq \mathbb{F}_p$ be a set, and $\mathsf{E}^{\times}(S) \gtrsim |S|^3$. If $|S|^5 \leqslant \|1_S\|_{A(\mathbb{F}_p)}^2 p^3$, then*

$$\|1_S\|_{A(\mathbb{F}_p)} \gtrsim |S|^{1/4}. \tag{3}$$

Corollary above improves some results from [1], see Corollaries 2, 4, 5 from the last paper, where instead of (3) several inequalities of the form $|S|^{\gamma}$, $\gamma < 1/2$ were obtained.

The family of sets with small Wiener norm contains a subfamily of symmetric sets with small negative Fourier coefficients (see, say, Lemma 13 below). It is a difficult problem to obtain a good lower bound for maximal modulus of negative Fourier coefficient, see [14, 15]. For example, in [14] Sanders obtained the following result.

Theorem 3 *Let p be a prime number, and $S \subseteq \mathbb{Z}/p\mathbb{Z}$ be a symmetric set. Suppose that $|S| \leqslant p/2$, $|S| := \delta p$. Then there are some functions $c(\delta) > 0$, $d(\delta) > 0$ such that*

$$- \min_{x \neq 0} \widehat{1}_S(x) \geqslant c(\delta) \cdot |S|^{d(\delta)} .$$

Here the functions $c(\delta)$ and $d(\delta)$ tends to zero as $\delta \to 0$ ($d(\delta)$ turns out to be a linear on δ). In particular case $\delta = 1/2$ Sanders calculated the number $d(\delta)$ which is turned out to be $1/3$. It is easy to see, that the upper bound here is $1/2$. We show that, in contrary, if one wants to estimate from below the maximal value of *positive* nonzero Fourier coefficient, then it can be done rather easily.

Theorem 4 *Let $S \subseteq \mathbb{F}_p$ be a symmetric set, $|S| \leqslant p/4$. Then*

$$\max_{x \neq 0} \widehat{1}_S(x) \gg |S|^{1/3} . \tag{4}$$

A similar result holds for any abelian group. A connection of sets with small nonzero positive/negative Fourier coefficients and its multiplicative energy is discussed in Section 4, see Corollary 14.

The author is grateful to the reviewer for valuable suggestions and remarks.

2 Definitions

Let $\mathbf{G}$ be an abelian group. If $\mathbf{G}$ is finite, then denote by N the cardinality of $\mathbf{G}$. For example, $N = p$ if $\mathbf{G} = \mathbb{F}_p$ and $N = p - 1$ if $\mathbf{G} = \mathbb{F}_p^*$. It is well–known [16] that the dual group $\widehat{\mathbf{G}}$ is isomorphic to $\mathbf{G}$ in this case. Let f be a function from $\mathbf{G}$ to $\mathbb{C}$. We denote the Fourier transform of f by $\widehat{f}$,

$$\widehat{f}(\xi) = \sum_{x \in \mathbf{G}} f(x) e(-\xi \cdot x) , \tag{5}$$

where $e(x) = e^{2\pi i x}$ and $\xi \in \widehat{\mathbf{G}}$. We rely on the following basic identities

$$\sum_{x \in \mathbf{G}} |f(x)|^2 = \frac{1}{N} \sum_{\xi \in \widehat{\mathbf{G}}} |\widehat{f}(\xi)|^2. \tag{6}$$

$$\langle f, g \rangle := \sum_{x \in \mathbf{G}} f(x)\overline{g(x)} = \frac{1}{N} \sum_{\xi \in \widehat{\mathbf{G}}} \widehat{f}(\xi)\overline{\widehat{g}(\xi)} = \frac{1}{N} \langle \widehat{f}, \widehat{g} \rangle. \tag{7}$$

$$\sum_{y \in \mathbf{G}} \left| \sum_{x \in \mathbf{G}} f(x)g(y - x) \right|^2 = \frac{1}{N} \sum_{\xi \in \widehat{\mathbf{G}}} |\widehat{f}(\xi)|^2 |\widehat{g}(\xi)|^2. \tag{8}$$

and

$$f(x) = \frac{1}{N} \sum_{\xi \in \widehat{\mathbf{G}}} \widehat{f}(\xi)e(\xi \cdot x). \tag{9}$$

If

$$(f * g)(x) := \sum_{y \in \mathbf{G}} f(y)g(x - y) \quad \text{and} \quad (f \circ g)(x) := \sum_{y \in \mathbf{G}} f(y)g(y + x)$$

then

$$\widehat{f * g} = \widehat{f}\widehat{g} \quad \text{and} \quad \widehat{f \circ g} = \widehat{f}^c \widehat{g} = \overline{\widehat{f}\widehat{g}}, \tag{10}$$

where for a function $f : \mathbf{G} \to \mathbb{C}$ we put $f^c(x) := f(-x)$. Clearly, $(f * g)(x) = (g * f)(x)$ and $(f \circ g)(x) = (g \circ f)(-x)$, $x \in \mathbf{G}$. Write $\mathsf{E}^+(A, B)$ for the *additive energy* of two sets $A, B \subseteq \mathbf{G}$ (see, e.g., [10]), that is,

$$\mathsf{E}^+(A, B) = |\{a_1 + b_1 = a_2 + b_2 \ : \ a_1, a_2 \in A, \ b_1, b_2 \in B\}|.$$

If $A = B$ we simply write $\mathsf{E}^+(A)$ instead of $\mathsf{E}^+(A, A)$. More generally, we write

$$\mathsf{T}_k^+(A) := |\{(a_1, \ldots, a_k, a_1', \ldots, a_k') \in A^{2k} \ : \ a_1 + \ldots a_k = a_1' + \ldots a_k'\}|.$$

In the same way one can define the *multiplicative energy* of two sets $A, B \subseteq \mathbb{F}_p$ as

$$\mathsf{E}^{\times}(A, B) = |\{a_1 b_1 = a_2 b_2 \ : \ a_1, a_2 \in A, \ b_1, b_2 \in B\}|.$$

Further, by (10), clearly,

$$\mathsf{E}^+(A, B) = \sum_x (A * B)(x)^2 = \sum_x (A \circ B)(x)^2 = \sum_x (A \circ A)(x)(B \circ B)(x). \tag{11}$$

and by (8),

$$\mathsf{E}^+(A, B) = \frac{1}{N} \sum_{\xi} |\widehat{A}(\xi)|^2 |\widehat{B}(\xi)|^2 . \tag{12}$$

We write $\mathbb{F}_p^*$ for $\mathbb{F}_p \setminus \{0\}$.

3 On the Multiplicative Energy of Sets with Small Wiener Norm

The sum-product decomposition of a set by energy was initiated by Balog and Wooley in [13]. We need modern form of these results from [17, Proposition 2, Corollary 1] which develop the method from [12].

Lemma 5 *Let $S \subseteq \mathbb{F}_p$ be a set, $|S|^6 \leqslant p^2 \mathsf{E}^\times(S)$. Then there is a set $S_1 \subseteq S$, $|S_1|^2 \gtrsim \mathsf{E}^\times(S)/|S|$ and*

$$\mathsf{E}^+(S_1)^2 \mathsf{E}^\times(S)^3 \lesssim |S_1|^{11} |S|^3 . \tag{13}$$

The same result holds if one swaps $+$ with $\times$ and vice versa.

Lemma 6 *Let $S \subseteq \mathbb{F}_p$ be a set, $|S|^6 \leqslant p^2 \mathsf{E}^\times(S)$. Then there is a set $S' \subseteq S$, $|S'|^3 \gtrsim \mathsf{E}^\times(S)$ and*

$$\mathsf{E}^+(S')^2 \mathsf{E}^\times(S)^3 \lesssim |S|^{14} . \tag{14}$$

The same result holds if one swaps $+$ with $\times$ and vice versa.

We need in a result on the energy of subsets of sets with small Wiener norm (also, see [3, Lemma 4]).

Lemma 7 *Let $S_1 \subseteq S \subseteq \mathbf{G}$ be sets and $K := \|1_S\|_{A(\mathbf{G})}$. Then*

$$\mathsf{E}^+(S, S_1) \geqslant \frac{|S_1|^3}{K^2} ,$$

and for any $k \geqslant 2$ the following holds

$$\mathsf{T}_k^+(S_1) \geqslant \frac{|S_1|^{2k}}{|S| K^{2k-2}} .$$

Proof By formula (7), we have

$$|S_1| N = \sum_{\xi} \widehat{1_{S_1}}(\xi) \widehat{1_S}(\xi) . \tag{15}$$

Using the Hölder inequality twice as well as identity (6) and (12), we get

$$|S_1|^4 \leqslant \left(KN^{-1} \sum_\xi |\widehat{1_{S_1}}(\xi)|^2 |\widehat{1_S}(\xi)| \right)^2$$

$$\leqslant K^2 |S_1| N^{-1} \sum_\xi |\widehat{1_{S_1}}(\xi)|^2 |\widehat{1_S}(\xi)|^2 = K^2 |S_1| \mathsf{E}^+(S, S_1).$$

Similarly, returning to (15) and applying the Hölder inequality again, we obtain

$$(|S_1|N)^{2k} \leqslant \sum_\xi |\widehat{1_{S_1}}(\xi)|^{2k} \cdot \left(\sum_\xi |\widehat{1_S}(\xi)|^{2k/(2k-1)} \right)^{2k-1}$$

$$\leqslant \sum_\xi |\widehat{1_{S_1}}(\xi)|^{2k} \cdot \sum_\xi |\widehat{1_S}(\xi)|^2 (NK)^{2k-2} =$$

$$= N^{2k} K^{2k-2} \mathsf{T}_k^+(S_1)|S|.$$

This completes the proof. $\qquad\qquad\qquad\qquad\qquad\qquad\qquad\qquad\qquad\qquad\square$

Take $k \geqslant 0$ and consider an arbitrary vector $\vec{s} = (s_1, \dots, s_k) \in (\mathbb{F}_p^*)^k$. For any set $Q \subseteq \mathbb{F}_p$ and $\vec{s}$ put $Q_{\vec{s}}^\times(x) := 1_Q(x)1_Q(s_1 x)\dots 1_Q(s_k x)$. In particular, $Q_{\vec{s}}^\times \subseteq Q$. We can easily bound the multiplicative Fourier coefficients $\tilde{Q}_{\vec{s}}^\times$ of the sets $Q_{\vec{s}}^\times$. Recall that the multiplicative Fourier transform of a function $f : \mathbb{F}_p^* \to \mathbb{C}$ is defined as

$$\tilde{f}(\chi) = \sum_{x \in \mathbb{F}_p^*} f(x)\overline{\chi(x)}, \qquad\qquad (16)$$

where χ is any multiplicative character.

Lemma 8 *Let $Q \subseteq \mathbb{F}_p$ be a set, $k \geqslant 0$ be an integer, and $s = (s_1, \dots, s_k) \in (\mathbb{F}_p^*)^k$ be a vector. Then for any non–principal multiplicative character χ the following holds*

$$|\tilde{Q}_{\vec{s}}^\times(\chi)| \leqslant \sqrt{p} \cdot \|1_Q\|_{A(\mathbb{F}_p)}^{k+1}. \qquad\qquad (17)$$

Proof By the inversion formula, we get

$$\widehat{Q}_{\vec{s}}^\times(r) = \sum_x 1_Q(x)1_Q(s_1 x)\dots 1_Q(s_k x)e(-xr) =$$

$$= \frac{1}{p^{k+1}} \sum_{z_0,z_1,\dots,z_k} \widehat{1_Q}(z_0)\widehat{1_Q}(z_1)\dots \widehat{1_Q}(z_k)e(z_0 x + z_1 s_1 x + \dots + z_k s_k x - xr) =$$

$$= \frac{1}{p^k} \sum_{z_0, z_1, \ldots, z_k \, : \, z_0 + z_1 s_1 + \cdots + z_k s_k = r} \widehat{Q}(z_0) \widehat{Q}(z_1) \ldots \widehat{Q}(z_k) . \tag{18}$$

Further, using the definition of the multiplicative Fourier transform (16) and formula (7) for the additive Fourier transform, we have for any multiplicative character χ

$$\widetilde{Q}_{\vec{s}}^{\times}(\chi) = \sum_x Q_{\vec{s}}^{\times}(x) \overline{\chi(x)} = \frac{1}{p} \sum_r \widehat{Q}_{\vec{s}}^{\times}(r) \overline{G(\chi, r)} ,$$

where $G(\chi, r)$ is the Gauss sum (Q can contain zero or not here because $\chi(0) = 0$ by the definition). Thus, by formula (18) and the well–known estimate for the absolute value of $G(\chi, r)$, we obtain

$$|\widetilde{Q}_{\vec{s}}^{\times}(\chi)| \leqslant \frac{\sqrt{p}}{p^{k+1}} \sum_r \sum_{z_1, \ldots, z_k} |\widehat{Q}(r - z_1 s_1 + \cdots + z_k s_k)| |\widehat{Q}(z_1)| \ldots |\widehat{Q}(z_k)| \leqslant$$

$$\leqslant \sqrt{p} \cdot p^{-(k+1)} \|\widehat{Q}\|_1^{k+1} = \sqrt{p} \|1_Q\|_{A(\mathbb{F}_p)}^{k+1} .$$

as required. $\qquad \square$

Using Lemmas 5 and 8 we can prove Theorem 1 from the Introduction.

Theorem 9 *Let $S \subseteq \mathbb{F}_p$ be a set, and $K := \|1_S\|_{A(\mathbb{F}_p)}$. We have*

$$\mathsf{E}^{\times}(S) \leqslant \frac{|S|^4}{p - 1} + \|1_S\|_{A(\mathbb{F}_p)}^2 |S| p . \tag{19}$$

Further, if $|S|^5 \leqslant K^2 p^3$, then

$$\mathsf{E}^{\times}(S) \lesssim K^{4/3} |S|^{8/3} , \tag{20}$$

and if $|S|^{27} \leqslant K^6 p^{17}$, then

$$\mathsf{E}^{\times}(S) \lesssim K^{12/17} |S|^{48/17} . \tag{21}$$

Proof First of all, notice that by formula (12) in the multiplicative form, combining with Parseval identity (6) as well as Lemma 8 with $k = 0$, we see that

$$\mathsf{E}^{\times}(S) = \frac{1}{p - 1} \sum_{\chi} |\widetilde{S}(\chi)|^4 \leqslant \frac{|S|^4}{p - 1} + \frac{1}{p - 1} \cdot p \|1_S\|_{A(\mathbb{F}_p)}^2 (p - 1) |S|$$

$$= \frac{|S|^4}{p - 1} + p \|1_S\|_{A(\mathbb{F}_p)}^2 |S| .$$

Further, if (20) takes place, then there is nothing to prove. Otherwise, thanks to the assumption $|S|^5 \leqslant K^2 p^3$ we see that the condition of Lemma 5 takes place, namely, $|S|^6 \leqslant p^2 \mathsf{E}^\times(S)$. By Lemma 5, we find $S_1 \subseteq S$ such that $|S_1|^2 \gtrsim \mathsf{E}^\times(S)/|S|$ and combining this result with Lemma 7, we get

$$\frac{|S_1|^8}{K^4 |S|^2} \mathsf{E}^\times(S)^3 \leqslant \mathsf{E}^+(S_1)^2 \mathsf{E}^\times(S)^3 \lesssim |S_1|^{11} |S|^3 .$$

and hence

$$\mathsf{E}^\times(S)^3 \lesssim K^4 |S_1|^3 |S|^5 \leqslant K^4 |S|^8$$

as required. If one applies Lemma 6, combining with Lemma 7, then for a set $S' \subseteq S$, $|S'|^3 \gtrsim \mathsf{E}^\times(S)$ the following holds

$$\frac{\mathsf{E}^\times(S)^{17/3}}{K^4 |S|^2} \lesssim \frac{|S'|^8}{K^4 |S|^2} \mathsf{E}^\times(S)^3 \leqslant \mathsf{E}^+(S')^2 \mathsf{E}^\times(S)^3 \lesssim |S|^{14} .$$

and it gives us

$$\mathsf{E}^\times(S) \lesssim |S|^{48/17} K^{12/17} .$$

Again if (21) takes place, then there is nothing to prove and otherwise Lemma 6 can be applied because of the condition $|S|^{27} \leqslant K^6 p^{17}$. This completes the proof. □

From the Parseval identity (6) one has $\|1_S\|_{A(G)} \leqslant |S|^{1/2}$ for any subset S of an abelian group $\mathbf{G}$. Theorem 9 gives us an interesting inverse inequality.

Corollary 10 *Let $S \subseteq \mathbb{F}_p$ be a set, and $\mathsf{E}^\times(S) \gtrsim |S|^3$. If $|S|^5 \leqslant \|1_S\|^2_{A(\mathbb{F}_p)} p^3$, then*

$$\|1_S\|_{A(\mathbb{F}_p)} \gtrsim |S|^{1/4} . \tag{22}$$

In any case $\|1_S\|_{A(\mathbb{F}_p)} \gg |S| p^{-1/2}$, provided that $|S| \leqslant p/2$.

It is interesting to notice that Corollary 10 does not hold in general fields, for example, if S is a subfield of size p in $\mathbb{F}_{p^2}$, then $\|1_S\|_{A(\mathbb{F}_{p^2})} = 1$ and $\mathsf{E}^\times(S) = |S|^3$.

Typical sets with $\mathsf{E}^\times(S) \gtrsim |S|^3$ are large subsets of multiplicative subgroups, large subsets of the sets of the form $\{1, g, \ldots, g^n\}$, where $g \in \mathbb{F}_p$ is a primitive root, see, e.g., [1] or, more generally, sets S with the small product set $SS := \{s_1 s_2 : s_1, s_2 \in S\}$. Let us remark another consequence of Theorem 9 which coincides with [1, Corollary 6] in the case of multiplicative subgroups (up to some logarithms).

Corollary 11 *Let $H \subseteq \mathbb{F}_p$ be a multiplicative subgroup, $|H| \leqslant p^{2/3}$ or $H = \{1, g, \ldots, g^n\}$, $n \leqslant p^{2/3}$. Then*

$$\|1_H\|_{A(\mathbb{F}_p)} \gtrsim |H|^{1/4}. \tag{23}$$

In any case $\|1_H\|_{A(\mathbb{F}_p)} \gg |H|p^{-1/2}$, *provided that* $|H| \leqslant p/2$.

Nevertheless, a simple application of Lemma 7, combining with [11, Theorem 3], namely, $\mathsf{E}^+(H) \ll |H|^{49/20} \log^{1/5} |H|$, $|H| \leqslant p^{1/2}$ in the case when H is a multiplicative subgroup, give us (another way to obtain the same is to use [11, Corollary 7])

Corollary 12 *Let* $H \subseteq \mathbb{F}_p$ *be a multiplicative subgroup,* $|H| \leqslant p^{1/2}$. *Then*

$$\|1_H\|_{A(\mathbb{F}_p)} \gtrsim |H|^{1/4+1/40}. \tag{24}$$

4 On the Quantity M$_+$

Let $S \subseteq \mathbf{G}$ be a symmetric sets. Then Fourier coefficients of $\widehat{1}_S$ are real and we put

$$\mathsf{M}_-^{\mathbf{G}}(S) = \mathsf{M}_-(S) = \max_{x \neq 0}(-\widehat{1}_S(x)),$$

and

$$\mathsf{M}_+^{\mathbf{G}}(S) = \mathsf{M}_+(S) = \max_{x \neq 0} \widehat{1}_S(x).$$

In a similar way we consider $\mathsf{M}_-^{\mathbf{G}}(\alpha)$, $\mathsf{M}_+^{\mathbf{G}}(\alpha)$ of a symmetric real function α.

As we noticed in the Introduction there is a connection of the problem of estimating M_+ with the Littlewood conjecture. Let us formulate a simple lemma about some bounds for $\|1_S\|_{A(\mathbf{G})}$ in terms of $\mathsf{M}_+(S)$.

Lemma 13 *Let* $\alpha : \mathbf{G} \to \mathbb{R}^+$ *be a symmetric function. Then*

$$\sum_x |\widehat{\alpha}(x)| \leqslant 2(\|\alpha\|_1 + N\mathsf{M}_+(\alpha)) - N\alpha(0), \tag{25}$$

and

$$\sum_x |\widehat{\alpha}(x)| \leqslant 2N\mathsf{M}_-(\alpha) + N\alpha(0). \tag{26}$$

Proof Let us prove (25), another bound follows similarly. By formula (9), we have

$$\sum_x \widehat{\alpha}(x) = \widehat{\alpha}(0) + \sum_{x \neq 0 \,:\, \widehat{\alpha}(x) > 0} \widehat{\alpha}(x) + \sum_{x \neq 0 \,:\, \widehat{\alpha}(x) < 0} \widehat{\alpha}(x) = N\alpha(0). \tag{27}$$

Thus

$$\sum_x |\widehat{\alpha}(x)| = \widehat{\alpha}(0) + \sum_{x \neq 0 \,:\, \widehat{\alpha}(x)>0} \widehat{\alpha}(x) - \sum_{x \neq 0 \,:\, \widehat{\alpha}(x)<0} \widehat{\alpha}(x) = 2\widehat{\alpha}(0) + 2 \sum_{x \neq 0 \,:\, \widehat{\alpha}(x)>0} \widehat{\alpha}(x) - N\alpha(0)$$

$$\leqslant 2(\|\alpha\|_1 + N\mathsf{M}_+(\alpha)) - N\alpha(0)$$

and we are done.   □

Corollary 14 *Let* $S \subseteq \mathbb{F}_p$ *be a symmetric set with* $\mathsf{E}^\times(S) \gtrsim |S|^3$. *Then* $\mathsf{M}_+(S), \mathsf{M}_-(S) \gtrsim |S|^{1/4}$.

Now we are ready to prove our second main result and we use the arguments from [14]. It gives us an unconditional lower bound for $\mathsf{M}_+(S)$.

Theorem 15 *Let* $S \subseteq \mathbf{G}$ *be a symmetric set,* $\zeta \in (0, 1/2)$ *be a real number, and* $|S| = (1/2 - \zeta)N$. *Then*

$$\mathsf{M}_+(S) \geqslant 2^{-2} \min\{\zeta^{1/3}|S|^{1/3}, \zeta|S|^{1/2}\} \,. \tag{28}$$

In particular, if $|S| \leqslant N/4$, *then*

$$\mathsf{M}_+(S) \gg |S|^{1/3} \,. \tag{29}$$

Proof In view of the Parseval identity, we get

$$|S|N = |S|^2 + \sum_{x \,:\, \widehat{1_S}(x)>0} \widehat{1_S^2}(x) + \sum_{x \,:\, \widehat{1_S}(x)<0} |\widehat{1_S}(x)|^2 \,. \tag{30}$$

From this formula it follows that $\mathsf{M}_-(S) > 0$ because otherwise by inequality (26) of Lemma 13, we obtain

$$|S|N/2 \leqslant |S|(N - |S|) \leqslant \mathsf{M}_+(S)N$$

and hence $\mathsf{M}_+(S) \geqslant |S|/2$. Further, by formula (10), we have

$$\widehat{1_S}(x) = \frac{1}{N} \sum_z \widehat{1_S}(x - z)\widehat{1_S}(z) = \frac{2|S|\widehat{1_S}(x)}{N} + \frac{1}{N} \sum_{z \neq 0,\, z \neq x} \widehat{1_S}(x - z)\widehat{1_S}(z) \,.$$

Now take x such that $\widehat{1_S}(x) = -\mathsf{M}_-(S)$. Then

$$-2\zeta\mathsf{M}_-(S) = \frac{1}{N} \sum_{z \neq 0,\, z \neq x} \widehat{1_S}(x - z)\widehat{1_S}(z) \,. \tag{31}$$

We see that the negativity of the left–hand side implies that one of the elements in the product $\widehat{1}_S(x-z)\widehat{1}_S(z)$ must be positive. Hence as in Lemma 13, we have

$$\zeta M_-(S) \leqslant M_+(S)\,(2M_+(S)+1)\,. \tag{32}$$

Because of $|S| = (1/2-\zeta)N$, we get in view of (30)

$$|S|N/2 \leqslant |S|N(1/2+\zeta) \leqslant M_+^2(S)N + \sum_{x\,:\,\widehat{1}_S(x)<0} |\widehat{1}_S^2(x)|\,. \tag{33}$$

If $M_+^2(S) \geqslant |S|/4$, then we are done, otherwise (33) implies that

$$|S|/4 \leqslant M_-(S)\cdot N^{-1} \sum_{x\,:\,\widehat{1}_S(x)<0} |\widehat{1}_S(x)| \leqslant M_-^2(S) \tag{34}$$

and, similarly,

$$|S|/4 \leqslant M_-(S)N^{-1}\left(\sum_{x\,:\,\widehat{1}_S(x)>0} |\widehat{1}_S(x)| + |A|\right) \leqslant 2M_-(S)M_+(S)\,. \tag{35}$$

Here we have assumed that $M_+(S) \geqslant 1/2$. Returning to (32) and using Lemma 13 as well as bound (35), we obtain

$$\zeta|S| \leqslant 8M_+^2(S)\,(2M_+(S)+1) \leqslant 32M_+^3(S)$$

as required. Again, we have assumed that $M_+(S) \geqslant 1/2$. If not, then (34) and (32) give us

$$M_+(S) \geqslant \zeta|S|^{1/2}/4\,.$$

This completes the proof. $\qquad\qquad\qquad\qquad\qquad\qquad\qquad\qquad\qquad\square$

Remark 16 The condition $|S| < N/2$ of Theorem 15 is quite natural. If $|S| = N/2$, then $M_+(S)$ can be zero. Indeed, let $\mathbf{G} = \mathbb{F}_2^n$ and A be an affine subspace of codimension 1 such that $0 \notin S$. Then $\widehat{S}(0) = |S|$, further, there is the only Fourier coefficient equals $(-|S|)$ and all other Fourier coefficients of S vanish. Thus, $M_+(S) = 0$. Moreover, if we delete from S a random subset Ω of cardinality ζN, then we get with high probability $M_+(S') \gg (\zeta|S'|)^{1/2}$ for this new set $S' = S \setminus \Omega$.

References

1. V.C. Garcia, The finite Littlewood problem in $\mathbb{F}_p$. **47**(85), 1–14 (2018). https://doi.org/10.1007/s11139-018-0038-3
2. B.J. Green, S.V. Konyagin, On the Littlewood problem modulo a prime. Canad. J. Math. **61**(1), 141–164 (2009)
3. S.V. Konyagin, I.D. Shkredov, Quantitative version of Beurling–Helson theorem. Funct. Anal. Its Appl. **49**(2), 110–121 (2015)
4. S.V. Konyagin, I.D. Shkredov, On the Wiener norm of subsets of Z/pZ of medium size. J. Math. Sci. **218**(5), 599–608 (2016)
5. T. Sanders, The Littlewood–Gowers problem. J. Anal. Math. **101**, 123–162 (2007)
6. T. Sanders, Boolean functions with small spectral norm, revisited. arXiv:1804.04050v1 [math.CA] 11 Apr 2018
7. T. Schoen, On the Littlewood conjecture in $\mathbb{Z}/p\mathbb{Z}$. MJCNT **7**(3), 66–72 (2017)
8. S.V. Konyagin, On a problem of Littlewood. Izvestiya Russ. Acad. Sci. **45**(2), 243–265 (1981)
9. O.C. McGehee, L. Pigno, B. Smith, Hardy's inequality and the L^1 norm of exponential sums. Ann. Math. **113**, 613–618 (1981)
10. T. Tao, V. Vu, *Additive Combinatorics* (Cambridge University Press, Cambridge, 2006)
11. B. Murphy, M. Rudnev, I.D. Shkredov, Y.N. Shteinikov, On the few products, many sums problem. arXiv:1712.00410v1 [math.CO] 1 Dec 2017
12. M. Rudnev, I.D. Shkredov, S. Stevens, On the energy variant of the sum–product conjecture. arXiv:1607.05053
13. A. Balog, T.D. Wooley, A low-energy decomposition theorem. Q. J. Math. **68**(1), 207–226 (2017)
14. T. Sanders, Chowla's cosine problem. Isr. J. Math. **179**(1), 1–28 (2010)
15. I. Ruzsa, Negative values of cosine sums. Acta Arithmetica **111**, 179–186 (2004)
16. W. Rudin, *Fourier Analysis on Groups* (Wiley, New York, 1962)
17. I.D. Shkredov, An application of the sum–product phenomenon to sets avoiding several linear equations. Sb. Math. **209**(4), 580–603 (2018)
18. T. Sanders, The coset and stability rings. arXiv:1810.10461v1 [math.CO] 24 Oct 2018

Order Estimates of Best Orthogonal Trigonometric Approximations of Classes of Infinitely Differentiable Functions

Tetiana A. Stepanyuk

Abstract In this paper we establish exact order estimates for the best uniform orthogonal trigonometric approximations of the classes of 2π-periodic functions, whose (ψ, β)–derivatives belong to unit balls of spaces L_p, $1 \leq p < \infty$, in the case, when the sequence $\psi(k)$ tends to zero faster, than any power function, but slower than geometric progression. Similar estimates are also established in the L_s-metric, $1 < s \leq \infty$, for the classes of differentiable functions, which (ψ, β)–derivatives belong to unit ball of space L_1.

Keywords Fourier series · Best orthogonal trigonometric approximation · Classes of infinitely differentiable functions · $(\psi \cdot \beta)$-derivative

1 Introduction

Let C be a space of 2π–periodic continuous functions with the following norm: $\|f\|_C := \max_{t \in [0,2\pi)} |f(t)|$; L_∞ be the space of 2π–periodic functions f, which are Lebesgue measurable and essentially bounded with the norm $\|f\|_\infty := \operatorname{ess\,sup}_t |f(t)|$ and L_p, $1 \leq p < \infty$, be the space of 2π–periodic functions f summable to the power p on $[0, 2\pi)$, with the norm $\|f\|_p := \left(\int_0^{2\pi} |f(t)|^p dt \right)^{\frac{1}{p}}$.

T. A. Stepanyuk (✉)
Institute of Analysis and Number Theory, Graz University of Technology, Graz, Austria

Johann Radon Institute for Computational and Applied Mathematics (RICAM), Austrian Academy of Sciences, Linz, Austria

Institute of Mathematics of NAS of Ukraine, Kyiv, Ukraine
e-mail: tania_stepaniuk@ukr.net

© Springer Nature Switzerland AG 2020
A. Raigorodskii, M. T. Rassias (eds.), *Trigonometric Sums and Their Applications*,
https://doi.org/10.1007/978-3-030-37904-9_13

Let $f : \mathbb{R} \to \mathbb{R}$ be the function from L_1, whose Fourier series is given by

$$\sum_{k=-\infty}^{\infty} \hat{f}(k)e^{ikx},$$

where $\hat{f}(k) = \frac{1}{2\pi} \int\limits_{-\pi}^{\pi} f(t)e^{-ikt}dt$ are the Fourier coefficients of the function f, $\psi(k)$ is an arbitrary fixed sequence of real numbers and β is a fixed real number. Then, if the series

$$\sum_{k \in \mathbb{Z}\setminus\{0\}} \frac{\hat{f}(k)}{\psi(|k|)} e^{i(kx+\frac{\beta\pi}{2}\mathrm{sign}k)}$$

is the Fourier series of some function φ from L_1, then this function is called the (ψ, β)–derivative of the function f and is denoted by f_β^ψ. A set of functions f, whose (ψ, β)–derivatives exist, is denoted by L_β^ψ (see [16]).

Let

$$B_p^0 := \left\{ \varphi \in L_p : \|\varphi\|_p \leq 1, \ \varphi \perp 1 \right\}, \quad 1 \leq p \leq \infty.$$

If $f \in L_\beta^\psi$, and, at the same time $f_\beta^\psi \in B_p^0$, then we say that the function f belongs to the class $L_{\beta,p}^\psi$.

Denote $C_\beta^\psi = C \cap L_\beta^\psi$ and $C_{\beta,p}^\psi = C \cap L_{\beta,p}^\psi$.

By $\mathfrak{M}$ we denote the set of all convex (downward) continuous functions $\psi(t)$, $t \geq 1$, such that $\lim\limits_{t \to \infty} \psi(t) = 0$. Assume that the sequence $\psi(k)$, $k \in \mathbb{N}$, specifying the class $L_{\beta,p}^\psi$, $1 \leq p \leq \infty$, is the restriction of the functions $\psi(t)$ from $\mathfrak{M}$ to the set of natural numbers.

Following Stepanets (see, e.g., [16]), by using the characteristic $\mu(\psi; t)$ of functions ψ from $\in \mathfrak{M}$ of the form

$$\mu(t) = \mu(\psi; t) := \frac{t}{\eta(t) - t}, \tag{1}$$

where $\eta(t) = \eta(\psi; t) := \psi^{-1}(\psi(t)/2)$, ψ^{-1} is the function inverse to ψ, we select the following subsets of the set $\mathfrak{M}$:

$$\mathfrak{M}_\infty^+ = \left\{ \psi \in \mathfrak{M} : \ \mu(\psi; t) \uparrow \infty \right\}.$$

$$\mathfrak{M}_\infty'' = \left\{ \psi \in \mathfrak{M}_\infty^+ : \ \exists K > 0 \ \eta(\psi; t) - t \geq K \ t \geq 1 \right\}.$$

The functions $\psi_{r,\alpha}(t) = \exp(-\alpha t^r)$ are typical representatives of the set $\mathfrak{M}_\infty^+$. Moreover, if $r \in (0, 1]$, then $\psi_{r,\alpha} \in \mathfrak{M}_\infty''$. The classes $L_{\beta,p}^\psi$, generated by the functions $\psi = \psi_{r,\alpha}$ are denoted by $L_{\beta,p}^{\alpha,r}$.

If $\psi \in \mathfrak{M}_\infty^+$ then (see, e.g., [15, p. 97]) the function $\psi(t)$ vanishes faster than any power function, i.e.,

$$\lim_{t \to \infty} t^r \psi(t) = 0 \quad \forall r \in \mathbb{R}.$$

This implies that, under the condition $\psi \in \mathfrak{M}_\infty^+$, the Fourier series of any function f from $C_{\beta,p}^\psi$, $\beta \in \mathbb{R}$, can be differentiated infinitely many times and, as a result, we get uniformly convergent series. Hence, the classes $C_{\beta,p}^\psi$ with $\psi \in \mathfrak{M}_\infty^+$ consist of infinitely differentiable functions. On the other hand, as shown in [17, p. 1692], for any infinitely differentiable 2π–periodic function f, one can indicate a function from the set $\mathfrak{M}_\infty^+$ such that $f \in C_\beta^\psi$ for any $\beta \in \mathbb{R}$.

For functions f from classes $L_{\beta,p}^\psi$ we consider: L_s–norms of deviations of the functions f from their partial Fourier sums of order $n - 1$, i.e., the quantities

$$\|\rho_n(f; \cdot)\|_s = \|f(\cdot) - S_{n-1}(f; \cdot)\|_s, \quad 1 \le s \le \infty, \tag{2}$$

where

$$S_{n-1}(f; x) = \sum_{k=-n+1}^{n-1} \hat{f}(k)e^{ikx};$$

and the best orthogonal trigonometric approximations of the functions f in metric of space L_s, i.e., the quantities of the form

$$e_m^\perp(f)_s = \inf_{\gamma_m} \|f(\cdot) - S_{\gamma_m}(f; \cdot)\|_s, \quad 1 \le s \le \infty, \tag{3}$$

where γ_m, $m \in \mathbb{N}$, is an arbitrary collection of m integer numbers, and

$$S_{\gamma_m}(f; x) = \sum_{k \in \gamma_m} \hat{f}(k)e^{ikx}.$$

We set

$$\mathscr{E}_n(L_{\beta,p}^\psi)_s = \sup_{f \in L_{\beta,p}^\psi} \|\rho_n(f; \cdot)\|_s, \quad 1 \le p, s \le \infty, \tag{4}$$

$$e_n^\perp(L_{\beta,p}^\psi)_s = \sup_{f \in L_{\beta,p}^\psi} e_n^\perp(f)_s, \quad 1 \le p, s \le \infty. \tag{5}$$

It is clear, that if $f \in C$, then

$$\|\rho_n(f; x)\|_\infty = \|\rho_n(f; x)\|_C, \quad e_m^\perp(f)_C = e_m^\perp(f)_\infty.$$

That is why

$$\mathscr{E}_n(C_{\beta,p}^{\psi})_C = \mathscr{E}_n(C_{\beta,p}^{\psi})_\infty, \quad e_m^{\perp}(C_{\beta,p}^{\psi})_C = e_m^{\perp}(C_{\beta,p}^{\psi})_\infty.$$

The following inequalities follow from given above definitions (4) and (5)

$$e_{2n}^{\perp}(L_{\beta,p}^{\psi})_s \leq e_{2n-1}^{\perp}(L_{\beta,p}^{\psi})_s \leq \mathscr{E}_n(L_{\beta,p}^{\psi})_s, \quad 1 \leq p, s \leq \infty. \tag{6}$$

In the case when $\psi(k) = k^{-r}$, $r > 0$, the classes $L_{\beta,p}^{\psi}$, $1 \leq p \leq \infty$, $\beta \in \mathbb{R}$, are well–known Weyl–Nagy classes $W_{\beta,p}^r$. For these classes, the order estimates of quantities $e_n^{\perp}(L_{\beta,p}^{\psi})_s$ are known for $1 < p, s < \infty$ (see [4, 5]), for $1 \leq p < \infty$, $s = \infty, r > \frac{1}{p}$ and also for $p = 1$, $1 < s < \infty$, $r > \frac{1}{s'}$, $\frac{1}{s} + \frac{1}{s'} = 1$ (see [5, 6]).

In the case, when $\psi(k)$ tends to zero not faster than some power function, order estimates for quantities (5) were established in [1, 10, 12–14]. In the case, when $\psi(k)$ tends to zero not slower than geometric progression, exact order estimates for $e_n^{\perp}(L_{\beta,p}^{\psi})_s$ were found in [11] for all $1 \leq p, s \leq \infty$.

Our aim is to establish the exact-order estimates of $e_n^{\perp}(L_{\beta,p}^{\psi})_\infty$, $1 \leq p < \infty$, and $e_n^{\perp}(L_{\beta,1}^{\psi})_s$, $1 < s < \infty$, in the case, when ψ decreases faster than any power function, but slower than geometric progression ($\psi \in \mathfrak{M}_\infty''$).

2 Best Orthogonal Trigonometric Approximations of the Classes $L_{\beta,p}^{\psi}$, $1 < p < \infty$, in the Metric of Space L_∞

We write $a_n \asymp b_n$ to mean that there exist positive constants C_1 and C_2 independent of n such that $C_1 a_n \leq b_n \leq C_2 a_n$ for all n.

Theorem 1 *Let $1 < p < \infty$, $\psi \in \mathfrak{M}_\infty''$ and the function $\frac{\psi(t)}{|\psi'(t)|} \uparrow \infty$ as $t \to \infty$. Then, for all $\beta \in \mathbb{R}$ the following order estimates hold*

$$e_{2n-1}^{\perp}(L_{\beta,p}^{\psi})_\infty \asymp e_{2n}^{\perp}(L_{\beta,p}^{\psi})_\infty \asymp \psi(n)(\eta(n) - n)^{\frac{1}{p}}. \tag{7}$$

In Theorem 1 and further we will assume $\psi'(t) := \psi'(t+0)$.

Proof According to Theorem 1 from [8] under conditions $\psi \in \mathfrak{M}_\infty^+$, $\beta \in \mathbb{R}$, $1 \leq p < \infty$, for $n \in \mathbb{N}$, such that $\eta(n) - n \geq a > 2$, $\mu(n) \geq b > 2$ the following estimate is true

$$\mathscr{E}_n(L_{\beta,p}^{\psi})_\infty \leq K_{a,b}(2p)^{1-\frac{1}{p}}\psi(n)(\eta(n) - n)^{\frac{1}{p}}, \tag{8}$$

where

$$K_{a,b} = \frac{1}{\pi} \max \left\{ \frac{2b}{b-2} + \frac{1}{a}, \; 2\pi \right\}.$$

It should be noticed, that condition $\frac{\psi(t)}{|\psi'(t)|} \to \infty$ as $t \to \infty$ implies that always exists $n_0 \in \mathbb{N}$, such that for all $n > n_0$, $n \in \mathbb{N}$ the following inequalities take place: $\eta(\psi, n) - n \geq a > 2$ and $\mu(\psi, n) \geq b > 2$. This fact follows from Remark 3.13.1 [16], which says, that for every $\psi \in \mathfrak{M}_\infty^+$

$$K_1(\eta(t) - t) \leq \frac{\psi(t)}{|\psi'(t)|} \leq K_2(\eta(t) - t), \; K_1, \; K_2 > 0, \tag{9}$$

and from the definition of quantity $\mu(\psi, t)$.

Using inequalities (6) and (8), we obtain

$$e_{2n}^\perp(L_{\beta,p}^\psi)_\infty \leq e_{2n-1}^\perp(L_{\beta,p}^\psi)_\infty \leq K_{a,b} \, (2p)^{1-\frac{1}{p}} \psi(n)(\eta(n) - n)^{\frac{1}{p}}. \tag{10}$$

Let us find the lower estimate for the quantity $e_{2n}^\perp(L_{\beta,p}^\psi)_\infty$. With this purpose we construct the function

$$f_{p,n}^*(t) = f_{p,n}^*(\psi; t) := \frac{\lambda_p}{\psi(n)(\eta(n) - n)^{\frac{1}{p'}}} \left(\frac{1}{2}\psi(1)\psi(2n) + \right.$$

$$\left. + \sum_{k=1}^{n-1} \psi(k)\psi(2n - k) \cos kt + \sum_{k=n}^{2n} \psi^2(k) \cos kt \right), \; \frac{1}{p} + \frac{1}{p'} = 1. \tag{11}$$

Let us show that $f_{p,n}^* \in L_{\beta,p}^\psi$. The definition of (ψ, β)–derivative yields

$$(f_{p,n}^*(t))_\beta^\psi = \frac{\lambda_p}{\psi(n)(\eta(n) - n)^{\frac{1}{p'}}} \left(\sum_{k=1}^{n-1} \psi(2n - k) \cos \left(kt + \frac{\beta\pi}{2} \right) \right.$$

$$\left. + \sum_{k=n}^{2n} \psi(k) \cos \left(kt + \frac{\beta\pi}{2} \right) \right). \tag{12}$$

Obviously

$$\left| (f_{p,n}^*(t))_\beta^\psi \right| \leq \frac{\lambda_p}{\psi(n)(\eta(n) - n)^{\frac{1}{p'}}} \left(\sum_{k=1}^{n-1} \psi(2n - k) + \sum_{k=n}^{2n} \psi(k) \right) <$$

$$< \frac{2\lambda_p}{\psi(n)(\eta(n) - n)^{\frac{1}{p'}}} \sum_{k=n}^{2n} \psi(k) \leq \frac{2\lambda_p}{\psi(n)(\eta(n) - n)^{\frac{1}{p'}}} \left(\psi(n) + \int_n^\infty \psi(u)du \right). \tag{13}$$

To estimate the integral from the right part of formula (13), we use the following statement [7, p. 500].

Proposition 1 *If $\psi \in \mathfrak{M}_\infty^+$, then for arbitrary $m \in \mathbb{N}$, such that $\mu(\psi, m) > 2$ the following condition holds*

$$\int_m^\infty \psi(u)du \le \frac{2}{1 - \frac{2}{\mu(m)}}\psi(m)(\eta(m) - m). \tag{14}$$

Formulas (13) and (14) imply that

$$\left|(f_{p,n}^*(t))_\beta^\psi\right| \le \frac{2\lambda_p}{\psi(n)(\eta(n) - n)^{\frac{1}{p'}}}\left(\psi(n) + \frac{2b}{b-2}\psi(n)(\eta(n) - n)\right) <$$

$$< \frac{5\lambda_p b}{b-2}(\eta(n) - n)^{\frac{1}{p}}. \tag{15}$$

We denote

$$D_{k,\beta}(t) := \frac{1}{2}\cos\frac{\beta\pi}{2} + \sum_{j=1}^k \cos\left(jt + \frac{\beta\pi}{2}\right). \tag{16}$$

Applying Abel transform, we have

$$\sum_{k=1}^{n-1}\psi(2n - k)\cos\left(kt + \frac{\beta\pi}{2}\right) = \sum_{k=1}^{n-2}(\psi(2n - k) - \psi(2n - k + 1))D_{k,\beta}(t)$$

$$+ \psi(n + 1)D_{n-1,\beta}(t) - \psi(2n - 1)\frac{1}{2}\cos\frac{\beta\pi}{2} \tag{17}$$

and

$$\sum_{k=n}^{2n}\psi(k)\cos\left(kt + \frac{\beta\pi}{2}\right) = \sum_{k=n}^{2n-1}(\psi(k) - \psi(k + 1))D_{k,\beta}(t)$$

$$+ \psi(2n)D_{2n,\beta}(t) - \psi(n)D_{n-1,\beta}(t). \tag{18}$$

Since

$$\sum_{k=0}^{N-1}\sin(\gamma + kt) = \sin\left(\gamma + \frac{N-1}{2}t\right)\sin\frac{Nt}{2}\frac{1}{\sin\frac{t}{2}} \tag{19}$$

(see, e.g., [2, p.43]), for $N = k + 1$, $\gamma = (\beta - 1)\frac{\pi}{2}$, the following inequality holds

$$
|D_{k,\beta}(t)| = \left| \frac{\cos\left(\frac{kt}{2} + \frac{\beta\pi}{2}\right) \sin\frac{k+1}{2}t}{\sin\frac{t}{2}} - \frac{1}{2}\cos\frac{\beta\pi}{2} \right|
$$

$$
= \left| \frac{\sin\left((k + \frac{1}{2})t + \frac{\beta\pi}{2}\right) - \cos\frac{t}{2}\sin\frac{\beta\pi}{2}}{2\sin\frac{t}{2}} \right| \le \frac{\pi}{t}, \quad 0 < |t| \le \pi. \tag{20}
$$

According to (12), (17), (18) and (20), we obtain

$$
\left| (f_{p,n}^*(t))_\beta^\psi \right| \le \frac{\lambda_p}{\psi(n)(\eta(n)-n)^{\frac{1}{p'}}} \frac{\pi}{|t|} \left(\sum_{k=1}^{n-2} |\psi(2n-k)-\psi(2n-k-1)| + \psi(n+1) \right.
$$

$$
+ \psi(2n-1) + \sum_{k=n}^{2n-1} |\psi(k) - \psi(k+1)| + \psi(2n) + \psi(n) \Bigg)
$$

$$
= \frac{\lambda_p}{\psi(n)(\eta(n)-n)^{\frac{1}{p'}}} \frac{2\pi}{|t|}(\psi(n+1) + \psi(n)) \le \frac{4\pi\lambda_p}{(\eta(n)-n)^{\frac{1}{p'}}} \frac{1}{|t|}. \tag{21}
$$

So, (15) and (21) imply

$$
\left\| (f_{p,n}^*(t))_\beta^\psi \right\|_p
$$

$$
\le \lambda_p \max\left\{ \frac{5b}{b-2}, 4\pi \right\} \left(\int_{|t| \le \frac{1}{\eta(n)-n}} (\eta(n)-n)dt + \frac{1}{(\eta(n)-n)^{\frac{p}{p'}}} \int_{\frac{1}{\eta(n)-n} \le |t| \le \pi} \frac{dt}{|t|^p} \right)^{\frac{1}{p}}
$$

$$
\le 2\lambda_p \max\left\{ \frac{5b}{b-2}, 4\pi \right\} \left(1 + \frac{1}{p-1} \right)^{\frac{1}{p}} = 2\lambda_p \max\left\{ \frac{5b}{b-2}, 4\pi \right\} (p')^{\frac{1}{p}}.
$$

Hence, for

$$
\lambda_p = \frac{1}{2(p')^{\frac{1}{p}} \max\left\{ \frac{5b}{b-2}, 4\pi \right\}}
$$

the embedding $f_{p,n}^* \in L_{\beta,p}^\psi$ is true.

Let us consider the quantity

$$
I_1 := \inf_{\gamma_{2n}} \left| \int_{-\pi}^{\pi} (f_{p,n}^*(t) - S_{\gamma_{2n}}(f_{p,n}^*; t))V_{2n}(t)dt \right|, \tag{22}
$$

where V_{2n} are de la Vallée-Poisson kernels of the form

$$V_m(t) := \frac{1}{2} + \sum_{k=1}^{m} \cos kt + 2 \sum_{k=m+1}^{2m-1} \left(1 - \frac{k}{2m}\right) \cos kt, \ m \in \mathbb{N}. \tag{23}$$

Proposition A1.1 from [3] implies

$$I_1 \leq \inf_{\gamma_{2n}} \|f_p^*(t) - S_{\gamma_{2n}}(f_{p,n}^*; t)\|_\infty \|V_{2n}\|_1 = e_{2n}^{\perp}(f_{p,n}^*)_\infty \|V_{2n}\|_1. \tag{24}$$

Since (see, e.g., [18, p.247])

$$\|V_m\|_1 \leq 3\pi, \ m \in \mathbb{N}, \tag{25}$$

from (24) and (25) we can write down the estimate

$$e_{2n}^{\perp}(f_{p,n}^*)_\infty \geq \frac{1}{3\pi} I_1. \tag{26}$$

Notice, that

$$f_{p,n}^*(t) - S_{\gamma_{2n}}(f_{p,n}^*; t)$$
$$= \frac{\lambda_p}{2\psi(n)(\eta(n) - n)^{\frac{1}{p'}}} \left(\sum_{\substack{|k| \leq n-1, \\ k \notin \gamma_{2n}}} \psi(|k|)\psi(2n - |k|)e^{ikt} + \sum_{\substack{n \leq |k| \leq 2n, \\ k \notin \gamma_{2n}}} \psi^2(|k|)e^{ikt} \right), \tag{27}$$

where $\psi(0) := \psi(1)$
   Whereas

$$\int_{-\pi}^{\pi} e^{ikt} e^{imt} dt = \begin{cases} 0, & k + m \neq 0, \\ 2\pi, & k + m = 0, \end{cases} \ k, m \in \mathbb{Z}, \tag{28}$$

and taking into account (23), we obtain

$$\int_{-\pi}^{\pi} (f_{p,n}^*(t) - S_{\gamma_{2n}}(f_{p,n}^*; t))V_{2n}(t)dt \tag{29}$$

$$= \frac{\lambda_p}{4\psi(n)(\eta(n)-n)^{\frac{1}{p'}}} \int_{-\pi}^{\pi} \left(\sum_{\substack{0 \leq k \leq n-1, \\ k \notin \gamma_{2n}}} \psi(k)\psi(2n - k)e^{ikt} + \sum_{\substack{-n+1 \leq k \leq -1, \\ k \notin \gamma_{2n}}} \psi(|k|)\psi(2n-|k|)e^{ikt} \right)$$

$$
+ \sum_{\substack{n \le k \le 2n, \\ k \notin \gamma_{2n}}} \psi^2(k) e^{ikt} + \sum_{\substack{-2n \le k \le -n, \\ k \notin \gamma_{2n}}} \psi^2(|k|) e^{ikt} \Bigg) \times
$$

$$
\times \Bigg(\sum_{0 \le k \le 2n} e^{ikt} + \sum_{-2n \le k \le -1} e^{ikt} + 2 \sum_{2n+1 \le |k| \le 4n-1} \Big(1 - \frac{|k|}{4n} \Big) e^{ikt} \Bigg) dt \tag{30}
$$

$$
= \frac{\lambda_p \pi}{2\psi(n)(\eta(n) - n)^{\frac{1}{p'}}} \Bigg(\sum_{\substack{|k| \le n-1, \\ k \notin \gamma_{2n}}} \psi(|k|)\psi(2n - |k|) + \sum_{\substack{n \le |k| \le 2n, \\ k \notin \gamma_{2n}}} \psi^2(|k|) \Bigg). \tag{31}
$$

The function $\phi_n(t) := \psi(t)\psi(2n - t)$ decreases for $t \in [1, n]$. Indeed

$$
\phi_n'(t) = |\psi'(t)||\psi'(2n - t)| \Big(\frac{\psi(t)}{|\psi'(t)|} - \frac{\psi(2n - t)}{|\psi'(2n - t)|} \Big) \le 0,
$$

because $\frac{\psi(t)}{|\psi'(t)|} \uparrow \infty$ for large n.

Thus, the monotonicity of function $\phi_n(t)$ and (31) imply

$$
I_1 = \frac{\pi \lambda_p}{2\psi(n)(\eta(n) - n)^{\frac{1}{p'}}} \Bigg(\psi^2(n) + \sum_{n+1 \le |k| \le 2n} \psi^2(|k|) \Bigg)
$$

$$
> \frac{\pi \lambda_p}{2\psi(n)(\eta(n) - n)^{\frac{1}{p'}}} \sum_{k=n}^{2n} \psi^2(k) \ge \frac{\pi \lambda_p}{2\psi(n)(\eta(n) - n)^{\frac{1}{p'}}} \int_n^{\eta(n)} \psi^2(t) dt
$$

$$
> \frac{\pi \lambda_p}{2\psi(n)(\eta(n) - n)^{\frac{1}{p'}}} \psi^2(\eta(n))(\eta(n) - n) = \frac{\pi \lambda_p}{8} \psi(n)(\eta(n) - n)^{\frac{1}{p}}. \tag{32}
$$

By considering (26) and (32) we can write

$$
e_{2n}^{\perp}(L_{\beta,p}^{\psi})_\infty \ge e_{2n}^{\perp}(f_{p,n}^*)_\infty \ge \frac{1}{3\pi} I_1 \ge \frac{\lambda_p}{24} \psi(n)(\eta(n) - n)^{\frac{1}{p}}. \tag{33}
$$

Theorem 1 is proved.

In fact in the proof of Theorem 1 we obtained estimates with constants in explicit form.

Proposition 2 Let $\psi \in \mathfrak{M}_\infty^+$, $\beta \in \mathbb{R}$, $1 < p < \infty$, $\frac{1}{p} + \frac{1}{p'} = 1$, and the function $\frac{\psi(t)}{|\psi'(t)|}$ increases monotonically. Then for $n \in \mathbb{N}$, such that $\mu(\psi, n) \ge b > 2$ and $\eta(\psi, n) - n \ge a > 2$, the following estimates hold

$$
K_{b,p} \psi(n)(\eta(n) - n)^{\frac{1}{p}} \le e_{2n}^{\perp}(L_{\beta,p}^{\psi})_\infty \le e_{2n-1}^{\perp}(L_{\beta,p}^{\psi})_\infty \le K_{a,b,p} \psi(n)(\eta(n) - n)^{\frac{1}{p}}, \tag{34}
$$

where

$$K_{a,b,p} = \frac{1}{\pi} \max \left\{ \frac{2b}{b-2} + \frac{1}{a}, \ 2\pi \right\} (2p)^{\frac{1}{p'}}. \tag{35}$$

$$K_{b,p} = \frac{1}{48 \max \left\{ \frac{5b}{b-2}, \ 4\pi \right\} (p')^{\frac{1}{p}}}. \tag{36}$$

3 Best Orthogonal Trigonometric Approximations of the Classes $L_{\beta,1}^{\psi}$ in the Metric of Space L_{∞}

Theorem 2 *Let $\psi \in \mathfrak{M}_{\infty}^{+}$. Then for all $\beta \in \mathbb{R}$ order estimates are true*

$$e_{2n-1}^{\perp}(L_{\beta,1}^{\psi})_{\infty} \asymp e_{2n}^{\perp}(L_{\beta,1}^{\psi})_{\infty} \asymp \psi(n)(\eta(n) - n). \tag{37}$$

Proof According to formula (48) from [18] under conditions $\psi \in \mathfrak{M}$, $\sum\limits_{k=1}^{\infty} \psi(k) < \infty$, $\beta \in \mathbb{R}$, for all $n \in \mathbb{N}$ the following estimate holds

$$\mathscr{E}_n(L_{\beta,1}^{\psi})_{\infty} \le \frac{1}{\pi} \sum_{k=n}^{\infty} \psi(k). \tag{38}$$

Using Proposition 1, we have

$$e_{2n}^{\perp}(L_{\beta,1}^{\psi})_{\infty} \le e_{2n-1}^{\perp}(L_{\beta,1}^{\psi})_{\infty} \le \mathscr{E}_n(L_{\beta,1}^{\psi})_{\infty} \le \frac{1}{\pi} \sum_{k=n}^{\infty} \psi(k)$$

$$\le \frac{1}{\pi} \left(\psi(n) + \int\limits_{n}^{\infty} \psi(u)du \right) \le \frac{\psi(n)}{\pi} \left(1 + \frac{b}{b-2}(\eta(n) - n) \right). \tag{39}$$

Let us find the lower estimate for the quantity $e_{2n}^{\perp}(L_{\beta,1}^{\psi})_{\infty}$.
We consider the quantity

$$I_2 := \inf_{\gamma_{2n}} \left| \int\limits_{-\pi}^{\pi} (f_{2n}^{*}(t) - S_{\gamma_{2n}}(f_{2n}^{*}; t)) V_{2n}(t)dt \right|, \tag{40}$$

where V_m are de la Vallée-Poisson kernels of the form (23), and

$$f_m^*(t) = f_m^*(\psi; t) := \frac{1}{5\pi m}\left(\frac{1}{2}\psi(1) + \sum_{k=1}^{m} k\psi(k)\cos kt + \sum_{k=m+1}^{2m}(2m+1-k)\psi(k)\cos kt\right).$$

$$(41)$$

In [18, p. 263–265] it was shown that $\|(f_m^*)_\beta^\psi\|_1 \leq 1$, i.e., f_m^* belongs to the class $L_{\beta,1}^\psi$ for all $m \in \mathbb{N}$.

Using Proposition A1.1 from [3] and inequality (25), we have

$$I_2 \leq \inf_{\gamma_{2n}}\|f_{2n}^*(t) - S_{\gamma_{2n}}(f_{2n}^*; t)\|_\infty \|V_{2n}\|_1 \leq 3\pi\, e_{2n}^\perp(f_{2n}^*)_\infty. \tag{42}$$

Assuming again $\psi(0) := \psi(1)$, from (23) and (41), we derive

$$I_2 = \frac{1}{20\pi n}\inf_{\gamma_{2n}}\left|\int_{-\pi}^{\pi}\left(\sum_{\substack{|k|\leq 2n,\\ k\notin\gamma_{2n}}}|k|\psi(|k|)e^{ikt} + \sum_{\substack{2n+1\leq|k|\leq 4n,\\ k\notin\gamma_{2n}}}(4n+1-|k|)\psi(|k|)e^{ikt}\right)\times\right.$$

$$\left.\times\left(\sum_{|k|\leq 2n}e^{ikt} + 2\sum_{2n+1\leq|k|\leq 4n-1}\left(1 - \frac{|k|}{4n}\right)e^{ikt}\right)dt\right|$$

$$= \frac{1}{10n}\inf_{\gamma_{2n}}\left(\sum_{\substack{|k|\leq 2n,\\ k\notin\gamma_{2n}}}|k|\psi(|k|) + \sum_{\substack{2n+1\leq|k|\leq 4n,\\ k\notin\gamma_{2n}}}\left(1 - \frac{|k|}{4n}\right)(4n+1-|k|)\psi(|k|)\right)$$

$$> \frac{1}{10n}\inf_{\gamma_{2n}}\sum_{\substack{|k|\leq 2n,\\ k\notin\gamma_{2n}}}|k|\psi(|k|) = \frac{1}{10n}\left(n\psi(n) + 2\sum_{k=n+1}^{2n}k\psi(k)\right)$$

$$> \frac{1}{10}\sum_{k=n}^{2n}\psi(k) > \frac{1}{10}\int_{n}^{\eta(n)}\psi(t)dt > \frac{1}{20}\psi(n)(\eta(n)-n), \tag{43}$$

where we have used, that function $t\psi(t)$ decreases monotonically from some number t_0. Indeed,

$$(t\psi(t))' = |\psi'(t)|t\left(\frac{\psi(t)}{|\psi'(t)|} - 1\right),$$

and relations (1) and (9) yield

$$\frac{\psi(t)}{|\psi'(t)|} \asymp \frac{\eta(t)-t}{t} = \frac{1}{\mu(t)} \to 0, \quad \text{as } t \to \infty \text{ for } \psi \in \mathfrak{M}_\infty^+.$$

Formulas (42) and (43) imply, that for $n > t_0$

$$e_{2n}^{\perp}(L_{\beta,1}^{\psi})_{\infty} \geq e_{2n}^{\perp}(f_{2n}^{*})_{\infty} \geq \frac{1}{3\pi}I_2 > \frac{1}{60\pi}\psi(n)(\eta(n) - n).$$

Theorem 2 is proved.

Corollary 1 *Let $r \in (0, 1)$, $\alpha > 0$, $1 \leq p < \infty$ and $\beta \in \mathbb{R}$. Then for all $n \in \mathbb{N}$ the following estimates are true*

$$e_n^{\perp}(L_{\beta,p}^{\alpha,r})_{\infty} \asymp \exp(-\alpha n^r)n^{\frac{1-r}{p}}. \tag{44}$$

4 Best Orthogonal Trigonometric Approximations of the Classes $L_{\beta,1}^{\psi}$ in the Metric of Spaces L_s, $1 < s < \infty$

Theorem 3 *Let $1 < s < \infty$, $\psi \in \mathfrak{M}_{\infty}''$ and function $\frac{\psi(t)}{|\psi'(t)|} \uparrow \infty$ as $t \to \infty$. Then for all $\beta \in \mathbb{R}$ order estimates hold*

$$e_{2n-1}^{\perp}(L_{\beta,1}^{\psi})_s \asymp e_{2n}^{\perp}(L_{\beta,1}^{\psi})_s \asymp \psi(n)(\eta(n) - n)^{\frac{1}{s'}}, \quad \frac{1}{s} + \frac{1}{s'} = 1. \tag{45}$$

Proof According to Theorem 2 from [8] under conditions $\psi \in \mathfrak{M}_{\infty}^{+}$, $\beta \in \mathbb{R}$, $1 < s \leq \infty$ for $n \in \mathbb{N}$, such that $\eta(n) - n \geq a > 2$, $\mu(n) \geq b > 2$ the following estimate holds

$$\mathscr{E}_n(L_{\beta,1}^{\psi})_s \leq K_{a,b}(2s')^{\frac{1}{s}}\psi(n)(\eta(n) - n)^{\frac{1}{s'}}. \tag{46}$$

Since, $\frac{\psi(t)}{|\psi'(t)|} \uparrow \infty$, then as it was noticed in the proof of Theorem 1, exists number n_0, such that for all $n > n_0$ inequalities $\eta(n) - n \geq a > 2$, $\mu(n) \geq b > 2$ hold.

Using inequalities (6) and (46), we get

$$e_{2n}^{\perp}(L_{\beta,1}^{\psi})_s \leq e_{2n-1}^{\perp}(L_{\beta,1}^{\psi})_s \leq K_{a,b,s'}(2s')^{\frac{1}{s}}\psi(n)(\eta(n) - n)^{\frac{1}{s'}}. \tag{47}$$

Let us find the lower estimate of the quantity $e_{2n}^{\perp}(L_{\beta,1}^{\psi})_s$.
We consider the quantity

$$I_3 := \inf_{\gamma_{2n}}\left| \int_{-\pi}^{\pi}(f_{2n}^{**}(t) - S_{\gamma_{2n}}(f_{2n}^{**}; t))f_{s',n}^{*}(t)dt \right|, \tag{48}$$

where

$$f_m^{**}(t) = \frac{1}{3\pi} V_m(t),$$

and $f_{s',n}^*$ is defined by formula (11).

On the basis of Proposition A1.1 from [3] we derive

$$I_3 \leq \inf_{\gamma_{2n}} \| f_{2n}^{**}(t) - S_{\gamma_{2n}}(f_{2n}^{**}; t) \|_s \| f_{s'}^* \|_{s'} \leq e_{2n}^{\perp}(f_{2n}^{**})_s. \tag{49}$$

On other hand, using formulas (28), we write

$$I_3 = \frac{\lambda_{s'}}{12\pi \psi(n)(\eta(n)-n)^{\frac{1}{s}}} \inf_{\gamma_{2n}} \left| \int_{-\pi}^{\pi} \left(\sum_{\substack{|k|\leq 2n, \\ k\notin\gamma_{2n}}} e^{ikt} + 2 \sum_{\substack{2n+1\leq|k|\leq 4n-1, \\ k\notin\gamma_{2n}}} \left(1 - \frac{|k|}{4n}\right) e^{ikt} \right) \times \right.$$

$$\left. \times \left(\sum_{|k|\leq n-1} \psi(|k|)\psi(2n-|k|) e^{ikt} + \sum_{n\leq|k|\leq 2n} \psi^2(|k|) e^{ikt} \right) dt \right|$$

$$= \frac{\lambda_{s'}}{6\psi(n)(\eta(n)-n)^{\frac{1}{s}}} \inf_{\gamma_{2n}} \left(\sum_{\substack{|k|\leq n-1, \\ k\notin\gamma_{2n}}} \psi(|k|)\psi(2n-|k|) + \sum_{\substack{n\leq|k|\leq 2n, \\ k\notin\gamma_{2n}}} \psi^2(|k|) \right)$$

$$= \frac{\lambda_{s'}}{6\psi(n)(\eta(n)-n)^{\frac{1}{s}}} \left(\psi^2(n) + 2 \sum_{k=n+1}^{2n} \psi^2(k) \right) > \frac{\lambda}{6\pi\psi(n)(\eta(n)-n)^{\frac{1}{s}}} \sum_{k=n}^{2n} \psi^2(k)$$

$$> \frac{\lambda_{s'}}{6\psi(n)(\eta(n)-n)^{\frac{1}{s}}} \int_{n}^{\eta(n)} \psi^2(t) dt > \frac{\lambda_{s'}}{24} \psi(n)(\eta(n)-n)^{\frac{1}{s'}}. \tag{50}$$

Hence, formulas (49) and (50) imply

$$e_{2n}^{\perp}(L_{\beta,1}^{\psi})_s \geq e_{2n}^{\perp}(f_{s'}^{**})_s \geq I_3 \geq \frac{\lambda_{s'}}{24} \psi(n)(\eta(n)-n)^{\frac{1}{s'}}. \tag{51}$$

Theorem 3 is proved.

In fact in the proof of Theorem 3 we obtained estimates with constants in explicit form.

Proposition 3 *Let $\psi \in \mathfrak{M}_\infty^+$, $\beta \in \mathbb{R}$, $1 \leq p < \infty$ and function $\frac{\psi(t)}{|\psi'(t)|}$ increases monotonically. Then for all $n \in \mathbb{N}$, such that $\mu(\psi, n) \geq b > 2$ and $\eta(\psi, n) - n \geq a > 2$, the following estimates are true*

$$K_{b,s'}\psi(n)(\eta(n)-n)^{\frac{1}{s'}} \le e_{2n}^{\perp}(L_{\beta,1}^{\psi})_s \le e_{2n-1}^{\perp}(L_{\beta,1}^{\psi})_s$$

$$\le K_{a,b,s'}\psi(n)(\eta(n)-n)^{\frac{1}{s'}},$$

where $K_{a,b,s'}$ and $K_{b,s'}$ are defined by formulas (35) and (36) respectively.

Corollary 2 Let $r \in (0,1)$, $\alpha > 0$, $1 < s < \infty$ and $\beta \in \mathbb{R}$. Then for all $n \in \mathbb{N}$ the following estimates are true

$$e_n^{\perp}(L_{\beta,1}^{\alpha,r})_s \asymp \exp(-\alpha n^r)n^{\frac{1-r}{s'}}, \quad \frac{1}{s} + \frac{1}{s'} = 1. \tag{52}$$

Note, that functions

(1) $e^{-\alpha t^r}t^\gamma$, $\alpha > 0$, $r \in (0,1]$, $\gamma \le 0$;
(2) $e^{-\alpha t^r}\ln^\gamma(t+K)$, $\alpha > 0$, $r \in (0,1]$, $\gamma \le 0$, $K > e-1$,

etc., can be regarded as examples of functions ψ, which satisfy the conditions of Theorems 1 and 3.

Remark 1 It should be noticed, that from Theorem 1–3 it follows that the orders of quantities $e_n^{\perp}(L_{\beta,p}^{\psi})_s$ for $1 \le p < \infty$, $s = \infty$ and $p = 1$, $1 < s < \infty$, coincide with orders of the best approximations $E_n(L_{\beta,p}^{\psi})_s$ (see [9]).

Acknowledgements The author is supported by the Austrian Science Fund FWF projects F5503 and F5506-N26 (part of the Special Research Program (SFB) "Quasi-Monte Carlo Methods: Theory and Applications") and partially is supported by grant of NAS of Ukraine for groups of young scientists (project No16-10/2018).

References

1. A.S. Fedorenko, On the best m-term trigonometric and orthogonal trigonometric approximations of functions from the classes $L_{\beta,p}^{\psi}$. Ukr. Math. J. **51**(12), 1945–1949 (1999)
2. I.S. Gradshtein, I.M. Ryzhik, *Tables of Integrals, Sums, Series, and Products* [in Russian] (Fizmatgiz, Moscow, 1963)
3. N.P. Korneichuk, *Exact Constants in Approximation Theory*, vol. 38 (Cambridge University Press, Cambridge, New York, 1990)
4. A.S. Romanyuk, Approximation of classes of periodic functions of many variables. Mat. Zametki **71**(1), 109–121 (2002)
5. A.S. Romanyuk, Best trigonometric approximations of the classes of periodic functions of many variables in a uniform metric. Mat. Zametki **81**(2), 247–261 (2007)
6. A.S. Romanyuk, *Approximate Characteristics of Classes of Periodic Functions of Many Variables* [in Russian] (Institute of Mathematics, Ukrainian National Academy of Sciences, Kyiv, 2012)
7. A.S. Serdyuk, Approximation by interpolation trigonometric polynomials on classes of periodic analytic functions. Ukr. Mat. Zh. **64**(5), 698–712 (2012); English translation: Ukr. Math. J. **64**(5), 797–815 (2012)

8. A.S. Serdyuk, T.A. Stepaniuk, Order estimates for the best approximation and approximation by Fourier sums of classes of infinitely differentiable functions. Zb. Pr. Inst. Mat. NAN Ukr. **10**(1), 255–282 (2013). [in Ukrainian]

9. A.S. Serdyuk, T.A. Stepanyuk, Estimates for the best approximations of the classes of innately differentiable functions in uniform and integral metrics. Ukr. Mat. Zh. **66**(9), 1244–1256 (2014)

10. A.S. Serdyuk, T.A. Stepaniuk, Order estimates for the best orthogonal trigonometric approximations of the classes of convolutions of periodic functions of low smoothness. Ukr. Math. J. **67**(7), 1–24 (2015)

11. A.S. Serdyuk, T.A. Stepaniuk, Estimates of the best m-term trigonometric approximations of classes of analytic functions. Dopov. Nats. Akad. Nauk Ukr. Mat. Pryr. Tekh. Nauky No. 2 32–37 (2015). [in Ukrainian]

12. A.S. Serdyuk, T.A. Stepanyuk, Estimates for approximations by Fourier sums, best approximations and best orthogonal trigonometric approximations of the classes of (ψ, β)–*differentiable functions*. Bull. Soc. Sci. Lettres Lodz. Ser. Rech. Deform. **66**(2), 35–43 (2016)

13. V.V. Shkapa, Estimates of the best M-term and orthogonal trigonometric approximations of functions from the classes $L^{\psi}_{\beta,p}$ in a uniform metric. *Differential Equations and Related Problems* [in Ukrainian]. vol. 11, issue 2. (Institute of Mathematics, Ukrainian National Academy of Sciences, Kyiv, 2014), pp. 305–317

14. V.V. Shkapa, Best orthogonal trigonometric approximations of functions from the classes $L^{\psi}_{\beta,1}$, Approximation Theory of Functions and Related Problems [in Ukrainian], vol. 11, issue 3 (Institute of Mathematics, Ukrainian National Academy of Sciences, Kyiv, 2014), pp. 315–329

15. A.I. Stepanets, *Classification and Approximation of Periodic Functions* [in Russian] (Naukova Dumka, Kiev, 1987)

16. A.I. Stepanets, *Methods of Approximation Theory* (VSP: Leiden/Boston, 2005)

17. A.I. Stepanets, A.S. Serdyuk, A.L. Shidlich, Classification of infinitely differentiable periodic functions. Ukr. Mat. Zh. **60**(12), 1686–1708 (2008)

18. T.A. Stepaniuk, Estimates of the best approximations and approximations of Fourier sums of classes of convolutions of periodic functions of not high smoothness in integral metrics, [in Ukrainian]. Zb. Pr. Inst. Mat. NAN Ukr. **11**(3), 241–269 (2014)

Equivalent Conditions of a Reverse Hilbert-Type Integral Inequality with the Kernel of Hyperbolic Cotangent Function Related to the Riemann Zeta Function

Bicheng Yang

Abstract By the use of techniques of real analysis and weight functions, we study some equivalent conditions of a reverse Hilbert-type integral inequality with the non-homogeneous kernel of hyperbolic cotangent function, related to the Riemann zeta function. Some equivalent conditions of a reverse Hilbert-type integral inequality with the homogeneous kernel are deduced. We also consider some particular cases.

Keywords Reverse Hilbert-type integral inequality · Weight function · Equivalent form · Homogeneous kernel

2000 Mathematics Subject Classification 26D15

1 Introduction

If $0 < \int_0^\infty f^2(x)dx < \infty$ and $0 < \int_0^\infty g^2(y)dy < \infty$, then we have the following Hilbert integral inequality (cf. [1]):

$$\int_0^\infty \int_0^\infty \frac{f(x)g(y)}{x+y}dxdy < \pi \left(\int_0^\infty f^2(x)dx \int_0^\infty g^2(y)dy \right)^{\frac{1}{2}}, \qquad (1)$$

where the constant factor π is the best possible.

In 1925, Hardy [2] gave an extension of (1) as follows:

For $p > 1$, $\frac{1}{p} + \frac{1}{q} = 1$, $f(x), g(y) \geq 0$, $0 < \int_0^\infty f^p(x)dx < \infty$ and $0 < \int_0^\infty g^q(y)dy < \infty$, the following Hardy-Hilbert inequality holds:

B. Yang (✉)
Department of Mathematics, Guangdong University of Education, Guangdong, Guangzhou, P. R. China
e-mail: bcyang@gdei.edu.cn; bcyang818@163.com

$$\int_0^\infty \int_0^\infty \frac{f(x)g(y)}{x+y}dxdy < \frac{\pi}{\sin(\pi/p)}\left(\int_0^\infty f^p(x)dx\right)^{\frac{1}{p}}\left(\int_0^\infty g^q(y)dy\right)^{\frac{1}{q}},$$

(2)

where, the constant factor $\frac{\pi}{\sin(\pi/p)}$ is the best possible.

Inequalities (1) and (2) are important in analysis and its applications (cf. [3, 4]).

In 1934, Hardy et al. gave an extension of (2) as follows: If $p > 1, \frac{1}{p} + \frac{1}{q} = 1, k_1(x, y)$ is a non-negative homogeneous function of degree -1,

$$k_p = \int_0^\infty k_1(u, 1)u^{\frac{-1}{p}} du \in \mathbf{R}_+ = (0, \infty),$$

then we have the following Hardy-Hilbert-type integral inequality with the best possible constant k_p:

$$\int_0^\infty \int_0^\infty k_1(x, y)f(x)g(y)dxdy < k_p\left(\int_0^\infty f^p(x)dx\right)^{\frac{1}{p}}\left(\int_0^\infty g^q(y)dy\right)^{\frac{1}{q}};$$

(3)

for $0 < p < 1, \frac{1}{p} + \frac{1}{q} = 1$, the reverse of (2) follows (cf. [3], Theorem 319, Theorem 336). Also a Hilbert-type integral inequality with the non-homogeneous kernel is proved as follows: If $p > 1, \frac{1}{p} + \frac{1}{q} = 1, h(u) > 0, \phi(\sigma) = \int_0^\infty h(u)u^{\sigma-1}du \in \mathbf{R}_+$, then

$$\int_0^\infty \int_0^\infty h(xy)f(x)g(y)dxdy$$

$$< \phi\left(\frac{1}{p}\right)\left(\int_0^\infty x^{p-2}f^p(x)dx\right)^{\frac{1}{p}}\left(\int_0^\infty g^q(y)dy\right)^{\frac{1}{q}},$$

(4)

where, the constant factor $\phi(\frac{1}{p})$ is the best possible (cf. [3], Theorem 350).

In 1998, by introducing an independent parameter $\lambda > 0$, Yang gave an extension of (1) with the kernel $\frac{1}{(x+y)^\lambda}$ (cf. [5, 6]). In 2004, by introducing another pair conjugate exponents (r, s), Yang [7] gave an extension of (2) as follows: If $\lambda > 0, p, r > 1, \frac{1}{p} + \frac{1}{q} = \frac{1}{r} + \frac{1}{s} = 1, f(x), g(y) \geq 0, 0 < \int_0^\infty x^{p(1-\frac{\lambda}{r})-1}f^p(x)dx < \infty$ and $0 < \int_0^\infty y^{q(1-\frac{\lambda}{s})-1}g^q(y)dy < \infty$, then

$$\int_0^\infty \int_0^\infty \frac{f(x)g(y)}{x^\lambda + y^\lambda}dxdy$$

$$< \frac{\pi}{\lambda\sin(\pi/r)}\left[\int_0^\infty x^{p(1-\frac{\lambda}{r})-1}f^p(x)dx\right]^{\frac{1}{p}}\left[\int_0^\infty y^{q(1-\frac{\lambda}{s})-1}g^q(y)dy\right]^{\frac{1}{q}}, \quad (5)$$

where, the constant factor $\frac{\pi}{\lambda\sin(\pi/r)}$ is the best possible. In 2005, [8] also gave an extension of (2) as follows:

$$\int_0^\infty \int_0^\infty \frac{f(x)g(y)}{(x+y)^\lambda} dxdy$$

$$< B(\frac{\lambda}{r}, \frac{\lambda}{s}) \left[\int_0^\infty x^{p(1-\frac{\lambda}{r})-1} f^P(x)dx \right]^{\frac{1}{p}} \left[\int_0^\infty y^{q(1-\frac{\lambda}{s})-1} g^q(y)dy \right]^{\frac{1}{q}}, \quad (6)$$

where, the constant factor $B(\frac{\lambda}{r}, \frac{\lambda}{s})(\lambda > 0)$ is the best possible. Krnić et al. [9–16] provided some extensions and particular cases of (2), (3) and (4) with parameters.

In 2009, Yang gave an extension of (3), (5) and (6) as follows (cf. [17, 18]): If $\lambda_1 + \lambda_2 = \lambda \in \mathbf{R} = (-\infty, \infty)$, $k_\lambda(x, y)$ is a non-negative homogeneous function of degree $-\lambda$, satisfying

$$k_\lambda(ux, uy) = u^{-\lambda} k_\lambda(x, y)(u, x, y > 0),$$

$k(\lambda_1) = \int_0^\infty k_\lambda(u, 1) u^{\lambda_1-1} du \in \mathbf{R}_+ = (0, \infty)$, then for $p > 1, \frac{1}{p} + \frac{1}{q} = 1$, we have

$$\int_0^\infty \int_0^\infty k_\lambda(x, y) f(x)g(y)dxdy$$

$$< k(\lambda_1) \left(\int_0^\infty x^{p(1-\lambda_1)-1} f^P(x)dx \right)^{\frac{1}{p}} \left(\int_0^\infty y^{q(1-\lambda_2)-1} g^q(y)dy \right)^{\frac{1}{q}}, \quad (7)$$

where, the constant factor $k(\lambda_1)$ is the best possible; for $0 < p < 1, \frac{1}{p} + \frac{1}{q} = 1$, the reverse of (7) follows. Also an extension of (4) was given as follows: For $p > 1, \frac{1}{p} + \frac{1}{q} = 1$, we have

$$\int_0^\infty \int_0^\infty h(xy) f(x)g(y)dxdy$$

$$< \phi(\sigma) \left(\int_0^\infty x^{p(1-\sigma)-1} f^P(x)dx \right)^{\frac{1}{p}} \left(\int_0^\infty y^{q(1-\sigma)-1} g^q(y)dy \right)^{\frac{1}{q}}, \quad (8)$$

where, the constant factor $\phi(\sigma)$ is the best possible; for $0 < p < 1, \frac{1}{p} + \frac{1}{q} = 1$, the reverse of (8) follows (cf. [19]).

Some equivalent inequalities of (7) and (8) are considered by [18]. In 2013, Yang [19] also studied the equivalency of (7) and (8). In 2017, Hong [20] studied a equivalent condition between (7) and the related parameters.

In this chapter, by the use of the way of real analysis and the weight functions, we consider some equivalent conditions of a reverse of (8) in the kernel $h(xy) = \coth(xy) - 1$ for $0 < p < 1$, related to the Riemann zeta function. Some equivalent conditions of the reverse of (7) in the kernel $k_0(x, y) = \cot h(x/y) - 1$ are deduced. We also consider some particular cases.

2 An Example and Two Lemmas

Example 1 Setting $h(u) = coth(u) - 1 = \frac{2}{e^{2u}-1}$ $(u > 0)$, where, $coth(u) = \frac{e^u + e^{-u}}{e^u - e^{-u}}$ is the hyperbolic cotangent function, then we find $coth(xy) - 1 = \frac{2}{e^{2xy}-1}$, $coth(\frac{x}{y}) - 1 = \frac{2}{e^{2x/y}-1}$ and for $\sigma > 1$,

$$
\begin{aligned}
k(\sigma) &= \int_0^\infty (coth(u) - 1) u^{\sigma-1} du \\
&= \int_0^\infty \frac{2u^{\sigma-1}}{e^{2u}-1} du = \int_0^\infty \frac{2u^{\sigma-1}e^{-2u}}{1 - e^{-2u}} du \\
&= 2\int_0^\infty u^{\sigma-1} \sum_{k=0}^\infty e^{-2(k+1)u} du = 2\sum_{k=1}^\infty \int_0^\infty u^{\sigma-1}e^{-2ku} du.
\end{aligned}
$$

Setting $v = 2ku$ in the above integral, we have

$$
k(\sigma) = 2^{1-\sigma} \int_0^\infty v^{\sigma-1}e^{-v} dv \sum_{k=1}^\infty \frac{1}{k^\sigma} = 2^{1-\sigma}\Gamma(\sigma)\zeta(\sigma) \in \mathbf{R}_+, \tag{9}
$$

where, $\Gamma(\sigma) := \int_0^\infty v^{\sigma-1}e^{-v} dv$ $(\sigma > 0)$ is the gamma function, and $\zeta(\sigma) := \sum_{k=1}^\infty \frac{1}{k^\sigma}$ $(\sigma > 1)$ is the Riemann zeta function.

Setting $\delta_0 = \frac{\sigma-1}{2} > 0, \sigma \pm \delta_0 \geq \sigma - \frac{\sigma-1}{2} = \frac{\sigma+1}{2} > 1$, we still have $k(\sigma \pm \delta_0) < \infty$,

In the following, we make appointment that $0 < p < 1, \frac{1}{p} + \frac{1}{q} = 1, \sigma > 1, \sigma_1 \in \mathbf{R}$.

For $n \in \mathbf{N} = \{1, 2, \dots\}$, we define the following two expressions:

$$
I_1 := \int_1^\infty \left[\int_0^1 (coth(xy) - 1)x^{\sigma+\frac{1}{pn}-1} dx \right] y^{\sigma_1 - \frac{1}{qn}-1} dy, \tag{10}
$$

$$
I_2 := \int_0^1 \left[\int_1^\infty (coth(xy) - 1)x^{\sigma-\frac{1}{pn}-1} dx \right] y^{\sigma_1 + \frac{1}{qn}-1} dy. \tag{11}
$$

Setting $u = xy$ in (10) and (11), we have

$$
\begin{aligned}
I_1 &= \int_1^\infty \left[\int_0^y (coth(u) - 1)\left(\frac{u}{y}\right)^{\sigma+\frac{1}{pn}-1} \frac{1}{y} du \right] y^{\sigma_1 - \frac{1}{qn}-1} dy \\
&= \int_1^\infty y^{(\sigma_1 - \sigma) - \frac{1}{n}-1} \left[\int_0^y (coth(u) - 1)u^{\sigma+\frac{1}{pn}-1} du \right] dy, \tag{12}
\end{aligned}
$$

$$I_2 = \int_0^1 \left[\int_y^\infty (coth(u) - 1) \left(\frac{u}{y} \right)^{\sigma - \frac{1}{pn} - 1} \frac{1}{y} du \right] y^{\sigma_1 + \frac{1}{qn} - 1} dy$$

$$= \int_0^1 y^{(\sigma_1 - \sigma) + \frac{1}{n} - 1} \left[\int_y^\infty (cot h(u) - 1) u^{\sigma - \frac{1}{pn} - 1} du \right] dy. \qquad (13)$$

Lemma 1 *If there exists a constant $M > 0$, such that for any non-negative measurable functions $f(x)$ and $g(y)$ in $(0, \infty)$, the following inequality*

$$I := \int_0^\infty \int_0^\infty (coth(xy) - 1) f(x)g(y)dxdy$$

$$\geq M \left[\int_0^\infty x^{p(1-\sigma)-1} f^p(x)dx \right]^{\frac{1}{p}} \left[\int_0^\infty y^{q(1-\sigma_1)-1} g^q(y)dy \right]^{\frac{1}{q}} \qquad (14)$$

holds true, then we have $\sigma_1 = \sigma$ and $M \leq k(\sigma)$.

Proof If $\sigma_1 > \sigma$, then for $n > \frac{1}{\delta_0 p} (n \in \mathbf{N}, 0 < p < 1)$, we set the following two functions:

$$f_n(x) := \begin{cases} 0, 0 < x < 1 \\ x^{\sigma - \frac{1}{pn} - 1}, x \geq 1 \end{cases},$$

$$g_n(y) := \begin{cases} y^{\sigma_1 + \frac{1}{qn} - 1}, 0 < y \leq 1 \\ 0, y > 1 \end{cases}.$$

We find

$$J_2 := \left[\int_0^\infty x^{p(1-\sigma)-1} f_n^p(x)dx \right]^{\frac{1}{p}} \left[\int_0^\infty y^{q(1-\sigma_1)-1} g_n^q(y)dy \right]^{\frac{1}{q}}$$

$$= \left(\int_1^\infty x^{-\frac{1}{n} - 1} dx \right)^{\frac{1}{p}} \left(\int_0^1 y^{\frac{1}{n} - 1} dy \right)^{\frac{1}{q}} = n.$$

By (13), we have

$$I_2 \leq \int_0^1 y^{(\sigma_1 - \sigma) + \frac{1}{n} - 1} dy \int_0^\infty (coth(u) - 1) u^{\sigma - \frac{1}{pn} - 1} du$$

$$= \frac{1}{\sigma_1 - \sigma + \frac{1}{n}} \left[\int_0^1 (cot h(u) - 1) u^{\sigma - \frac{1}{pn} - 1} du \right.$$

$$+ \int_1^\infty (cot h(u) - 1) u^{\sigma - \frac{1}{pn} - 1} du \right]$$

$$\leq \frac{1}{\sigma_1 - \sigma} \left[\int_0^1 (\cot h(u) - 1) u^{(\sigma - \delta_0) - 1} du \right.$$

$$\left. + \int_1^\infty (\cot h(u) - 1) u^{\sigma - 1} du \right]$$

$$\leq \frac{1}{\sigma_1 - \sigma} (k(\sigma - \delta_0) + k(\sigma)),$$

and then by (14), it follows that

$$\frac{1}{\sigma_1 - \sigma} (k(\sigma - \delta_0) + k(\sigma))$$

$$\geq I_2 = \int_0^\infty \int_0^\infty (\cot h(xy) - 1) f_n(x) g_n(y) dx dy \geq M J_2 = Mn. \quad (15)$$

By (15), in view of $\sigma_1 - \sigma > 0$, $0 \leq k(\sigma - \delta_0) + k(\sigma) < \infty$, for $n \to \infty$, we find

$$\infty > \frac{1}{\sigma_1 - \sigma} (k(\sigma - \delta_0) + k(\sigma)) \geq \infty,$$

which is a contradiction.

If $\sigma_1 < \sigma$, then for $n \in \mathbf{N}$, $n > \frac{1}{\delta_0 p}$, we set the following two functions:

$$\widetilde{f}_n(x) := \begin{cases} x^{\sigma + \frac{1}{pn} - 1}, & 0 < x \leq 1 \\ 0, & x > 1 \end{cases},$$

$$\widetilde{g}_n(y) := \begin{cases} 0, & 0 < y < 1 \\ y^{\sigma_1 - \frac{1}{qn} - 1}, & y \geq 1 \end{cases}.$$

We find

$$\widetilde{J}_2 := \left[\int_0^\infty x^{p(1-\sigma) - 1} \widetilde{f}_n^p(x) dx \right]^{\frac{1}{p}} \left[\int_0^\infty y^{q(1-\sigma_1) - 1} \widetilde{g}_n^q(y) dy \right]^{\frac{1}{q}}$$

$$= \left(\int_0^1 x^{\frac{1}{n} - 1} dx \right)^{\frac{1}{p}} \left(\int_1^\infty y^{-\frac{1}{n} - 1} dy \right)^{\frac{1}{q}} = n.$$

By (12), we have

$$I_1 \leq \int_1^\infty y^{(\sigma_1 - \sigma) - \frac{1}{n} - 1} dy \int_0^\infty (\coth(u) - 1) u^{\sigma + \frac{1}{pn} - 1} du$$

$$= \frac{1}{\sigma - \sigma_1 + \frac{1}{n}} \left[\int_0^1 (\coth(u) - 1) u^{\sigma + \frac{1}{pn} - 1} du \right.$$

$$+ \int_1^\infty (\cot h(u) - 1)u^{\sigma + \frac{1}{pn} - 1} du \Bigg]$$

$$\leq \frac{1}{\sigma - \sigma_1} \Bigg[\int_0^1 (\cot h(u) - 1)u^{\sigma - 1} du$$

$$+ \int_1^\infty (coth(u) - 1)u^{\sigma + \delta_0 - 1} du \Bigg]$$

$$\leq \frac{1}{\sigma - \sigma_1} (k(\sigma) + k(\sigma + \delta_0)),$$

and then by (14), it follows that

$$\frac{1}{\sigma - \sigma_1} (k(\sigma) + k(\sigma + \delta_0))$$

$$\geq I_1 = \int_0^\infty \int_0^\infty (coth(xy) - 1) \tilde{f}_n(x) \tilde{g}_n(y) dx dy \geq M \tilde{J}_2 = Mn. \quad (16)$$

By (16), for $n \to \infty$, we still find that

$$\infty > \frac{1}{\sigma - \sigma_1} (k(\sigma) + k(\sigma + \delta_0)) \geq \infty,$$

which is a contradiction.

Hence, we conclude that $\sigma_1 = \sigma$.

For $\sigma_1 = \sigma$, we have $nM = M J_2 \leq I_2$ by using (14). Then we reduce (13) as follows:

$$M = \frac{1}{n} M J_2 \leq \frac{1}{n} I_2$$

$$= \frac{1}{n} \int_0^1 y^{\frac{1}{n} - 1} \Bigg[\int_y^\infty (coth(u) - 1)u^{\sigma - \frac{1}{pn} - 1} du \Bigg] dy$$

$$= \frac{1}{n} \int_0^1 y^{\frac{1}{n} - 1} \Bigg[\int_y^1 (\cot h(u) - 1)u^{\sigma - \frac{1}{pn} - 1} du \Bigg] dy$$

$$+ \int_1^\infty (coth(u) - 1)u^{\sigma - \frac{1}{pn} - 1} du$$

$$= \frac{1}{n} \int_0^1 \left(\int_0^u y^{\frac{1}{n} - 1} dy \right) (\cot h(u) - 1)u^{\sigma - \frac{1}{pn} - 1} du$$

$$+ \int_1^\infty (coth(u) - 1)u^{\sigma - \frac{1}{pn} - 1} du$$

$$\leq \int_0^1 (\cot h(u) - 1)u^{\sigma + \frac{1}{qn} - 1} du + \int_1^\infty (coth(u) - 1)u^{\sigma - 1} du. \quad (17)$$

Since for $n > \frac{1}{\delta_0|q|}$ $(n \in \mathbf{N})$, we have

$$(coth(u) - 1)u^{\sigma + \frac{1}{qn} - 1} \leq (coth(u) - 1)u^{\sigma - \delta_0 - 1}(0 < u \leq 1)$$

and

$$\int_0^1 (coth(u) - 1)u^{\sigma - \delta_0 - 1}du \leq k(\sigma - \delta_0) < \infty,$$

then by (17) and Lebesgue control convergence theorem (cf. [21]), we have

$$M \leq \lim_{n \to \infty} \left[\int_0^1 (\cot h(u) - 1)u^{\sigma + \frac{1}{qn} - 1}du + \int_1^\infty (coth(u) - 1)u^{\sigma - 1}du \right]$$

$$= \int_0^1 \lim_{n \to \infty} (coth(u) - 1)u^{\sigma + \frac{1}{qn} - 1}du + \int_1^\infty (\cot h(u) - 1)u^{\sigma - 1}du = k(\sigma).$$

The lemma is proved.

For $\sigma_1 = \sigma$, by Lemma 1, we still have

Lemma 2 *If there exists a constant $M > 0$, such that for any non-negative measurable functions $f(x)$ and $g(y)$ in $(0, \infty)$, the following inequality*

$$I := \int_0^\infty \int_0^\infty (coth(xy) - 1)f(x)g(y)dxdy$$

$$\geq M \left[\int_0^\infty x^{p(1-\sigma)-1} f^p(x)dx \right]^{\frac{1}{p}} \left[\int_0^\infty y^{q(1-\sigma)-1} g^q(y)dy \right]^{\frac{1}{q}} \quad (18)$$

holds true, then we have $M \leq k(\sigma)$.

3 Main Results

Theorem 1 *The following conditions are equivalent:*

(i) There exists a constant $M > 0$, such that for any $f(x) \geq 0$,

$$0 < \int_0^\infty x^{p(1-\sigma)-1} f^p(x)dx < \infty,$$

we have the following inequality:

$$J := \left\{ \int_0^\infty y^{p\sigma_1 - 1} \left[\int_0^\infty (coth(xy) - 1) f(x) dx \right]^p dy \right\}^{\frac{1}{p}}$$

$$> M \left[\int_0^\infty x^{p(1-\sigma)-1} f^p(x) dx \right]^{\frac{1}{p}}; \tag{19}$$

(ii) there exist a constant $M > 0$, such that for any $g(y) \geq 0$,

$$0 < \int_0^\infty y^{q(1-\sigma_1)-1} g^q(y) dy < \infty,$$

we have the following inequality:

$$K := \left\{ \int_0^\infty x^{q\sigma - 1} \left[\int_0^\infty (coth(xy) - 1) g(y) dy \right]^q dx \right\}^{\frac{1}{q}}$$

$$> M \left[\int_0^\infty y^{q(1-\sigma_1)-1} g^q(y) dy \right]^{\frac{1}{q}}; \tag{20}$$

(iii) there exists a constant $M > 0$, such that for any $f(x), g(y) \geq 0$,

$$0 < \int_0^\infty x^{p(1-\sigma)-1} f^p(x) dx < \infty,$$

and

$$0 < \int_0^\infty y^{q(1-\sigma_1)-1} g^q(y) dy < \infty,$$

we have the following inequality:

$$I = \int_0^\infty \int_0^\infty (coth(xy) - 1) f(x) g(y) dx dy$$

$$> M \left[\int_0^\infty x^{p(1-\sigma)-1} f^p(x) dx \right]^{\frac{1}{p}} \left[\int_0^\infty y^{q(1-\sigma_1)-1} g^q(y) dy \right]^{\frac{1}{q}}; \tag{21}$$

(iv) $\sigma_1 = \sigma$.

Proof "$(i) \implies (iii)$". By the reverse Hölder's inequality (cf. [22]), we have

$$I = \int_0^\infty \left[y^{\sigma_1 - \frac{1}{p}} \int_0^\infty (coth(xy) - 1) f(x) dx \right] \left(y^{\frac{1}{p} - \sigma_1} g(y) \right) dy$$

$$\geq J \left[\int_0^\infty y^{q(1-\sigma_1)-1} g^q(y) dy \right]^{\frac{1}{q}}. \tag{22}$$

Then by (19), we have (21).

"$(ii) => (iii)$". Still by the reverse Hölder's inequality, we have

$$I = \int_0^\infty \left(y^{\frac{1}{q}-\sigma} f(x) \right) \left[x^{\sigma-\frac{1}{q}} \int_0^\infty (coth(xy) - 1)g(y)dy \right] dx$$

$$\geq \left[\int_0^\infty x^{p(1-\sigma)-1} f^p(x)dx \right]^{\frac{1}{p}} K. \tag{23}$$

Then by (20), we have (21).

"$(iii) => (iv)$". By Lemma 1, we have $\sigma_1 = \sigma$.

"$(iv) => (i)$". Setting $u = xy$, we obtain the following weight function: For $y > 0$,

$$\omega(\sigma, y) : = y^\sigma \int_0^\infty (coth(xy) - 1)x^{\sigma-1}dx$$

$$= \int_0^\infty (coth(u) - 1)u^{\sigma-1}du = k(\sigma). \tag{24}$$

By the reverse Hölder's inequality with weight and (24), we have

$$\left[\int_0^\infty (coth(xy) - 1)f(x)dx \right]^p$$

$$= \left\{ \int_0^\infty (coth(xy) - 1) \left[\frac{y^{(\sigma-1)/p}}{x^{(\sigma-1)/q}} f(x) \right] \left[\frac{x^{(\sigma-1)/q}}{y^{(\sigma-1)/p}} \right] dx \right\}^p$$

$$\geq \int_0^\infty (coth(xy) - 1) \frac{y^{\sigma-1}}{x^{(\sigma-1)p/q}} f^p(x)dx$$

$$\times \left[\int_0^\infty (coth(xy) - 1) \frac{x^{\sigma-1}}{y^{(\sigma-1)q/p}} dx \right]^{p/q}$$

$$= \left[\omega(\sigma, y)y^{q(1-\sigma)-1} \right]^{p-1} \int_0^\infty (coth(xy) - 1) \frac{y^{\sigma-1}}{x^{(\sigma-1)p/q}} f^p(x)dx$$

$$= (k(\sigma))^{p-1} y^{-p\sigma+1} \int_0^\infty (coth(xy) - 1) \frac{y^{\sigma-1}}{x^{(\sigma-1)p/q}} f^p(x)dx \tag{25}$$

If (25) takes the form of equality for a $y \in (0, \infty)$, then (cf. [22]), there exists constants A and B, such that they are not all zero, and

$$A \frac{y^{\sigma-1}}{x^{(\sigma-1)p/q}} f^p(x) = B \frac{x^{\sigma-1}}{y^{(\sigma-1)q/p}} \quad a.e. \ in \ \mathbf{R}_+.$$

We suppose that $A \neq 0$ (otherwise $B = A = 0$). Then it follows that

$$x^{p(1-\sigma)-1} f^p(x) = y^{q(1-\sigma)} \frac{B}{Ax} \ a.e. \ in \ \mathbf{R}_+,$$

which contradicts the fact that $0 < \int_0^\infty x^{p(1-\sigma)-1} f^p(x) dx < \infty$. Hence, (25) takes the form of strict inequality.

For $\sigma_1 = \sigma$, by (25) and Fubini theorem, we have

$$J > (k(\sigma))^{\frac{1}{q}} \left[\int_0^\infty \int_0^\infty (coth(xy) - 1) \frac{y^{\sigma-1}}{x^{(\sigma-1)p/q}} f^p(x) dx dy \right]^{\frac{1}{p}}$$

$$= (k(\sigma))^{\frac{1}{q}} \left\{ \int_0^\infty \left[\int_0^\infty (coth(xy) - 1) \frac{y^{\sigma-1}}{x^{(\sigma-1)(p-1)}} dy \right] f^p(x) dx \right\}^{\frac{1}{p}}$$

$$= (k(\sigma))^{\frac{1}{q}} \left[\int_0^\infty \omega(\sigma, x) x^{p(1-\sigma)-1} f^p(x) dx \right]^{\frac{1}{p}}$$

$$= k(\sigma) \left[\int_0^\infty x^{p(1-\sigma)-1} f^p(x) dx \right]^{\frac{1}{p}}. \tag{26}$$

Setting $0 < M \leq k(\sigma)$, then (19) follows.

"$(iv) \Longrightarrow (ii)$". In the same way, we obtain (20).

Therefore, the conditions (i), (ii), (iii) and (iv) are equivalent.

For $\sigma_1 = \sigma$, we still have

Theorem 2 *The following conditions are equivalent:*

(i) For any $f(x) \geq 0$, $0 < \int_0^\infty x^{p(1-\sigma)-1} f^p(x) dx < \infty$, we have the following inequality:

$$\left\{ \int_0^\infty y^{p\sigma-1} \left[\int_0^\infty (coth(xy) - 1) f(x) dx \right]^p dy \right\}^{\frac{1}{p}}$$

$$> k(\sigma) \left[\int_0^\infty x^{p(1-\sigma)-1} f^p(x) dx \right]^{\frac{1}{p}}; \tag{27}$$

(ii) for any $g(y) \geq 0$, $0 < \int_0^\infty y^{q(1-\sigma)-1} g^q(y) dy < \infty$, we have the following inequality:

$$\left\{ \int_0^\infty x^{q\sigma-1} \left[\int_0^\infty (coth(xy) - 1) g(y) dy \right]^q dx \right\}^{\frac{1}{q}}$$

$$> k(\sigma) \left[\int_0^\infty y^{q(1-\sigma)-1} g^q(y) dy \right]^{\frac{1}{q}}; \tag{28}$$

(iii) for any $f(x), g(y) \geq 0, 0 < \int_0^\infty x^{p(1-\sigma)-1} f^p(x)dx < \infty$, and

$$0 < \int_0^\infty y^{q(1-\sigma)-1} g^q(y)dy < \infty,$$

we have the following inequality:

$$\int_0^\infty \int_0^\infty (coth(xy) - 1) f(x)g(y)dxdy$$

$$> k(\sigma) \left[\int_0^\infty x^{p(1-\sigma)-1} f^p(x)dx \right]^{\frac{1}{p}} \left[\int_0^\infty y^{q(1-\sigma)-1} g^q(y)dy \right]^{\frac{1}{q}}. \quad (29)$$

Moreover, the constant factor $k(\sigma)$ in (27), (28) and (29) is the best possible.

Proof For $\sigma_1 = \sigma$ in Theorem 1, since $0 < k(\sigma) < \infty$, setting $M = k(\sigma)$ in (19), (20) and (21), in the same way, we still can prove that the conditions (i), (ii) and (iii) are equivalent in Theorem 2. If there exists a constant $M \geq k(\sigma)$, such that (29) is valid, then by Lemma 2, we have $M \leq k(\sigma)$. Hence, the constant factor $M = k(\sigma)$ in (29) is the best possible. The constant factor $k(\sigma)$ in (27) ((28)) is still the best possible. Otherwise, by (22) (or (23)) for $\sigma_1 = \sigma$, we can conclude that the constant factor $M = k(\sigma)$ in (29) is not the best possible.

4   Some Corollaries

In particular, for $\sigma = \frac{1}{p}(> 1)$ in Theorem 2, we have

Corollary 1 *The following conditions are equivalent:*

(i) For any $f(x) \geq 0, 0 < \int_0^\infty x^{p-2} f^p(x)dx < \infty$, we have the following inequality:

$$\left[\int_0^\infty \left(\int_0^\infty (coth(xy) - 1) f(x)dx \right)^p dy \right]^{\frac{1}{p}} > k(\frac{1}{p}) \left(\int_0^\infty x^{p-2} f^p(x)dx \right)^{\frac{1}{p}};$$

$$(30)$$

(ii) for any $g(y) \geq 0, 0 < \int_0^\infty g^q(y)dy < \infty$, we have the following inequality:

$$\left[\int_0^\infty x^{q-2} \left(\int_0^\infty (coth(xy) - 1)g(y)dy \right)^q dx \right]^{\frac{1}{q}} > k(\frac{1}{p}) \left(\int_0^\infty g^q(y)dy \right)^{\frac{1}{q}};$$

$$(31)$$

(iii) for any $f(x), g(y) \geq 0, 0 < \int_0^\infty x^{p-2} f^p(x)dx < \infty$, and $0 < \int_0^\infty g^q(y)dy < \infty$, we have the following inequality:

$$\int_0^\infty \int_0^\infty (coth(xy) - 1) f(x)g(y)dxdy$$

$$> k(\frac{1}{p}) \left(\int_0^\infty x^{p-2} f^p(x)dx \right)^{\frac{1}{p}} \left(\int_0^\infty g^q(y)dy \right)^{\frac{1}{q}} ; \qquad (32)$$

Moreover, the constant factor

$$k(\frac{1}{p}) = 2^{1/q} \Gamma(\frac{1}{p}) \zeta(\frac{1}{p})$$

in (30), (31) and (32) is the best possible.

Setting $y = \frac{1}{Y}$, $G(Y) = g(\frac{1}{Y})\frac{1}{Y^2}$ in Theorem 1–2, then replacing Y $(G(Y))$ by y $(g(y))$, we have

Corollary 2 *The following conditions are equivalent:*

(i) There exists a constant $M > 0$, such that for any $f(x) \geq 0$,

$$0 < \int_0^\infty x^{p(1-\sigma)-1} f^p(x)dx < \infty,$$

we have the following inequality:

$$\left\{ \int_0^\infty y^{-p\sigma_1-1} \left[\int_0^\infty (coth(\frac{x}{y}) - 1) f(x)dx \right]^p dy \right\}^{\frac{1}{p}}$$

$$> M \left[\int_0^\infty x^{p(1-\sigma)-1} f^p(x)dx \right]^{\frac{1}{p}} ; \qquad (33)$$

(ii) there exists a constant $M > 0$, such that for any $g(y) \geq 0$,

$$0 < \int_0^\infty y^{q(1+\sigma_1)-1} g^q(y)dy < \infty,$$

we have the following inequality:

$$\left\{ \int_0^\infty x^{q\sigma-1} \left[\int_0^\infty (coth(\frac{x}{y}) - 1)g(y)dy \right]^q dx \right\}^{\frac{1}{q}}$$

$$> M \left[\int_0^\infty y^{q(1+\sigma_1)-1} g^q(y)dy \right]^{\frac{1}{q}} ; \qquad (34)$$

(iii) there exists a constant $M > 0$, such that for any $f(x), g(y) \geq 0$,

$$0 < \int_0^\infty x^{p(1-\sigma)-1} f^p(x)dx < \infty,$$

and

$$0 < \int_0^\infty y^{q(1+\sigma_1)-1} g^q(y)dy < \infty,$$

we have the following inequality:

$$\int_0^\infty \int_0^\infty (\coth(\frac{x}{y}) - 1) f(x)g(y)dxdy$$

$$> M \left[\int_0^\infty x^{p(1-\sigma)-1} f^p(x)dx \right]^{\frac{1}{p}} \left[\int_0^\infty y^{q(1+\sigma_1)-1} g^q(y)dy \right]^{\frac{1}{q}}; \quad (35)$$

(iv) $\sigma_1 = \sigma$.

Corollary 3 *The following conditions are equivalent:*

(i) For any $f(x) \geq 0$, $0 < \int_0^\infty x^{p(1-\sigma)-1} f^p(x)dx < \infty$, we have the following inequality:

$$\left\{ \int_0^\infty y^{-p\sigma-1} \left[\int_0^\infty (\coth(\frac{x}{y}) - 1) f(x)dx \right]^p dy \right\}^{\frac{1}{p}}$$

$$> k(\sigma) \left[\int_0^\infty x^{p(1-\sigma)-1} f^p(x)dx \right]^{\frac{1}{p}}; \quad (36)$$

(ii) for any $g(y) \geq 0$, $0 < \int_0^\infty y^{q(1+\sigma)-1} g^q(y)dy < \infty$, we have the following inequality:

$$\left\{ \int_0^\infty x^{q\sigma-1} \left[\int_0^\infty (\coth(\frac{x}{y}) - 1) g(y)dy \right]^q dx \right\}^{\frac{1}{q}}$$

$$> k(\sigma) \left[\int_0^\infty y^{q(1+\sigma)-1} g^q(y)dy \right]^{\frac{1}{q}}; \quad (37)$$

(iii) for any $f(x), g(y) \geq 0, 0 < \int_0^\infty x^{p(1-\sigma)-1} f^p(x)dx < \infty$, and

$$0 < \int_0^\infty y^{q(1+\sigma)-1} g^q(y)dy < \infty,$$

we have the following inequality:

$$\int_0^\infty \int_0^\infty (coth(\frac{x}{y}) - 1) f(x)g(y)dxdy$$

$$> k(\sigma) \left[\int_0^\infty x^{p(1-\sigma)-1} f^p(x)dx\right]^{\frac{1}{p}} \left[\int_0^\infty y^{q(1+\sigma)-1} g^q(y)dy\right]^{\frac{1}{q}}. \quad (38)$$

Moreover, the constant factor $k(\sigma)$ in (36), (37) and (38) is the best possible. In particular, for $\sigma = \frac{1}{p}(> 1)$ in Corollary 3, we have

Corollary 4 *The following conditions are equivalent:*

(i) For any $f(x) \geq 0, 0 < \int_0^\infty x^{p-2} f^p(x)dx < \infty$, we have the following inequality:

$$\left\{\int_0^\infty y^{-2} \left[\int_0^\infty (coth(\frac{x}{y}) - 1) f(x)dx\right]^p dy\right\}^{\frac{1}{p}}$$

$$> k(\frac{1}{p}) \left(\int_0^\infty x^{p-2} f^p(x)dx\right)^{\frac{1}{p}}; \quad (39)$$

(ii) for any $g(y) \geq 0, 0 < \int_0^\infty y^{2(q-1)} g^q(y)dy < \infty$, we have the following inequality:

$$\left\{\int_0^\infty x^{q-2} \left[\int_0^\infty (coth(\frac{x}{y}) - 1) g(y)dy\right]^q dx\right\}^{\frac{1}{q}}$$

$$> k(\frac{1}{p}) \left[\int_0^\infty y^{2(q-1)} g^q(y)dy\right]^{\frac{1}{q}}; \quad (40)$$

(iii) for any $f(x), g(y) \geq 0, 0 < \int_0^\infty x^{p-2} f^p(x)dx < \infty$, and $0 < \int_0^\infty y^{2(q-1)} g^q(y)dy < \infty$, we have the following inequality:

$$\int_0^\infty \int_0^\infty (coth(\frac{x}{y}) - 1) f(x)g(y)dxdy$$

$$> k(\frac{1}{p}) \left[\int_0^\infty x^{p-2} f^p(x)dx\right]^{\frac{1}{p}} \left[\int_0^\infty y^{2(q-1)} g^q(y)dy\right]^{\frac{1}{q}}. \quad (41)$$

Moreover, the constant factor

$$k(\frac{1}{p}) = 2^{1/q} \Gamma(\frac{1}{p}) \zeta(\frac{1}{p})$$

in (39), (40) and (41) is the best possible.

5 Conclusions

By the use of the way of real analysis and weight functions, we study some equivalent conditions of a reverse Hilbert-type integral inequality with the non-homogeneous kernel of the hyperbolic cotangent function, related to the Riemann zeta function in Theorems 1–2. Some equivalent conditions of a reverse Hilbert-type integral inequality with the homogeneous kernel are deduced in Corollary 2. We also consider some particular cases in Corollarys 1 and 3–4. The lemmas and theorems provide and extensive account of this type of inequalities.

Acknowledgements This work is supported by the National Natural Science Foundation (No. 61772140), and Science and Technology Planning Project Item of Guangzhou City (No. 201707010229). We are grateful for this help.

References

1. I. Schur, Bernerkungen sur Theorie der beschrankten Billnearformen mit unendlich vielen Veranderlichen. J. Math. **140**, 1–28 (1911)
2. G.H. Hardy, Note on a theorem of Hilbert concerning series of positive terms. Proc. Lond. Math. Soc. **23**(2), xlv–xlvi (1925). Records of Proc
3. G.H. Hardy, J.E. Littlewood, G. Pólya, *Inequalities* (Cambridge University Press, Cambridge, USA)
4. D.S. Mitrinović, J.E. Pecaric, A.M. Fink, *Inequalities Involving Functions and their Integrals and Deivatives* (Kluwer Academic, Boston, 1991)
5. B.C. Yang, On Hilbert's integral inequality. J. Math. Anal. Appl. **220**, 778–785 (1998)
6. B.C. Yang, A note on Hilbert's integral inequality. Chin. Q. J. Math. **13**(4), 83–86 (1998)
7. B.C. Yang, On an extension of Hilbert's integral inequality with some parameters. Aust. J. Math. Anal. Appl. **1**(1), 1–8 (2004), Art.11
8. B.C. Yang, I. Brnetic, M. Krnic, J.E. Pecaric, Generalization of Hilbert and Hardy-Hilbert integral inequalities. Math. Ineq. Appl. **8**(2), 259–272 (2005)
9. M. Krnic, J.E. Pecaric, Hilbert's inequalities and their reverses. Publ. Math. Debrecen **67**(3–4), 315–331 (2005)
10. Y. Hong, On Hardy-Hilbert integral inequalities with some parameters. J. Ineq. Pure Appl. Math. **6**(4), 1–10 (2005), Art. 92
11. B. Arpad, O. Choonghong, Best constant for certain multi linear integral operator. J. Inequal. Appl. **2006**(28582), 1–12 (2006)
12. Y.J. Li, B. He, On inequalities of Hilbert's type. Bull. Aust. Math. Soc. **76**(1), 1–13 (2007)
13. W.Y. Zhong, B.C. Yang, On multiple Hardy-Hilbert's integral inequality with kernel. J. Inequal. Appl. **2007**, 17, Art ID 27962
14. J.S. Xu, Hardy-Hilbert's Inequalities with two parameters. Adv. Math. **36**(2), 63–76 (2007)
15. M. Krnić, M.Z. Gao, J.E. Pečarić, et al., On the best constant in Hilbert's inequality. Math. Inequal. Appl. **8**(2), 317–329 (2005)
16. Y. Hong, On Hardy-type integral inequalities with some functions. Acta MathmaticaS inica **49**(1), 39–44 (2006)
17. B.C. Yang, *The Norm of Operator and Hilbert-Type Inequalities* (Science Press, Beijing, 2009)
18. B.C. Yang, *Hilbert-Type Integral Inequalities* (Bentham Science Publishers Ltd., The United Emirates, 2009)

19. B.C. Yang, On Hilbert-type integral inequalities and their operator expressions. J. Guangaong Univ. Edu. **33**(5), 1–17 (2013)
20. Y. Hong, On the structure character of Hilbert's type integral inequality with homogeneous kernal and applications. J. Jilin Univ. (Science Edition) **55**(2), 189–194 (2017)
21. J.C. Kuang, *Real and Functional Analysis* (Continuation)(second volume) (Higher Education Press, Beijing, 2015)
22. J.C. Kuang, *Applied Inequalities* (Shangdong Science and Technology Press, Jinan, 2004)

Index

© Springer Nature Switzerland AG 2020
A. Raigorodskii, M. T. Rassias (eds.), *Trigonometric Sums and Their Applications*,
https://doi.org/10.1007/978-3-030-37904-9

Printed in the United States
by Baker & Taylor Publisher Services